# *Robustness of Evolving Power Grids*

Tse and co-authors take a multidisciplinary approach to address the complex issues facing the sustainable and robust management of modern power grids—particularly the integration of renewable energy, artificial intelligence, and cyber networks—and the resulting necessity for enhanced resilience.

This book concentrates on three aspects central to robustness. First, it presents a methodology for modeling cascading failures in power grids using complex network theory that captures the propagation of failures triggered by initial events. Next, it proposes new evaluation metrics to accurately assess the robustness of complex power grids against major blackout incidents caused by cascading failures. Unlike traditional structure-based metrics, these metrics incorporate network characteristics, electrical variables, and learning-based methods. Lastly, it presents new strategies for enhancing the robustness of evolving power grids, focusing on both the system planning phase and the system control/protection phase. Throughout, this book offers comprehensive and in-depth analyses to ensure that readers gain a comprehensive understanding of the topic. It also draws from the latest advances in the field of complex circuits and systems and incorporates findings from academic studies, industry reports, and real-world case studies, providing a solid foundation for technical descriptions. The content is designed with practical applicability in mind; readers will benefit from actionable strategies directly achievable in real-world scenarios. Readers will acquire a holistic and multidisciplinary understanding of the robustness of evolving power grids and will be able to take a holistic perspective on modern power grid challenges and opportunities.

This is a vital and cutting-edge resource for research students, undergraduate students, and researchers in power engineering. Engineers and policymakers will also find valuable recommendations and methodologies to enhance robustness and mitigate risks.

# *Robustness of Evolving Power Grids*
## *Modeling and Analysis*

Dong Liu, Xi Zhang, and Chi Kong Tse

CRC Press
Taylor & Francis Group
Boca Raton London New York

CRC Press is an imprint of the
Taylor & Francis Group, an **informa** business

First edition published 2026
by CRC Press
2385 NW Executive Center Drive, Suite 320, Boca Raton FL 33431

and by CRC Press
4 Park Square, Milton Park, Abingdon, Oxon, OX14 4RN

*CRC Press is an imprint of Taylor & Francis Group, LLC*



ISBN: 9781041030744 (hbk)
ISBN: 9781041030768 (pbk)
ISBN: 9781003622130 (ebk)

DOI: 10.1201/9781003622130

Typeset in Times
by codeMantra

Publisher's Note: This book has been prepared from camera-ready copy provided by the author.

# Contents

# Preface

The increasing complexity of modern power grids and their pivotal role in sustaining economic and societal functions necessitate robust and resilient infrastructure. Power grids, the backbone of modern societies, ensure the continuous flow of energy that powers industries, businesses, and homes. As the energy landscape evolves with the integration of renewable energy sources, artificial intelligence, and complex connectivity with cyber coupling, the power grid faces unprecedented challenges, including cascading failures, cyber vulnerabilities, and the need for efficient resource management. These challenges demand systematic methodologies, advanced modeling techniques, and innovative strategies to ensure the reliability and stability of power systems under varying operating conditions. This book aims to address these pressing issues by offering a comprehensive and interdisciplinary exploration of power grid robustness, focusing on methodologies, evaluation metrics, and mitigation strategies.

The motivation for this book stems from the growing importance of developing robust power systems capable of withstanding disruptions in an increasingly interconnected and dynamic world. With the transition toward greener energy solutions, the integration of renewable energy sources, such as wind and solar, introduces variability and uncertainty into power grid operations. While these renewable sources are essential for achieving sustainability goals and combating climate change, their intermittent nature presents unique challenges for grid stability. Meanwhile, the coupling of power systems with cyber networks has transformed traditional grids into complex cyber-physical systems. While this digital transformation enhances grid monitoring, control, and efficiency, it also exposes power systems to new vulnerabilities, including cyberattacks, communication failures, and software glitches. These factors, combined with the ever-present risks of natural disasters, aging infrastructure, and equipment malfunctions, underscore the urgent need for advanced tools, frameworks, and strategies to assess and enhance power grid robustness.

The technical foundation of this book is built upon the principles of complex network theory, power system engineering, and artificial intelligence. By integrating these diverse perspectives, this book offers a multidisciplinary approach to understanding the intricate dynamics of power grids and their vulnerabilities. It focuses on three critical aspects of robustness: cascading failure modeling, robustness assessment, and strategies for enhancing grid resilience. Each of these topics is explored in depth, with an emphasis on practical applicability and theoretical rigor. By bridging the gap between theoretical insights and real-world challenges, this book provides a roadmap for developing solutions that ensure the reliability and stability of power grids in the face of evolving threats.

The evolution of this book was also inspired by the increasing attention to the broader societal impacts of power grid failures. Recent major disruptions, such as prolonged outages caused by hurricanes, wildfires, and cyberattacks, have highlighted the cascading consequences of power grid instability on public safety, economic activity, and national security. These events serve as stark reminders of the critical need to design grids that can withstand, adapt to, and recover from a wide range of disruptions. Our goal is to provide a comprehensive resource that not only equips readers with a deep understanding of these challenges but also empowers them with actionable tools to address them.

This book is structured around four major topics that form the core of its content:

1. The initial three chapters provide an overview of power network representation, power flow models, and essential protection strategies. These fundamentals serve to present the state of the art in power system analysis and control and establish the groundwork for understanding various issues associated with the robustness of emerging power grids to be presented in the subsequent chapters.

2. In Chapter 4, the methodology for *cascading failure modeling for power grids* is presented in detail using complex network theory. This methodology is based on the abstraction of the power grid as a network of interconnected nodes and links, enabling the simulation of failure propagation triggered by initial events. The methodology integrates stochastic methods and circuit-based power flow analysis to capture the partially deterministic, time-dependent, yet uncertain nature of cascading failures.

3. In Chapter 5, we introduce a suite of novel *metrics for evaluating power grid robustness*. These metrics go beyond traditional structure-based indices by incorporating electrical variables, network dynamics, and learning-based techniques. Through computer simulation of cascading failures, these metrics provide actionable insights into the grid's ability to withstand major disruptions, under the inevitable condition of the ever-increasing level of integration of power electronics equipment and renewable energy sources.

4. Lastly, in Chapters 6 and 7, we outline the *strategies for improving power grid resilience* in both the system planning and operational phases. Topics include critical component identification, network connection optimization, cyber-physical system integration, and AI-enabled mitigation strategies. Practical examples and case

studies are used to illustrate the implementation of these strategies in real-world scenarios. Chapter 8 concludes the book with a forward-looking perspective on power system robustness, emphasizing the complementary roles of researchers, industry practitioners, and policymakers in advancing resilient, intelligent, and sustainable power grids.

The target audience for this book includes researchers, engineers, and professionals in the fields of power systems, electrical engineering, and complex networks. Policymakers and industry practitioners will also find valuable insights into the challenges and opportunities associated with modern power grid operations. Additionally, graduate students and academics seeking to deepen their understanding of power grid robustness will benefit from this book's comprehensive analysis and rigorous research foundation. By addressing the needs of such a diverse audience, we aim to foster collaboration across disciplines and sectors to tackle the pressing challenges faced by modern power grids.

Readers can expect to gain a broad and nuanced understanding of the technical, operational, and strategic aspects of power grid robustness. By exploring cascading failure models, evaluation metrics, and mitigation strategies, this book equips readers with the tools and knowledge needed to address the multifaceted challenges of modern power grids. The incorporation of cutting-edge research, industry practices, and real-world case studies ensures that the content remains relevant and applicable to current and future developments in the field.

This book strives to strike a balance between depth of analysis and accessibility. Complex concepts are broken down into manageable components, and the use of examples, illustrations, and diagrams enhances clarity and comprehension. By bridging theory and practice, this book provides readers with a holistic perspective on power grid robustness, empowering them to develop innovative solutions to ensure the reliability and stability of critical energy systems. In an era where resilient energy systems are more crucial than ever, it is our hope that this book serves as a valuable resource for advancing the understanding and implementation of robust power grid technologies.

We would like to extend our heartfelt gratitude to Professor Ron Chen, Director of the Center for Complexity and Complex Systems at City University of Hong Kong, for introducing us to the fascinating field of complex networks. His valuable inspiration and continuous encouragement have been instrumental throughout our journey of applying complex network theory to power system analysis. We deeply appreciate his guidance and support.

We sincerely thank all those researchers, practitioners, and policymakers whose work has inspired and shaped this endeavor. It is through collective efforts and interdisciplinary collaboration that we can build power grids capable of meeting the challenges of the future. In particular, we have been fortunate to draw inspiration from members of our research groups in Hong Kong and Beijing, especially our former and current students and postdocs: Julian Fang, Meng Huang, Biwei Li, Jade Li, Shuo Li, Zhihai Rong, Jiajing Wu, Xingtang Wu, Zhenxi Wu, Yongxiang Xia, Jingxi Yang, Katie Yang, and Choujun Zhan. Their valuable contributions to previous related projects have laid the foundation for many of the ideas and methodologies presented in this work. Their unwavering support and dedication have been instrumental in bringing this book to fruition, and we are deeply grateful for their commitment and collaboration.

Finally, we hope that this book serves as a valuable guide for researchers, engineers, and professionals in the fields of power systems, electrical engineering, and complex networks to navigate the complexities of modern power grids. By addressing the technical, operational, and strategic challenges of power grid robustness, we aim to inspire multidisciplinary innovation and collaboration in building energy systems that are resilient, efficient, and capable of meeting the demands of the 21st century.

*Dong Liu*
*Xi Zhang*
*Chi Kong Tse*

# *About the Authors*

**Dong Liu** earned a B.Eng. (Hons) degree in Electronic Engineering with first-class honors from The Hong Kong Polytechnic University in 2014; a B.Eng. in Microelectronics from Sun Yat-sen University in Guangzhou, China, in 2014; and a Ph.D. in Electrical Engineering from the Hong Kong Polytechnic University in 2019. He previously held a postdoctoral position at the Institute of Textile and Clothing at the Hong Kong Polytechnic University and currently serves as a research fellow at the Centre for Complexity and Complex Networks and the Department of Electrical Engineering at City University of Hong Kong. He is currently a full member of the IEEE Power and Energy Circuits and Systems Technical Committee and a member of the IEEE Power and Energy Society. He is the author or co-author of 14 publications in peer-reviewed journals, including various *IEEE Transactions*, and has delivered over 10 invited presentations at international conferences. His research interests encompass cascading failure modeling and analysis, cyber-physical systems, and machine learning for assessing robustness in smart grids.

**Xi Zhang** earned a B.Eng. degree in Automation from Beijing Jiaotong University, Beijing, China, in 2013 and a Ph.D. in Electronic and Information Engineering from the Hong Kong Polytechnic University, Hong Kong, China, in 2017. He is currently Assistant Professor at the School of Automation, Beijing Institute of Technology, Beijing, China. His research interests focus on power system resilience, applications of AI technologies to the analysis and decision-making in power networks, etc. He served as Associate Editor of *IEEE Transactions on Circuits and Systems-II: Express Briefs* (2022–2023) and Guest Editor of *IEEE Journal on Emerging and Selected Topics in Circuits and Systems* (Q3 2023), where he received the Best Associate Editor Award.

**Chi Kong Tse** earned B.Eng. (Hons) and Ph.D. degrees in Electrical Engineering from the University of Melbourne. He is currently a Chair Professor of Electrical Engineering at City University of Hong Kong and is concurrently being appointed as Associate Vice President (Innovation). He is/was an honorary professor and distinguished fellow with several Australian, Canadian, and Chinese universities, including Melbourne University, RMIT University, University of Western Australia, University of Calgary, Huazhong University of Science and Technology (where he is Chang Jiang Scholars Chair), Northeastern University, among others. His research interests are in power electronics and nonlinear systems. He has been awarded numerous invention, research, and teaching prizes. Among them, the IEEE CASS Charles A. Desoer Technical Achievement Award 2022 recognized his "outstanding contributions and continued leadership in the development of research in complex behavior of power electronics and energy systems", where he was the Editor-in-Chief of the *IEEE Transactions on Circuits and Systems II* (2016–2019) and of the *IEEE Circuits and Systems Magazine* (2013–2016). He was elected an IEEE Fellow in 2005 for his contributions to power electronics systems.

# *Symbols*

| | |
|---|---|
| $\delta$ | Rotor angle (in context of power system) |
| $P_e$ | Electrical power in power generation |
| $P_m$ | Mechanical power provided by turbine |
| $H$ | Normalized inertia |
| $S_{\text{base}}$ | Machine rating |
| $V_i$ | Voltage of node $i$ |
| $B_{ik}$ | Susceptance of line $(i, k)$ |
| $G_{ik}$ | Conductance of line $(i, k)$ |
| $P_{ik}$ | Active power injected to node $i$ from node $k$ |
| $Q_{ik}$ | Reactive power injected to node $i$ from node $k$ |
| $T_{GTi}$ | Turbine-governor time constant for node $i$ |
| $R_i$ | Droop coefficient for node $i$ |
| $\boldsymbol{\theta}$ | Voltage phase vector |
| $\boldsymbol{P}$ | Injected power vector |
| $P_{G_i}$ | Power generated by generator $i$ |
| $P_{\text{demand}}$ | Power demanded in the whole system |
| $P_{\text{line},k}$ | Power flow in line $k$ |
| $P_{\text{limit},k}$ | Maximum power flow permitted in line $k$ |
| $V_{\text{bus},m}$ | Voltage of bus $m$ |
| $V_{\text{max}}$ | Maximum bus voltage |
| $V_{\text{min}}$ | Minimum bus voltage |
| $P_{\text{reserve}}$ | Total reserve power |
| $\langle k \rangle$ | Average node degree |
| $P(k)$ | Node degree distribution |
| $r$ | Degree-degree correlation |
| $L$ | Shortest path |
| $B_i$ | Betweenness centrality of node $i$ |
| $\Gamma_i$ | Clustering coefficient of node $i$ |
| $\Gamma$ | Average clustering coefficient |
| $R_{\text{e}}$ | Power grid resilience metric |
| $P(s)$ | Occurrence probability of failure set $s$ caused by a natural disaster |
| $E_{\text{loss}}(s)$ | Power supply loss associated with failure set $s$ |
| $P(s\|\eta)$ | Occurrence probability of failure set $s$ under a disaster mode |
| $P(\eta)$ | Probability of disaster mode $\eta$ |
| $E_{\text{loss}}(s)$ | Power supply loss |
| $E_{\text{initial}}$ | Initial amount of power supply |

| | |
|---|---|
| $E_{\text{remaining}}(s)$ | Remaining maximum amount of power supply of the power grid after the failure set $s$ occurred |
| $E_t(s)$ | Maximum power supply provided by the grid at time $t$ |
| $R_{\text{critical}}(i)$ | Component resilience-criticality metric |
| $\hat{\text{R}}\text{e}$ | Average value of $E_{loss}$ of all generated samples by Monte Carlo method |
| $\hat{R}_{\text{critical}}(i)$ | Estimated value of resilience-criticality metric of component $i$ |
| SSE | Sum of squared errors |
| $\omega$ | Family of failure sets that may occur in the power grid |
| $s$ | Initial damage event which is a failure set within this family |
| $\eta$ | Disaster mode, including the specific disaster type, level, location, and other characteristics (in context of damage model) |
| $m_\eta$ | Range of $\eta$ |
| $p_{i,\eta}$ | Damage probability of component $i$ under a disaster mode |
| $p_{j,\eta}$ | Damage probability of component $j$ under a disaster mode |
| $p_{k,\eta}$ | Damage probability of component $k$ under a disaster mode,<br>$k$ is one of the damaged components except for components $i$ |
| $E_i$ | Active power injected into the power grid from node $i$ |
| $G$ | Set of all component states under a certain failure condition |
| $\omega_i$ | Weight of load node $i$, representing the importance level of the load |
| $E_{ij}$ | Active power flow across link $(i, j)$ that connects node $i$ and node $j$ |
| $a_{ij}$ | Derivative of the edge connecting node $i$ and node $j$ |
| $\mathbf{A}$ | Vector of $a_{ij}$ |
| $\theta_i$ | Phase angle of node $i$ |
| $\theta_j$ | Phase angle of node $j$ |
| $\theta$ | Vector of $\theta_i$ |
| $C_{ij}$ | Power flow limit of the edge passing through node $i$ and node $j$ |
| $C_i$ | Power limit of node $i$ |
| $N_d$ | Set of load nodes |
| $N_g$ | Set of generator nodes |
| $\chi_i$ | Random number |
| $X_i$ | State of component $i$ |
| $\Omega_a$ | Set of random samples of damage events |
| $\|\Omega_a\|$ | Number of elements of $\Omega$ |
| $\Omega_{ai}$ | Set of all generated samples of damage events that contain component $i$ |

| | |
|---|---|
| $\|\Omega_{ai}\|$ | Number of elements in $\omega_{ai}$ |
| $C(i)$ | Cluster $i$ with several resilience-criticality metric |
| po | A resilience-criticality metric in $C(i)$ |
| $m_i$ | Centroid of $C(i)$ |
| $\epsilon$ | Average propagation latency |
| $\rho$ | Relative coupling correlation coefficient |
| $\Delta f_{\mathrm{pp}}$ | Peak-to-peak frequency deviation |
| $t_{\mathrm{att}}$ | Attack initiation time |
| $w_{\mathrm{PEG}}$ | Vulnerability factor |
| $r_{\mathrm{RE}}$ | Integration level of renewable energy source |
| $r_{\mathrm{outage}}$ | Power outage rate (proportion of generation loss) |
| ACE | Area control error |

# 1

# *Introduction*

Modern electrical power systems are vital infrastructures that support our economy and everyday lives. Yet, history has consistently demonstrated that even the most advanced grids remain vulnerable to catastrophic failures. From the Northeast U.S. blackouts in 1965 and 2003 to India's massive outage in 2012, Texas in 2021, and the Iberian Peninsula in 2025, localized issues have frequently escalated into cascading failures, triggering widespread blackouts. These events result in billions of dollars in economic losses, disrupt societies, and put lives at risk. Such incidents highlight the urgent need for grid robustness while continuing to maintain essential functionality [212]. Here, the term *robustness* refers to the ability to withstand, absorb, and adapt to disruptions such as equipment failures, extreme weather, cyberattacks, or fluctuations in renewable energy output. Our understanding of robustness has evolved dramatically over time. For decades, grid planners relied on static, deterministic checks to guard against single-point failures, such as the traditional "$N-1$ criterion". However, recent research has shifted toward dynamic simulations and risk-based assessments of cascading failures, providing a more comprehensive approach to capturing the complex chain reactions that lead to blackouts [251]. We are now transitioning toward a more comprehensive framework, one that accounts for multiple uncertainties (such as variable renewable sources, fluctuating loads, and random faults), cyber-physical interdependencies (particularly emerging cyber threats), and the distinct characteristics of modern grids, including low inertia, power-electronics dominance, distributed energy resources, and climate extremes. This evolution, from simplistic "single-failure" rules to integrated, uncertainty-aware modeling, reflects both a deeper understanding of robustness and an urgent response to the rapidly shifting realities of electric power systems. In the sections that follow, we delve into how advanced modeling and analytical techniques can ensure that future power grids remain resilient in the face of escalating challenges. This chapter examines the role of power systems and the lessons learned from major blackouts; defines and compares methods for assessing robustness; traces the grid's evolution under high renewable integration, interdependencies, and AI integration; and highlights critical robustness challenges, including low inertia, network topology, cybersecurity, and extreme weather.

DOI: 10.1201/9781003622130-1

## 1.1 Background

The secure and reliable operation of power systems is essential, as they serve as critical infrastructures supporting modern economies and societies. As a result, evaluating and mitigating the risks of power grid failures has long been a key focus for both academic research and industrial practice. The profound impact of such failures is starkly illustrated by historical large-scale blackouts, which underscore the diverse and complex risks confronting power systems. These risks encompass physical threats (such as equipment failures and natural disasters), operational challenges (including cascading failures and stability loss), and emerging risks (like cybersecurity vulnerabilities and the intermittency of renewable energy sources). In response to these significant challenges, major economies around the world have developed and continuously refined strategies to enhance system robustness and minimize the risk of failures.

### 1.1.1 Major Blackout Events

Historical records reveal that since the 21st century, large-scale blackouts affecting regions with populations exceeding 50 million have occurred, as depicted in Figure 1.1. These events highlight that, despite the continuous advancement of power systems, the risk of major blackouts remains significant. A closer examination of these blackouts shows that the risks faced by power systems can be broadly categorized as follows:

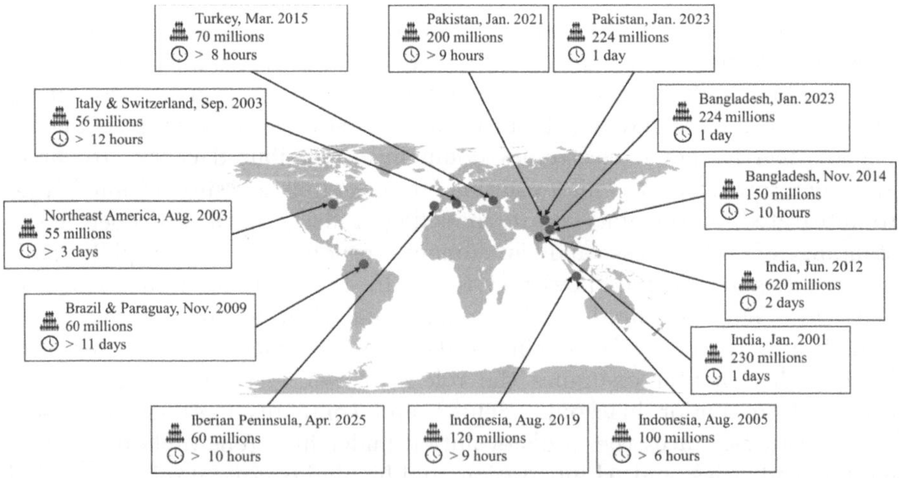


**FIGURE 1.1**
Large-scale blackouts impacting over 50 million people since the 21st century.

1. Physical Threats:
   (a) Equipment failure: Aging or sudden malfunctioning of generators, transformers, transmission lines, and/or other critical components (potentially causing localized outages or cascading failures).
   (b) Natural disasters: Equipment damage caused by typhoons, earthquakes, or ice storms (e.g., tsunami-induced damage contributing to the 2011 Fukushima nuclear accident in Japan).
   (c) Extreme weather events: Supply disruptions due to extreme cold; surging loads triggered by severe heatwaves (e.g., the 2021 Texas power crisis during a winter storm; load spikes during heatwaves).
2. Operational Threats:
   (a) Cascading failures: A single failure triggering power flow redistribution, leading to line/transformer overloads, protective tripping, and system separation/islanding (e.g., 2003 Northeast US-Canada blackout affecting 50 million people).
   (b) Stability violations: System collapse due to voltage instability (e.g., Tokyo blackout, 1987) or frequency instability (e.g., nationwide blackout in Turkey, 2015).
   (c) Renewable energy variability: Increased frequency regulation burden caused by the inherent stochasticity of increased renewable generation (e.g., intra-day power fluctuations exceeding 30 GW in Germany, 2020).
3. Emerging Threats:
   (a) Cyberattacks: Malicious activities targeting critical infrastructure control systems, such as False Data Injection (FDI) attacks against SCADA systems (e.g., the 2015 cyberattack on Ukraine's power grid).
   (b) Market mechanism risks: Unplanned withdrawal of generation resources driven by extreme electricity price volatility (e.g., contributing to the California energy crisis, 2000-2001).

### 1.1.2 Reducing Blackout Risk

Major economies have developed multi-layered approaches to reduce blackout risk. In North America, mandatory reliability standards are bolstered by advanced modeling techniques and emergency procedures that are designed to anticipate and isolate failures [220]. The region adheres to a rigorous North American Electric Reliability Corporation (NERC) framework, which has evolved from static "$N-1$ criteria" into a comprehensive set of binding standards encompassing both cyber-physical security and enforceable compliance penalties. Beyond these standards, NERC's long-term strategy emphasizes the use of advanced risk analytics and dynamic modeling, exemplified by initiatives like the North American Energy Resilience Model. These efforts aim to predict cascading failures and optimize mitigation measures, including black-start planning and enhanced transmission capacity [243]. Additionally, regulators and operators implement detailed emergency procedures, such as mandatory training on emergency operations, system restoration, and coordination across balancing authorities, to ensure rapid and controlled responses during extreme events.

China is making significant strides in enhancing the resilience of its power grid through substantial investments in real-time stability monitoring, demand-response programs, provincial pilot market reforms, and the expansion of ultra-high-voltage grids [236]. By modernizing its power infrastructure with advanced real-time monitoring technologies, China is able to detect instabilities early, allowing for preemptive interventions to avert cascading disturbances [190]. Complementing these technological advancements, the National Development and Reform Commission's 2024–2027 plan prioritizes expanding demand-response capacity to 5%–10% of peak load, upgrading flexible transmission and distribution systems, and integrating energy storage solutions to address the variability of renewable energy sources. Additionally, provincial pilot programs under recent market reforms are exploring innovative approaches, such as competitive dispatch, renewable energy trading, and digital substation deployments. These initiatives are developing scalable models that inform national policies, enhance resilience, and promote the diversification of resources and market mechanisms.

Europe has implemented the EU Risk-Preparedness Regulation, which encompasses national and regional crisis plans, coordinated outage planning, and short-term adequacy assessments, along with pan-European flexibility and stability roadmaps to ensure cross-border energy support [97]. Specifically, the EU's 2019/941 Risk-Preparedness Regulation mandates that all member states develop comprehensive national and regional crisis-management plans. These plans must outline preventive measures, mutual assistance mechanisms, and transparency protocols for cross-border support during electricity emergencies [90]. The European Network of Transmission System Operators for Electricity (ENTSO-E) enhances this framework through its outage planning coordination (OPC) and short-term adequacy (STA) processes. These initiatives harmonize maintenance schedules and assess supply adequacy across transmission system operators (TSOs), thereby minimizing the risks of simultaneous outages and ensuring the sharing of reserves where necessary [97]. Furthermore, strategic studies on system flexibility and inertia management provide a roadmap for integrating high shares of renewable energy. Recommendations include the installation of grid-forming inverters, the deployment of fast frequency response services, and the enhanced utilization of hydropower for rapid balancing during disturbances [87].

The Australian Energy Market Operator (AEMO) employs a range of strategies to navigate its energy transition, including System Restart and Local Black Start Procedures, minimum-system-load management, adaptations to under-frequency load shedding, off-market reserve procurement, and the rapid deployment of energy storage solutions such as pumped hydro and batteries. Specifically, AEMO's System Restart Plan, supported by Local Black Start Procedures, formalizes the procurement of black-start services and outlines system restoration pathways to re-energize regions following total shutdowns. To address low-demand risks stemming from high rooftop solar output, AEMO issues minimum system load notifications and can activate solar-curtailment programs. Additionally, under-frequency load shedding schemes have been adapted to incorporate distributed energy resources [18]. In anticipation of the gap left by retiring thermal plants, AEMO has tendered off-market reliability reserves and advocated for the accelerated deployment of batteries, pumped hydro, and fast-start gas units. In certain instances, the closure of coal plants has been deferred to maintain inertia and firm capacity during this transitional period [20].

## 1.2 Robustness Assessment in Power Grids

In this section, we present the fundamental principles, dual-dimensional framework, and core evaluation tasks essential for assessing power system robustness. First, we define robustness as the grid's ability to withstand both random disturbances and deliberate attacks, aiming to prevent the escalation of single faults into large-scale blackouts while maintaining critical functions such as voltage and frequency stability. We then break down robustness into structural and operational dimensions, illustrating how topology hardening and dynamic control collaborate to create a comprehensive static-dynamic defense. Finally, we explain how quantitative assessments, tailored to both planning and real-time operations, drive targeted improvements, effectively closing the loop between evaluation and optimization.

### 1.2.1 Power Grid Robustness

Power system robustness is defined as the ability of the electricity network to withstand both random perturbations, such as equipment failures, load fluctuations, and uncertainties in renewable generation, and intentional attacks, including cyber intrusions and physical destruction. The primary objective is to prevent a single fault from escalating into a widespread outage. Achieving this level of robustness involves inhibiting the propagation of successive failures and ensuring that, even under disturbance conditions, the grid continues to maintain essential services. This is particularly important for preserving voltage and frequency stability, thereby preventing local collapses from triggering global instability.

### 1.2.2 Network and Operational Robustness

*Network robustness* focuses on the grid's physical topology and its inherent tolerance to component loss. Enhancements include reinforcing critical nodes (e.g., major substations), adding redundant transmission paths, and adopting zonal partitioning to contain faults. Studies have also linked network metrics—such as average node degree and community structure—to robustness against desynchronization phenomena.

*Operational robustness* relies on dynamic control strategies designed to respond effectively to unfolding disturbances. Preventive controls, such as security-constrained dispatch and reserve allocation, are employed to keep the system within safe operating limits. In contrast, emergency controls, including under-frequency load shedding, rapid generator regulation, and intentional islanding, are activated post-contingency to quickly halt cascading failures and restore stability. Optimal control frameworks enhance these strategies by identifying the most effective hardening and response measures within dynamic constraints. Together, a synergistic defense is created, combining static hardening with dynamic agility, to form the foundation of a robust power system.

### 1.2.3 Robustness Assessment

The primary goal of robustness assessment is to establish quantitative frameworks applicable to both planning and operational contexts. During the planning stage, evaluations focus on measuring the static, end-state impacts of cascading failures, such as total load shed and the percentage of isolated regions. This helps identify topological weak points and informs network reinforcement strategies. In real-time operational analysis, the focus shifts to tracking the dynamic evolution of disturbances, including fault-propagation pathways, the decay rates of security margins, and the effectiveness of control actions. This enables operators to intervene promptly and interrupt cascading sequences. In practice, insights from both long-term planning studies and real-time monitoring feed directly into network design and operational controls. Through iterative cycles of performance evaluation, targeted infrastructure upgrades, policy adjustments, and reassessment, the system continuously learns and evolves, enhancing its capacity to absorb disturbances and prevent cascading failures.

### 1.2.4 Robustness, Resilience, and Reliability

Table 1.1 gives a concise comparison of robustness, resilience, and reliability. Robustness measures the system's ability to withstand disturbances prior to

**TABLE 1.1**
Comparison of Robustness, Reliability, and Resilience

| | **Robustness** | **Resilience** | **Reliability** |
|---|---|---|---|
| Time scale | Seconds ∼ Hours | Hours ∼ Days | Months ∼ Years |
| Core objective | Preventing fault escalation | End-to-end disturbance management | Avoiding random, low-probability failures |
| Intentional | Cascading failure | Extreme events | Random failures |
| Typical measurement | Load loss ratio | Recovery time | Duration of outages |

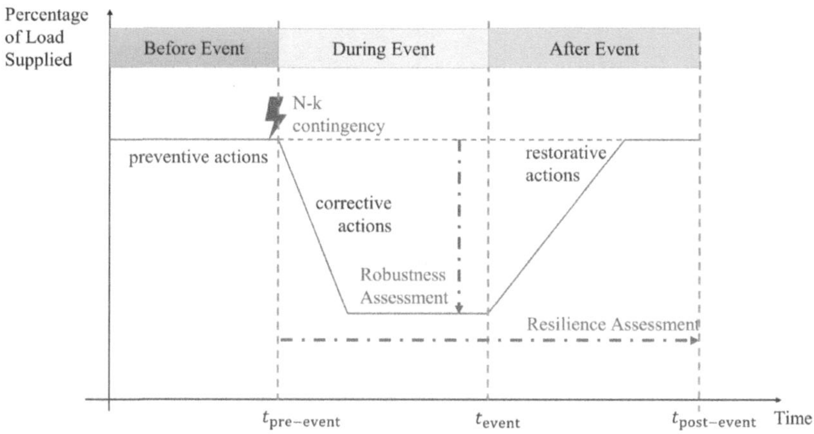


**FIGURE 1.2**
Robustness and resilience in a blackout event.

recovery, often quantified by the proportion of load lost before restoration. In contrast, resilience encompasses the entire "absorb-adapt-recover" cycle and is typically characterized by the time required to restore service following an event. Figure 1.2 shows the difference between the attainment of robustness and resilience in a blackout event. Reliability assesses the probability of uninterrupted service over extended observation periods and is quantified by indices such as SAIDI (System Average Interruption Duration Index) and SAIFI (System Average Interruption Frequency Index), which measure the frequency and duration of outages. It's important to note that vulnerability, being the inverse of robustness, indicates the sensitivity of system functions to specific threats. In subsequent chapters, we will use robustness and vulnerability interchangeably when discussing mitigation measures, as enhancing robustness inherently reduces vulnerability to targeted threats.

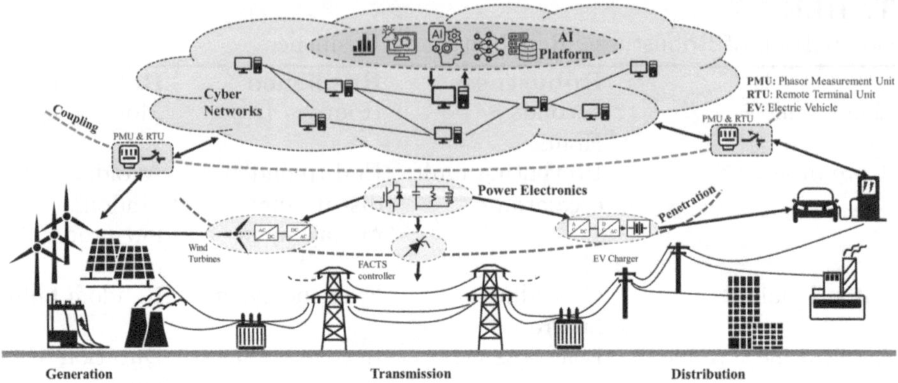


**FIGURE 1.3**
Evolving power grid.

## 1.3 Evolution of Power Grids

As illustrated in Figure 1.3, the power system is undergoing a historic transformation from a traditional electromechanical network into a power-electronics-driven, intelligent energy ecosystem enhanced by deep AI integration. This evolution, characterized by the widespread deployment of power-electronic interfaces and a tight cyber-physical convergence, positions the grid as a smart energy hub capable of self-sensing, autonomous decision-making, and self-healing. These capabilities prepare the grid to meet the demands of the carbon-neutral era, which include the increased use of renewable energy and the integration of diverse and pervasive loads.

### 1.3.1 Power-Electronics-Dominated Power Grids

Power electronics now permeate every layer of the power grid, from bulk generation and transmission to distribution and end-use, fundamentally reshaping its physical foundation:

1. Full-scale integration of power electronic converters: Today, nearly all new power generators are connected through inverter-based resources, such as photovoltaic inverters, wind turbine converters, Voltage Source Converter - High Voltage Direct Current (VSC-HVDC) links, as well as electric vehicle (EV) chargers and variable-frequency drives. The National Renewable Energy Laboratory (NREL) projects that wind and solar will contribute 60%–80% of the generation in a least-cost energy mix by 2035, highlighting the

widespread adoption of power electronics across the system, well beyond 60% of capacity [177].

2. Paradigm shift to low-inertia flexibility: The traditional grid, characterized by inertia-rich synchronous machines, has transitioned to a "low-inertia + multi-timescale-coupled" regime [143], introducing new stability challenges such as harmonic resonance and voltage flicker. This shift facilitates tighter, bidirectional interactions among energy sources, the network, loads, and storage systems. Advanced studies indicate that grid-forming converters, like virtual synchronous generators, can emulate inertia and support frequency regulation. However, their rapid dynamics necessitate careful stability design to ensure seamless integration.
3. Device-level reliability breakthroughs: Wide-bandgap semiconductors with higher thermal conductivity, larger breakdown fields, and lower switching losses, especially silicon carbide (SiC), are used in next-generation converters, providing dramatic gains in efficiency and reliability for high-power applications [35]. Alongside these hardware advances, intelligent "smart inverters" provide grid services (e.g., voltage and frequency regulation, synthetic inertia, ride-through, and anti-islanding) and embed resilience directly within power-electronic interfaces.

### 1.3.2 Cyber-Coupled Power Networks

The integration of advanced sensing, decentralized control, and digital platforms has created a highly interconnected cyber-physical power system. Ultra-fast, high-resolution measurement networks support collaborative control schemes that span cloud, edge, and device levels. Meanwhile, real-time digital twins and blockchain governance establish an intelligent agent framework capable of sensing, decision-making, and self-healing.

Wide-area deployment of phasor measurement units (PMUs) now exceeds 2,500 sites across North American bulk power systems, capturing synchronized measurements at 30–120 samples per second compared to the Supervisory Control and Data Acquisition (SCADA)'s 2–4 s scan rate, which is over 100× finer resolution for dynamic state estimation [34]. Globally, increasingly more electricity smart meters have been installed, enabling near real-time load monitoring and two-way communication on distribution feeders. Emerging quantum sensors are also being piloted to provide sub-microsecond, point-level measurements, promising unprecedented precision for grid resilience and control applications [59].

Traditional centralized, hierarchical control systems are being replaced by cloud–edge–endpoint collaboration, where microgrid clusters employ consensus-based algorithms to achieve autonomous, multi-objective dispatch cycles in as little as 100 ms. Simultaneously, drone swarms equipped with high-resolution cameras and LiDAR are conducting routine inspections of transmission and distribution assets, significantly reducing inspection times and enabling predictive maintenance through real-time anomaly detection [71].

### 1.3.3 Artificial Intelligence-Enabled Power Systems

Artificial intelligence (AI) is driving a transformative leap in power system capabilities by overcoming traditional computational limitations and seamlessly integrating advanced AI techniques into control and forecasting processes. This integration paves the way for a future where every power-electronic device functions as an autonomous agent. AI is enabling ultra-fast simulations, highly accurate forecasting, millisecond-scale controls, and autonomous device operation, thereby establishing the foundation for a grid that is self-sensing, self-deciding, and self-healing.

Surrogate models accelerate complex power system simulations by orders of magnitude. Recent hydropower turbine optimization studies using GAN-based surrogates reported up to a thousand-fold reduction in computation time. Trained deep surrogate models can serve as high-fidelity replacements for physics-based dynamic simulators, achieving comparable accuracy while reducing run-times from minutes to milliseconds. Meanwhile, online deep reinforcement learning controllers have demonstrated end-to-end control loop execution within single-digit milliseconds, far surpassing traditional SCADA cycle times [125].

Federated learning frameworks enable cross-regional model collaboration without sharing raw grid data. For example, provincial peak-shaving prediction models are trained locally and aggregated centrally, preserving data privacy while improving forecast performance. Multimodal foundation models unify the interpretation of textual regulations, time-series sensor streams, and high-resolution equipment imagery, facilitating automated grid inspection workflows and protocol compliance checks. Embodied intelligent agents power next-generation inspection robots that leverage sub-micron optical sensors and advanced AI to detect defects invisible to the human eye, enabling continuous, high-precision asset monitoring [138].

Looking ahead, each power-electronic interface is envisioned as an autonomous "agent" within a distributed network framework. Systems such as multi-agent energy resources coordinate millions of devices to optimize system-wide objectives in real time. Within this paradigm, the grid transforms into a self-reconfiguring network capable of dynamically isolating faults, rerouting power flows, and restoring service without human intervention, ultimately realizing a resilient energy internet for a carbon-neutral future [244].

## 1.4 Challenges in Power Grid Robustness

Emerging trends in modern grids, characterized by a widespread adoption of power-electronic interfaces, shifting network topologies and generation patterns, heightened cyber-physical interdependencies, and more frequent extreme weather, are placing new strains on system robustness. The high integration of inverter-based resources reduces inertia and introduces resonance and control challenges that demand advanced converter controls and stability analyses. At the same time, the arrangement of lines, generators, and loads critically influences how faults propagate and whether islanding strategies can contain failures. Cyber vulnerabilities, from denial-of-service attacks on control networks to firmware exploits in smart inverters, now pose an existential risk to grid operations, as underscored by both real-world inverter shutdowns and evolving regulatory mandates. Finally, the intensification of extreme weather events, such as hurricanes, wildfires, and ice storms, stresses aging infrastructure and complicates restoration efforts, making robust hardening and adaptive response strategies more essential than ever.

### 1.4.1 Impact of Power Electronics Integration

Research has shown that the power blackouts in the U.K. in 2019 and in Australia in 2016 were closely linked to the widespread deployment of power electronics in their respective power grids. Evaluating the robustness of grids dominated by power electronics involves addressing three interrelated challenges: (1) capturing multi-timescale dynamics, ranging from microsecond converter transients to seasonal load variations; (2) developing comprehensive models that accurately represent synchronous machines, inverter-based resources, storage systems, controls, and network topology; and (3) defining and calculating robustness metrics, particularly power-loss probability distributions through cascading-failure simulations, to quantify system resilience.

#### 1.4.1.1 Key Challenge 1: Multi-Timescale Dynamics

Modern power grids exhibit complex, multi-timescale behavior by integrating fast converter transients (microseconds to milliseconds), conventional synchronous-machine electromechanical swings (tens to hundreds of milliseconds), cascading failure propagation (seconds to minutes), daily demand fluctuations (hours), and seasonal trends (months to years). This complexity challenges single-scale analysis. To maintain tractability, studies often separate these time scales, averaging faster dynamics and treating slower ones as quasi-static. However, it remains essential to account for cross-scale influences to ensure alignment with measurements and real-world responses [208].

#### 1.4.1.2 Key Challenge 2: Comprehensive System Modeling

A robust model must encompass the full range of components, including synchronous generators (modeled using swing equations), grid-connected converters (utilizing electromagnetic transient (EMT) or averaged models), battery energy storage systems, power-electronics-controlled loads, and power-flow regulators, along with their interactions. By partitioning the model according to time scales, such as fast transients versus slow demand dynamics, targeted analyses can be conducted. For example, examining local stability of converters at the millisecond scale, assessing cascading failure risks over seconds to minutes, and conducting seasonal planning. This approach maintains coherence across subsystems. Additionally, the model must incorporate topology flexibility, such as clustered AC subnetworks connected by VSC-HVDC lines, to accurately reflect real-world configurations (e.g., China's interconnected regional clusters) and facilitate "what-if" studies under various converter integration scenarios.

#### 1.4.1.3 Key Challenge 3: Robustness Metrics and Simulation Platforms

Evaluating robustness requires a cascading-failure simulation environment that recalculates power flows and balances after each component's failure, capturing the probabilistic spread of disturbances. By running large ensembles of disturbance scenarios, one can derive the cumulative distribution of the total load shed, i.e., the probability that power loss stays below a threshold, which serves as a quantitative robustness indicator and reflects the grid's ability to maintain service under varying operating conditions. Comparing these distributions across different levels of converter integration then reveals how increasing power-electronics equipment affects the likelihood and severity of cascading events, guiding both planning and operational resilience strategies.

### 1.4.2 Impact of Topological and Dynamical Evolution

Modern power networks increasingly incorporate significant shares of inverter-based generation, widespread distributed energy resources, and active two-way demand-side participation. These developments complicate robustness assessment in several key ways: first, rapid fluctuations in renewable energy output and diverse geographical deployment patterns affect stability margins; second, real-time cascading failures necessitate the immediate identification of critical lines to protect and prevent failure propagation; and third, stochastic and strategic load behaviors, such as electric vehicle charging and price-responsive demand, challenge traditional deterministic models [202]. To address these challenges, assessment methods must integrate high-resolution temporal modeling, spatially explicit network representations, and probabilistic load-response analyses.

#### 1.4.2.1 Key Challenge 4: Quantifying the Impact of Generation Variability and Spatial Distribution

Traditional robustness metrics frequently fall short in addressing the dynamic characteristics of inverter-based generation. Unlike conventional synchronous generators, inverter-based resources are subject to rapid fluctuations in output, primarily driven by variable renewable energy sources [239]. These fluctuations, combined with the spatial distribution of generation—ranging from centralized wind farms to decentralized rooftop photovoltaics—can profoundly impact system stability and resilience. To effectively assess the robustness of these systems, it is essential to utilize models that incorporate both the variability and spatial diversity of generation sources. This approach allows for more accurate predictions of system behavior across various scenarios.

#### 1.4.2.2 Key Challenge 5: Dynamic Identification of Critical Transmission Lines

Identifying critical transmission lines is essential for preventing and mitigating cascading failures. Traditional static analyses, such as $N - k$ contingency assessments, often fall short in capturing the dynamic interactions between system components during fault conditions. Recent studies have introduced dynamic models that take into account the time-varying nature of power flows and the interdependencies among various system elements. For instance, employing dynamic interaction graphs can facilitate the real-time identification of vulnerable transmission lines, enabling timely interventions to avert widespread outages. Additionally, the significant increase in operating modes driven by renewable energy sources and flexible loads makes exhaustive contingency enumeration impractical, necessitating the use of approximation techniques that strike a balance between speed and accuracy.

#### 1.4.2.3 Key Challenge 6: Addressing Load Uncertainty and Demand Response Dynamics

The growing integration of demand-side resources, such as electric vehicles and smart appliances, introduces new uncertainties into power system operations. These resources can exhibit stochastic behaviors influenced by user preferences and external factors, such as pricing signals. Additionally, demand response programs elicit varied customer reactions to these price signals; while some consumers may choose to defer usage, others might suddenly increase their load, adding a layer of unpredictability akin to game theory dynamics. Traditional deterministic models often struggle to capture the complexities associated with demand response and its effects on system robustness. To address these challenges, advanced probabilistic and optimization-based approaches are essential for modeling the uncertain nature of demand-side resources and their interactions with the grid. By incorporating these considerations, system operators can better anticipate potential risks and develop strategies to enhance grid robustness [237].

### 1.4.3 Impact of Cyber Coupling

The transformation of the traditional grid into a fully digitalized cyber-physical system, where physical assets, Information and Communications Technology/Operational Technology (ICT/OT) platforms, data management layers, and applications are intricately interconnected, introduces new classes of vulnerabilities. These include hardware failures, IT/OT exploits, data integrity lapses, and service/application flaws, which legacy robustness metrics are ill-equipped to address. A notable example occurred on December 23, 2015, when coordinated cyberattacks on three Ukrainian distribution companies resulted in power outages affecting over 230,000 customers for durations ranging from 1 to 6 hours, with some estimates suggesting as many as 1.4 million users were impacted. This incident marked the first documented case of a cyberattack directly leading to a blackout. To effectively tackle these challenges, there is a pressing need for next-generation robustness frameworks that seamlessly integrate cyber-physical coupling, probabilistic threat propagation, and adaptive security dynamics. Such frameworks will ensure that digitalized power systems can be assessed and fortified against both traditional faults and sophisticated cyber-physical attacks [106].

#### 1.4.3.1 Key Challenge 7: Modeling Cyber-Physical Interdependencies

Smart grids are inherently interdependent networks that integrate electrical and communication infrastructures, with failure modes capable of propagating across both domains. Current robustness models often decouple these layers, treating cyber threats as isolated, static events and physical faults as independent random or malicious outages. This separation fails to account for the potential for failures to transfer between information nodes and power nodes, as well as the link-level propagation of issues. For instance, a malware injection into a substation Remote Terminal Unit (RTU) could disable relay logic and trigger a cascading failure [49]. Without a unified modeling framework that encompasses both coupling modes (cyber-to-power and power-to-cyber) and failure objects (nodes and links), it becomes impossible to accurately quantify real attack chains and their effects on overall grid stability.

#### 1.4.3.2 Key Challenge 8: Capturing Adaptive Attack Defense Interactions

Cyber-physical security in smart grids resembles a continuous arms race, where attackers seek new exploits while operators implement countermeasures. However, most analyses tend to assume static attacker objectives and fixed defense placements, treating attack entry points and hardened assets as unchanging. This approach overlooks the reality that adversaries adapt their strategies in response to defensive measures. Recent research in game-theoretic defense allocation highlights the importance of modeling this co-evolution; for instance, it emphasizes the need to allocate security resources to areas where attackers are likely to shift their targets once initial protections are established [49]. Additionally, multi-stage attack-defense simulations demonstrate how dynamic defenses, such as moving-target techniques and on-the-fly firewall reconfigurations, can effectively disrupt sophisticated intrusion chains. Without these adaptive models, risk trajectories remain unexamined, leaving operators without the insights necessary to fine-tune defenses against evolving threats.

### 1.4.4 Impact of Extreme Weather Events

Climate change has heightened the frequency and severity of extreme weather events, such as hurricanes, heavy rainfall, and ice storms, posing significant threats to grid stability and security. Power outages linked to severe weather have become increasingly common, occurring far more frequently than outages unrelated to such events. In the United States alone, 80% of major outages from 2000 to 2023 were directly associated with weather-related incidents [74]. These disasters cause direct physical damage, including downed power lines, collapsed poles, and failures of wind turbines and photovoltaic (PV) arrays, which disrupt supply and can trigger cascading failures in today's tightly coupled grids. Furthermore, structural shifts, such as high renewable energy integration leading to low system inertia, exacerbate instability risks under rapidly changing weather conditions. Traditional robustness analyses have yet to incorporate meteorological data into failure-mode modeling, resulting in an inadequate understanding of the disaster chain in which weather events lead to equipment damage, subsequently causing renewable variability and cascading failures. By developing high-resolution weather-to-failure models that capture the dual disturbance effects on renewable sources, model spatially coupled cascades, and reconcile defense resources across multiple timescales, future robustness frameworks can better equip modern power grids to confront the escalating challenges posed by extreme weather.

#### 1.4.4.1 Key Challenge 9: Weather-to-Failure Mapping Gaps

Most robustness models fall short in accurately translating meteorological data into equipment failure probabilities. While grid-scale weather models operate on kilometer-scale grids, the vulnerability of assets requires a much finer, meter-scale detail. Additionally, the damage mechanisms vary by asset type: for example, typhoon winds may buckle steel lattice towers, erode the tips of wind turbine blades, or crack photovoltaic (PV) glass [118]. Without data-driven mapping functions, such as deep learning models that link high-resolution weather forecasts to component failure rates, assessments would fail to capture the true risk chain from storm cells to system collapse. This gap in modeling hinders the ability to anticipate and mitigate the impacts of extreme weather on grid infrastructure effectively.

#### 1.4.4.2 Key Challenge 10: Dual Renewable-Disturbance Effects

Extreme weather events simultaneously lead to deficits in centralized generation and collapses in distributed output. For instance, wind farms often trip offline at cut-out speeds of approximately 20–25 m/s, resulting in significant bulk power shortfalls. Similarly, solar photovoltaic (PV) arrays can experience near-zero output during heavy cloud cover or hail events. These synchronous disturbances exacerbate power imbalances and strain reserve margins. However, current robustness metrics rarely account for the compounded effects of simultaneous failures in both centralized and distributed renewable sources [51]. This oversight limits the ability to effectively assess and enhance grid resilience in the face of extreme weather challenges.

#### 1.4.4.3 Key Challenge 11: Regional Cascade Amplification

Modern grids utilize long-distance, ultra-high-voltage AC/DC links to transport bulk power across regions. While these strong couplings enhance efficiency, they also have the potential to transform a localized weather-induced fault into a wide-area collapse. For example, damage in Region A can redirect substantial power flows into Region B, leading to overloaded lines and subsequent outages. Traditional $N - k$ studies often overlook this spatially propagated risk, as they typically analyze outages in isolation rather than considering them as part of a geographically spreading cascade [197]. This limitation hinders the ability to fully understand and mitigate the interconnected vulnerabilities within the grid.

## 1.5 Summary

In this introductory chapter, we review the major blackout events that occurred in the past few decades and briefly analyze their immediate causes. We introduce the concept of robustness of power grids and the various factors affecting it. Assessment methods for the robustness of power grids are also reviewed. The recent trends in the grid evolution are discussed in terms of the dominance of power electronics equipment, the coupling with cyber networks, and the use of AI methods. The key challenges in maintaining the robustness of power grids in view of these growing trends are discussed.

# 2

# *Power Network Representation and Power Flow Models*

Complex network theory provides a valuable framework for analyzing the robustness of power systems by representing the grid as a graph composed of nodes and edges. In this representation, generation plants and load centers are depicted as network nodes, while transmission and distribution lines are illustrated as edges. Topological metrics, such as node degree and betweenness centrality, can then be employed to identify critical vulnerabilities within cascading-failure scenarios. Hybrid approaches enhance purely topological models by integrating electrical engineering concepts, such as electrical betweenness and network efficiency, thereby establishing a more direct connection between graph properties and actual power flow behavior. From a dynamic perspective, any alteration in the network structure, such as the removal of a node or line, or variations in power injections, necessitates a complete redistribution of flows. Accurately computing these redistributions is essential for predicting system responses and potential cascades. DC power flow models, which linearize the full AC equations under small-angle assumptions, offer a fast and efficient method for estimating real power flows and are commonly used for initial robustness screenings. In contrast, AC power flow models account for voltage magnitudes and reactive power interactions, allowing for a more in-depth analysis of voltage stability and reactive margins-critical components for comprehensive robustness evaluations. In this chapter, we will illustrate how to represent power systems as complex networks and demonstrate the implementation of both DC and AC power flow models for assessing robustness and vulnerability. This foundational work will set the stage for further discussions on cascading failures and potential mitigation strategies.

## 2.1 Power Network Representation

The electric power system is a complex infrastructure made up of multiple interdependent components that work together to facilitate the generation, transmission, and distribution of electrical energy. At the generation stage,

DOI: 10.1201/9781003622130-2 

various primary energy sources are converted into electrical power through different types of generating units. Transmission facilities, including high-voltage transmission lines, transport this energy over long distances and across regions, enabling large-scale power exchanges. Transformers play a crucial role in adjusting voltage levels, either stepping up voltages for efficient long-distance transmission or stepping down voltages for safe delivery to end users. Distribution systems further reduce voltage levels to meet the specific requirements of residential, commercial, and industrial consumers, completing the energy delivery process. In this section, we formalize the abstraction of this physical power system into a complex network model, where equipment and substations are represented as network nodes, and physical connections, such as transmission and distribution lines, are modeled as edges. This network representation serves as a foundation for applying complex network theory, allowing for a systematic analysis and evaluation of the structural characteristics and robustness of modern power systems. By leveraging this framework, we can better understand the vulnerabilities and interdependencies within the grid, ultimately enhancing our ability to assess and improve its resilience.

### 2.1.1 Fundamentals

In real-world applications, a power grid serves as a critical infrastructure responsible for converting various forms of primary energy from nature into electrical energy and delivering it to locations where electricity is needed in human society. The core physical components of a power grid include the following:

- Generation units: Large-scale thermal power plants, hydroelectric stations, nuclear power plants, and other facilities that convert primary energy into electrical power.
- Substations: Facilities containing high-voltage equipment such as circuit breakers, disconnect switches, current/voltage transformers, busbars, power transformers, and associated protection and control devices.
- Transmission lines: Overhead conductors (typically three-phase) supported by towers or poles, responsible for long-distance, high-capacity power delivery.
- Loads: The aggregate of all electricity-consuming devices connected to substations (through step-down distribution transformers), including those in urban, industrial, commercial, and residential areas.

Figure 2.1 illustrates the geographical layout of the actual high-voltage transmission network in the New England region of the United States [201]. For power system simulations, such as load flow analysis, short-circuit calculations, and transient stability studies, the complex physical network must be abstracted into an equivalent circuit model, often referred to as the bus-branch

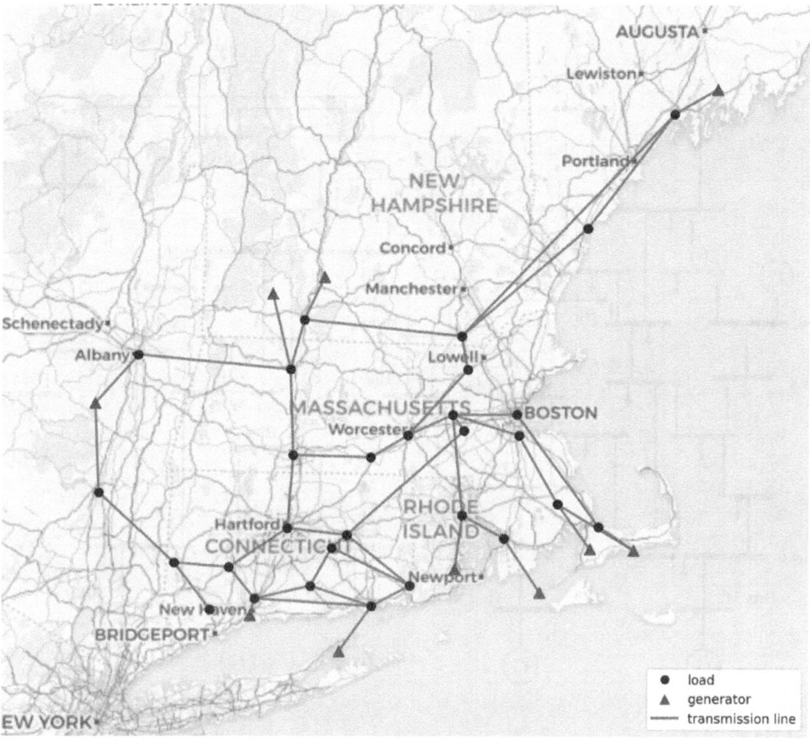


**FIGURE 2.1**
Geographical layout of the actual high-voltage transmission network in the New England region of the United States.

model. The IEEE 39 Bus system serves as a highly simplified representation of the actual New England high-voltage transmission network. In this model, each major substation high-voltage busbar, or its electrically equivalent connection point, is represented as a node (bus). The IEEE 39 Bus system consists of 39 such nodes, each labeled with a unique identifier (e.g., Bus 30, Bus 39), as shown in Figure 2.2. Transmission lines and transformers that connect two substations are modeled as branches, linking pairs of nodes. The IEEE 39 Bus system includes a total of 46 branches, comprising 34 transmission lines and 12 transformers. Additionally, generators, loads, and shunt compensation devices (such as capacitors or reactors) are connected to designated nodes within this framework. This abstraction allows for a more manageable analysis of the power system while retaining essential characteristics necessary for evaluating its performance and stability.

- Bus data: Bus number, type (slack/reference bus, PV bus, generator bus, PQ bus, load bus), nominal voltage (kV), active power injection

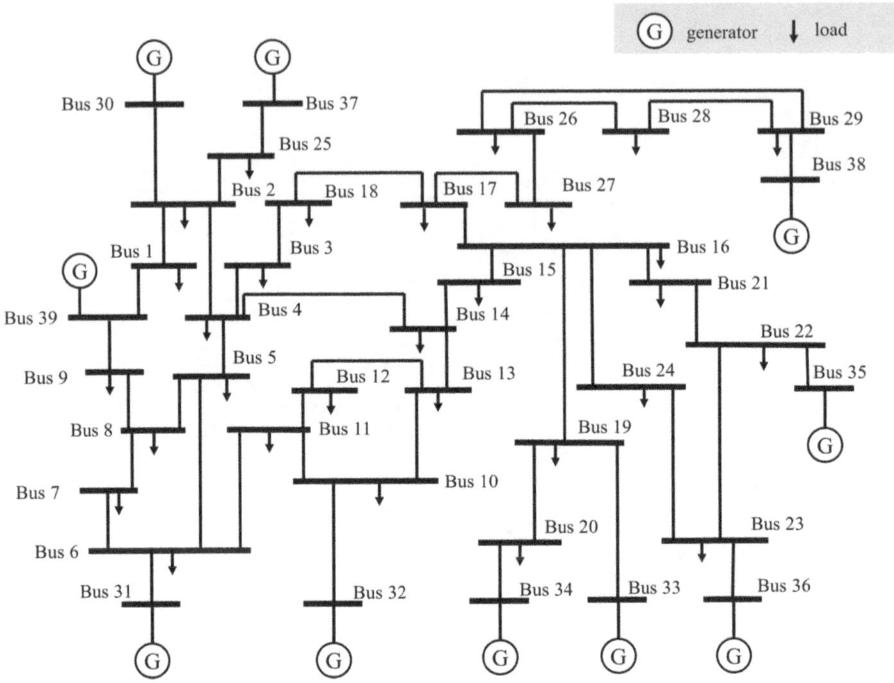


**FIGURE 2.2**
IEEE 39 Bus system, a simplified bus-branch model of the New England high-voltage transmission network. Each numbered node represents a major substation bus, while lines and transformers are represented as branches connecting the buses.

$P$(MW)/reactive power injection $Q$ (MVAR) (negative for loads, positive for generators, $P/Q$ of the slack bus to be solved), shunt conductance/susceptance.

- Branch data: From-bus, to-bus, resistance $R$ (p.u.), reactance $X$ (p.u.), line charging susceptance $B/2$ (p.u.) for transmission lines, transformer tap ratio, impedance, and tap position (if applicable).
- Generator data: Rated capacity (MVA), transient reactance $X'_d$, inertia constant $H$, excitation system parameters, and governor parameters.
- Load data: $P$, $Q$ demand (optionally including static characteristic coefficients).

For network structural analyses, such as connectivity studies, vulnerability assessment, network partitioning, and computation of topological metrics, the focus has shifted from detailed electrical parameters to the underlying connectivity structure. This leads to the construction of a pure topological graph

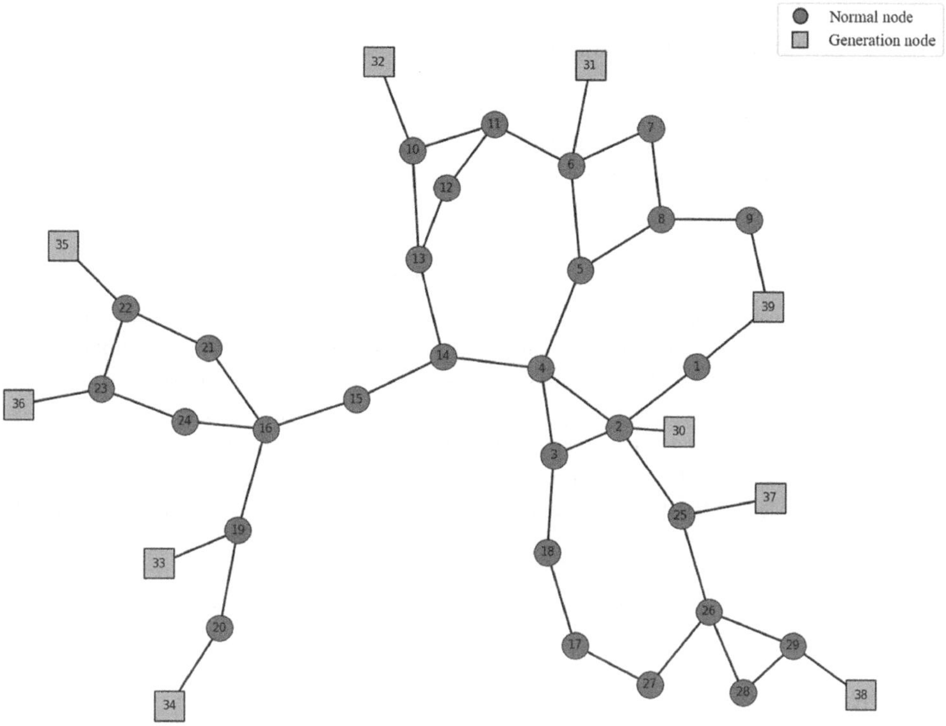


**FIGURE 2.3**
Pure topological graph representation of the IEEE 39 Bus system. All buses are retained as nodes, and physical connections (transmission lines and transformers) are represented as undirected edges, with electrical parameters omitted to emphasize network connectivity.

(Figure 2.3). The process involves two main steps: (1) all buses are retained as nodes in the graph, and (2) all physical connections, whether transmission lines or transformers, are represented as edges.

The key distinctions of the topological graph from the bus-branch model are:

1. Electrical parameters such as $R$, $X$, $B$, and transformer tap ratios are disregarded. Only the existence of a physical connection between two nodes is considered.

2. The transmission network is generally modeled as an undirected graph, as electrical power can flow in either direction under normal operating conditions. Directed graphs are used only in studies involving specific power flow directions or specialized equipment, such as back-to-back HVDC links.

3. Where multiple parallel circuits exist between two substations (a common feature in real-world networks, though less frequent in the IEEE 39 Bus system), they are typically represented as a single edge in the topological graph. This simplification reflects the fact that connectivity analysis concerns itself with whether two nodes are connected, rather than the number of parallel paths.

### 2.1.2 Power Grid: A Complex Network Perspective

Complex network theory has emerged as a widely utilized approach for modeling real-world systems, offering valuable insights into the relationships and interactions within these systems. The primary advantage of applying this theory lies in its ability to provide a robust analytical foundation for examining how network structure influences the functionality and behaviors of systems. This section will present the essential network topology measures, models of networked systems, and the concept of interdependent networks.

To explore the topological characteristics of complex networks, several important measures are commonly used, including node degree, degree distribution, shortest path length, and clustering coefficient.

#### 2.1.2.1 Node Degree

A network consists of nodes connected by edges. The *degree* is a fundamental metric that measures the topological importance of a node in the network. In an undirected network, where edges have no direction, the degree $k_i$ of node $i$ refers to the number of edges incident to it. In a directed network, the *out-degree* refers to the number of outgoing edges, and the *in-degree* refers to the number of incoming edges to a node. The degree is essential for analyzing the structure of real-world networks. Nodes with higher degrees, often referred to as *hubs*, are more critical in terms of network vulnerability. Degree-based attacks, which involve removing high-degree nodes, have been shown to significantly disrupt network connectivity, as removing highly connected nodes fragments the network more effectively than removing low-degree nodes.

#### 2.1.2.2 Average Node Degree

The *average node degree* $\langle k \rangle$ provides a statistical overview of the connection density in a network. It is the mean degree across all nodes, calculated as:

$$\langle k \rangle = \frac{1}{N} \sum_{i=1}^{N} k_i$$

where $N$ is the total number of nodes in the network. A higher $\langle k \rangle$ indicates a denser network with more connections among nodes.

#### 2.1.2.3 Node Degree Distribution

The *node degree distribution* $P(k)$ describes the probability that a randomly selected node has degree $k$. It can be expressed as:

$$P(k) = \frac{N(k)}{N}$$

where $N(k)$ is the number of nodes with degree $k$. Networks typically exhibit either a Poisson distribution or a power-law distribution. The Poisson distribution is characteristic of homogeneous networks, while the power-law distribution indicates a heterogeneous network, where most nodes have few connections, but a small number of nodes are highly connected (hubs).

#### 2.1.2.4 Degree-Degree Correlation

The *degree-degree correlation* measures how nodes with different degrees are connected. The *assortativity coefficient* $r$ quantifies this relationship and is defined as:

$$r = \frac{\mathfrak{m}^{-1}\sum_{(i,j)\in M} k_i k_j - \left[\mathfrak{m}^{-1}\sum_{(i,j)\in M}\frac{1}{2}(k_i + k_j)\right]^2}{\mathfrak{m}^{-1}\sum_{(i,j)\in M}\frac{1}{2}(k_i^2 + k_j^2) - \left[\mathfrak{m}^{-1}\sum_{(i,j)\in M}\frac{1}{2}(k_i + k_j)\right]^2} \tag{2.1}$$

where $k_i$ and $k_j$ are the degrees of nodes $i$ and $j$, respectively, $M$ is the set of edges in the network, $\mathfrak{m}$ is the number of edges in $M$. Positive assortativity ($r > 0$) indicates that high-degree nodes tend to connect with high-degree nodes, a characteristic of social networks. Negative assortativity ($r < 0$) is common in technical networks, such as power grids, where high-degree nodes often connect to low-degree nodes.

#### 2.1.2.5 Shortest Path

The *shortest path* between nodes $i$ and $j$ is the path with the fewest edges. The distance $d_{ij}$ between these nodes is the number of edges in the shortest path. The *average shortest path length* $L$ is the mean distance between all pairs of nodes in the network, defined as

$$L = \frac{1}{N(N-1)}\sum_{i\neq j} d_{ij} \tag{2.2}$$

This metric is essential for characterizing the overall connectivity of a network.

#### 2.1.2.6 Betweenness Centrality

Betweenness centrality is a measure of the importance of a node or edge in terms of its involvement in the shortest paths between pairs of nodes. The betweenness of node $i$ is given by:

$$B_i = \sum_{s\neq i\neq t}\frac{\sigma_{st}(i)}{\sigma_{st}} \tag{2.3}$$

where $\sigma_{st}$ is the total number of shortest paths from node $s$ to node $t$, and $\sigma_{st}(i)$ is the number of shortest paths passing through node $i$. Nodes with high betweenness centrality are critical for maintaining network connectivity, as they often act as bridges between different parts of the network.

#### 2.1.2.7 Clustering Coefficient

The *clustering coefficient* measures how connected a node's neighbors are to each other. The clustering coefficient of node $i$, denoted $\Gamma_i$, is given by:

$$\Gamma_i = \frac{2E_i}{k_i(k_i - 1)} \tag{2.4}$$

where $E_i$ is the number of edges between the neighbors of node $i$, and $k_i$ is the degree of node $i$. The *average clustering coefficient* of a network is the mean of $\Gamma_i$ over all nodes:

$$\Gamma = \frac{1}{N}\sum_{i=1}^{N}\Gamma_i \tag{2.5}$$

Empirical studies have shown that social networks tend to have a high clustering coefficient, whereas technical networks, like power grids, tend to have lower clustering due to their hierarchical nature [193].

Here, we introduce three prominent network models that categorize networks with similar topological properties: the Erdös-Rényi (ER) random network model [89], the Watts-Strogatz (WS) small-world network model [272], and the Barabási-Albert (BA) scale-free network model [23]. These models provide a framework for statistical analysis of network dynamics and structures.

#### 2.1.2.8 ER Random Network

The *Erdös-Rényi (ER)* random network model generates a random network where each pair of nodes is connected with a given probability. Proposed by Erdös and Rényi in 1959, the model follows two steps: first, $N$ isolated nodes are created; second, an edge is added between every pair of nodes with a probability $p$. The resulting network has $pN(N-1)/2$ edges, and the average degree of a node is $\langle k \rangle = Np$. The degree distribution in an ER network follows a Poisson distribution, i.e.,

$$P(k) = \frac{\langle k \rangle^k e^{-\langle k \rangle}}{k!} \tag{2.6}$$

which characterizes the network as homogeneous. The ER model typically exhibits a small average path length $L_{\text{random}} \sim \frac{\ln(N)}{\ln(\langle k \rangle)}$ and a low clustering coefficient $\Gamma_{\text{random}} \sim \frac{\langle k \rangle}{N}$. While useful for baseline comparisons, ER networks do not accurately represent real-world systems due to their overly simplistic structure.

#### 2.1.2.9 WS Small-World Network Model

The *Watts-Strogatz (WS)* small-world network model, introduced by Watts and Strogatz in 1998, is used to generate networks that exhibit the small-world effect. These networks feature short average path lengths and relatively high clustering coefficients, resembling real-world networks much more closely than ER or regular networks.

The WS model is based on a rewiring process that transforms a regular network (with large path lengths and high clustering) into a random network (with small path lengths and low clustering). Starting with a regular network of $N$ nodes with degree $k$, each edge is rewired with a probability $p$ by randomly selecting another node to replace one of its terminal nodes. As $p$ increases from 0 to 1, the network transitions from a regular structure to a random one. At $p = 0$, the network is regular, with high clustering and a large average path length. At $p = 1$, it becomes random, with low clustering and short average path lengths. The small-world network typically exhibits a clustering coefficient similar to the original regular network and a path length close to that of the random network.

WS small-world networks are a prominent model that captures the structural properties of many large-scale social networks. These networks are characterized by two key features: the presence of clusters of interconnected nodes and a short average path length among all nodes.

#### 2.1.2.10 BA Scale-free Network Model

Many real-world networks, such as social networks and the internet, follow a *scale-free* structure, where a small number of nodes (hubs) have a high degree, while most nodes have few connections. The *Barabási-Albert (BA)* model, introduced by Barabási and Albert in 1999, generates scale-free networks based on two main principles: *growth* and *preferential attachment.* In the BA model, new nodes are added to the network sequentially. At each step, a new node connects to existing nodes with a probability $\Pi_i$, i.e.,

$$\Pi_i = \frac{k_i}{\sum_j k_j} \tag{2.7}$$

where $k_i$ is the degree of node $i$. As the network grows, the degree distribution follows a power law, i.e.,

$$P(k) \sim k^{-3} \tag{2.8}$$

indicating that a few nodes acquire a disproportionately high number of connections, while most nodes have relatively few. This growth process mimics the development of many real-world networks, such as the Wikipedia citation network or protein-protein interaction networks, where new nodes are more likely to connect to existing, highly connected nodes.

The BA model effectively captures the emergence of hubs in networks, which is a key feature of many large-scale real-world networks. The preferential attachment mechanism explains why highly connected nodes attract more connections over time, leading to the formation of scale-free structures.

The topological properties of a network offer valuable insights into its structure and robustness. Key characteristics include the network's size (measured by the number of nodes and links), average node degree, clustering coefficient, and average shortest path length. Table 2.1 summarizes the fundamental topological properties of power networks across different regions. In comparison to other large-scale networks, such as the Internet, power networks are generally smaller in scale. The average node degree in power grids typically ranges from two to three, suggesting that power stations are not densely interconnected. Despite this relatively low average node degree, power networks still display distinctive structural characteristics that contribute to their functionality and resilience.

The clustering coefficient and average shortest path length presented in Table 2.1 are utilized to characterize the small-world properties of power networks. In a small-world network, the average path length is comparable to that of a random network of the same size, yet it features a significantly higher clustering coefficient. This unique structure facilitates efficient communication and ensures robust connectivity, even in the face of potential disruptions.

Power grids typically exhibit small-world characteristics when the clustering coefficient is significantly higher than that of a random network, while the average path length remains comparable to that of a random network. For instance, the power grids of Western America and France can be classified as small-world networks, demonstrating this distinctive structure. In contrast, regions such as North and Central China do not display the small-world property. This observation underscores the varying degrees of robustness present in the power grids of different regions.

One of the primary objectives of analyzing power grids through the lens of complex network theory is to identify structures that demonstrate high robustness. Research has indicated that power grids exhibiting small-world properties tend to be more resilient to random failures and targeted attacks compared to networks with more regular or random configurations. Additionally, previous studies have shown that networks featuring motifs (i.e., specific local patterns such as stars and triangles) are generally more susceptible to failures. Understanding these structural characteristics can help inform strategies for enhancing the resilience of power grids.

However, no single network topology has been universally applicable to real-world power grids. For example, the North American power grid displays an exponential degree distribution, while some studies suggest that its node degree follows a power-law distribution. Additionally, while certain power grids, such as those in Shanghai and parts of Europe, exhibit small-world properties, there has yet to be a consistent pattern that defines a "typical" power grid topology. This variability highlights the complexity and diversity of power grid structures across different regions.

**TABLE 2.1**
Basic Topological Properties of Different Power Networks, Including Number of Nodes, Number of Links, Average Node Degree ($\langle k \rangle$), Clustering Coefficient ($\Gamma$) and Average Shortest Path Length ($L$) of Different Power Networks

| **Region** | **Number of Nodes** | **Number of Links** | $\langle \mathbf{k} \rangle$ | $\Gamma$ | $\Gamma/\Gamma_{\mathbf{random}}$ | $\boldsymbol{L}$ | $\boldsymbol{L}/\boldsymbol{L}_{\mathbf{random}}$ |
|---|---|---|---|---|---|---|---|
| West America [14] | 4941 | 6594 | 2.67 | 0.080 | 148.045 | 18.7 | 2.159 |
| North China [248] | 8092 | 9018 | 2.23 | 0.002 | 6.169 | 32 | 2.852 |
| Center China [248] | 2379 | 2739 | 2.32 | 0.004 | 4.512 | 21.08 | 2.282 |
| Italy [221] | 127 | 171 | 2.69 | 0.156 | 7.365 | 8.47 | 1.730 |
| France [221] | 146 | 223 | 3.05 | 0.279 | 13.355 | 6.61 | 1.479 |
| Spain [221] | 98 | 175 | 3.57 | 0.316 | 8.675 | 4.92 | 1.366 |

A crucial aspect of robustness assessment in power grids is identifying critical nodes and links. These elements are deemed critical because their removal could lead to significant disruptions within the system. To evaluate their importance, vulnerability analyses are conducted, which involve the sequential or simultaneous removal of nodes or links to observe the impact on the network's performance. Research by Motter et al. [186] and Albert et al. [7] has demonstrated that nodes with higher degrees or loads tend to be more susceptible to attacks. Notably, intentional attacks on high-degree nodes can result in greater disruptions than random node removals. These findings indicate that relying solely on node degree and load may not fully capture the criticality of certain components within power networks, highlighting the need for a more nuanced approach to assessing vulnerability.

In addition to topological metrics, *electrical metrics* have been proposed to more effectively identify critical elements within power grids. These metrics consider not only the network's topology but also the inherent power flow parameters. Research by Arianos et al. [37] has demonstrated that electrical properties, such as net-ability and extended betweenness, provide a more comprehensive assessment of component criticality. These metrics are instrumental in pinpointing critical lines and nodes within a power grid, thereby enhancing robustness evaluations. However, the correlation between electrical metrics and actual malfunction data remains limited, indicating that further refinement is necessary to ensure their practical applicability in engineering contexts.

## 2.2 Cascading Failure

Cascading failure, which often leads to widespread power outages, is defined as a series of interconnected failures in which the initial failure of one or more components in a power system triggers subsequent failures of dependent components. These failures are typically interdependent, and the cascading effects can lead to significant system-wide disruptions [22]. The primary causes of the initial failures can vary, including human errors, natural disasters, or equipment malfunctions [290]. Once an initial failure occurs, the failure of additional components can follow due to various mechanisms, with overloading being one of the primary triggers. In these cases, power components are automatically disconnected by protection schemes when their loads exceed a specified threshold, leading to a chain reaction of failures throughout the system. Figure 2.4 illustrates the time sequence of a cascading failure process in a network, showing how the failure propagates through the system.

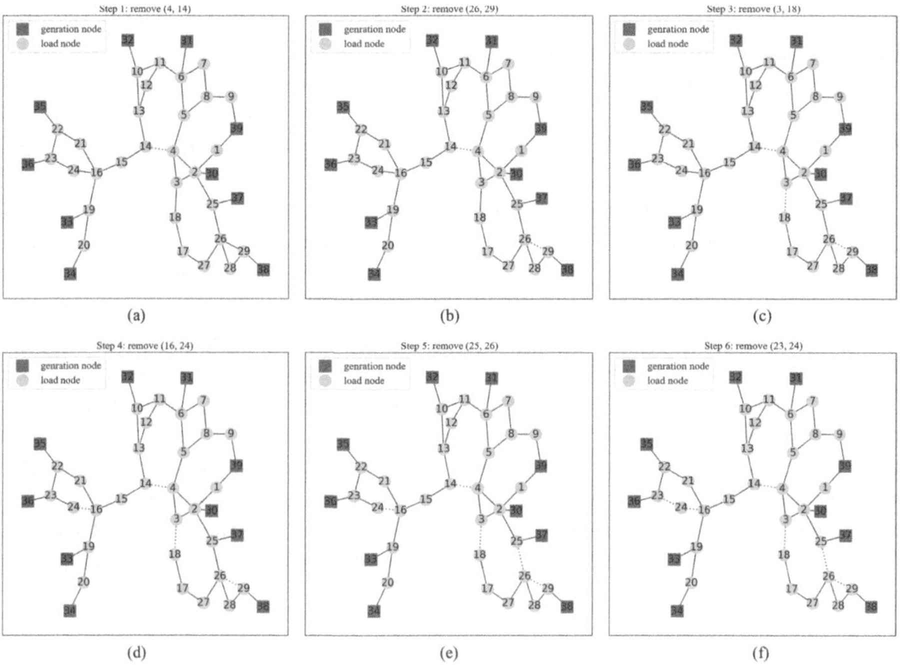


**FIGURE 2.4**
Sequential disconnection of network edges with fixed Kamada–Kawai layout. Panel (a–f) illustrate a stepwise removal of edges from the original network. In Step 1, only the edge (4, 14) is removed (dashed line). In each subsequent step (Steps 2–6), one additional edge is removed, cumulatively building on the previous configuration. Squares represent generator nodes, and Circles represent load nodes. Dashed lines indicate edges removed at or before the current step.

### 2.2.1 Basics of Cascading Failure Models

To model the propagation of cascading failure in power systems, Motter and Lai [186] proposed a step-by-step process that incorporates the overloading mechanism. In this model, it is assumed that energy flow follows the shortest path between each pair of nodes. The load at node $i$, denoted as $L_i$, is determined by counting the number of shortest paths passing through node $i$. The capacity of each node is defined as the maximum load it can carry. The capacity $C_i$ of node $i$ is assumed to be proportional to its initial load $L_i(0)$, given by

$$C_i = (1+\alpha)L_i(0), \quad j = 1, 2, \ldots, N \tag{2.9}$$

where $\alpha \geq 0$ is the tolerance parameter. When the load of a node exceeds its capacity at time $t$, i.e., $L_i(t) > C_i(t)$, the node is considered out of service and is removed from the network. The cascading failure process begins with the removal of a single node, which alters the network topology and redistributes the load along the new shortest paths. This redistribution may cause additional nodes to fail if their loads exceed their capacities. The process continues until there are no more overloaded nodes, at which point the cascading failure is considered complete. The extent of the failure is evaluated by the relative size of the largest connected component.

Simulation of the Western American power grid demonstrates that targeting the node with the largest load initially leads to a larger scale of failure compared to targeting the node with the largest degree or a random node.

In Wang and Rong's model [258], the initial load of a node is defined as a function of the product of the node's degree and the sum of the degrees of its neighboring nodes, rather than just using node degree or betweenness. When a node is removed, its load is redistributed proportionally to its neighbors, based on their initial loads. This model offers an intuitive understanding of how power flow is redistributed when nodes are removed, differing from the shortest-path-based flow redistribution in Motter and Lai's model [186].

However, models that solely rely on topological characteristics to redistribute power flow cannot adequately simulate cascading failures in power systems, as they overlook the actual power flow mechanisms governed by electrical principles, such as Kirchhoff's law. Omitting these physical laws may result in conclusions that are inconsistent with real-world power systems. Hines et al. [116] pointed out that, in a real power network, the failure of a new component may be distant from the current failure, which contradicts Wang and Rong's findings [258]. Therefore, power flow calculations must be integrated into cascading failure studies to obtain a more accurate and practical understanding of power system robustness.

### 2.2.2 Power Flow Models

The goal of using AC or DC power flow models is to calculate the steady-state solutions of power systems, including the voltage at each bus and the current through each transmission line. These electrical engineering approaches provide a reliable method for modeling power flow distribution in networks, offering a more realistic framework for studying cascading failures in power systems.

#### 2.2.2.1 AC Power Flow Model

The AC power flow model [105] is used to calculate the voltage and current for a power network at steady-state conditions. Given a power network having $N$ nodes, the voltage at node $i$ is given by $V_i = |V_i| \angle\theta_i$, where $|V_i|$ is the voltage amplitude and $\theta_i$ is the phase angle at node $i$. Likewise, for another node $k$, the voltage is denoted by $V_k = |V_k| \angle\theta_k$. The current passing through the transmission line $(i, k)$ between nodes $i$ and $k$ is calculated based on Ohm's Law, i.e., $I_{ik} = (V_i - V_k)(G_{ik} + jB_{ik})$, where $B_{ik}$ and $G_{ik}$ are the susceptance and the conductance of line $(i, k)$, respectively. Then, considering the power flow passing through transmission line $(i, k)$, the power $S_{ik}$ injected to node $i$ is given by

$$S_{ik} = V_k I_{ik} = P_{ik} + jQ_{ik} \tag{2.10}$$

where $P_{ik}$ and $Q_{ik}$ are the active and reactive power respectively. Here, $P_{ik}$ and $Q_{ik}$ can be further derived in the form of:

$$P_{ik} = |V_i|^2 G_{ik} - |V_i| |V_k| (G_{ik} \cos\theta_{ik} + B_{ik} \sin\theta_{ik}) \tag{2.11}$$

$$Q_{ik} = |V_i|^2 B_{ik} + |V_i| |V_k| (G_{ik} \sin\theta_{ik} - B_{ik} \cos\theta_{ik}) \tag{2.12}$$

The power externally injected at node $i$ is expressed by $S_i = P_i - jQ_i$, where $P_i$ is the active power and $Q_i$ is the reactive power. Since the sum of the power externally injected to node $i$ and the power injected to node $i$ through the transmission line is zero, the active and reactive power can be given as

$$P_i = -\sum_{k=1}^{N} P_{ik} = \sum_{i=1}^{N} |V_i| |V_k| (G_{ik} \cos\theta_{ik} + B_{ik} \sin\theta_{ik}) \tag{2.13}$$

$$Q_i = \sum_{k=1}^{N} Q_{ik} = \sum_{k=1}^{N} |V_i| |V_k| (G_{ik} \sin\theta_{ik} - B_{ik} \cos\theta_{ik}) \tag{2.14}$$

where $G_{ii} = -\sum_{k \neq i} G_{ik}$ and $B_{ii} = -\sum_{k \neq i} B_{ik}$. Here, $G_{ik}$ and $B_{ik}$ are both zero if there is no transmission line between nodes $i$ and $k$. Also, the phase difference between nodes $i$ and $k$ is denoted as $\theta_{ik} = \theta_i - \theta_k$. These two equations contain four variables: $P_i$, $Q_i$, $\theta_i$, and $V_i$. Thus, there are $2N$ equations and $4N$ variables for a power network with $N$ nodes. To solve these $2N$ equations, at least $2N$ variables need to be specified. In the AC power flow analysis, the nodes (buses) in a power network are classified into three types, i.e slack node, load node, and generator node. For each type of node, some variables are known, and the unknown variables are to be found by solving the equations.

These $2N$ equations are non-linear, with no closed-form solution. Thus, the solution must resort to suitable numerical iterative techniques including the Gauss-Seidel method, Newton-Raphson method, Fast-decoupled-load-flow method, and so on [105].

The propagation of cascading failures in power systems is closely tied to network topology. The rate at which failures spread can be modeled using a branching process, reflecting how the structure of the network influences the cascade. During a cascading failure, transmission lines are removed sequentially, redistributing power flow across the network. Using the AC power flow model, we calculate the load on each transmission line, and if the load exceeds the line's capacity, it trips. The time for a line to trip is faster with higher overload. Network topological features are correlated with the rate at which failures propagate.

To assess the vulnerability of power systems, identifying critical nodes is crucial. A network approach that combines topological and electrical characteristics helps pinpoint these key nodes. Node electrical centrality—which accounts for both the node's connectivity and the AC power flow—can identify these critical components. The removal of such nodes causes significant damage to the system. A vulnerability index, based on net-ability, can also evaluate how power system performance degrades under both normal and cascading failure conditions. Simulations confirm that nodes with high electrical centrality are vital, and their failure results in major disruptions.

The inclusion of AC power flow models has proven effective for studying cascading failures. Other models, such as the Manchester model and the AC Optimal Power Flow (OPA) model, also use this approach to analyze cascading dynamics in power grids.

#### 2.2.2.2 DC Power Flow Model

Due to the high complexity of the AC power flow model, substantial computational resources are required to obtain the numerical solution. For a large power network, the time taken for solving these nonlinear equations is unacceptably long. Thus, a DC power flow model is proposed as an alternative method to calculate the power flow in power networks. To derive the DC power flow equations, the following three assumptions are made:

1. The voltage magnitude at each bus is assigned to be 1 p.u.
2. The resistance is much smaller than the reactance of each transmission line, and hence it can be neglected.
3. The phase difference $\theta_{ik}$ between node $i$ and node $k$ of a transmission line $(i, k)$ is very small. Thus, $\sin\theta_{ik} \approx \theta_{ik}$ and $\cos\theta_{ik} \approx 0$.

Then, equation (2.11) can be rewritten as

$$P_{ik} \approx -B_{ik}\sin(\theta_i - \theta_k) \approx -B_{ik}(\theta_i - \theta_k) \tag{2.15}$$

The active power for each node is given by

$$P_i = -\sum P_{ik} = \sum_{k=1}^{N} B_{ik}(\theta_i - \theta_k) \tag{2.16}$$

For node $i$, equation (2.16) is regarded as a nodal equation. The nodal equations of all nodes in a power network can be expressed in matrix form, i.e.,

$$\mathbf{P} = \mathbf{B}\boldsymbol{\theta} \tag{2.17}$$

where $\mathbf{P} = \begin{bmatrix} P_1 & P_2 \cdots & P_N \end{bmatrix}^T$, $\boldsymbol{\theta} = \begin{bmatrix} \theta_1 & \theta_2 \cdots & \theta_N \end{bmatrix}^T$ and $\mathbf{B}$ is a $N \times N$ matrix in which each element represents the admittance of each transmission line of the power network.

The DC power flow model has become widely used in simulating cascading failures in power systems due to its simplicity and computational efficiency. This model simplifies the power flow equations, making it particularly effective for analyzing large-scale power grids and understanding the propagation of failures. In the context of cascading failures, the DC power flow model is employed to calculate the power flow across transmission lines, identifying when the load on a line exceeds its capacity. If the power flow surpasses the line's capacity, the line is tripped, triggering a chain of failures. This process allows for the identification of critical lines and nodes that are most likely to cause widespread disruption within the grid.

However, while the DC power flow model excels in simulating power flow and network topology changes, it does not account for the transient stability of the power system, which can be crucial in more complex fault scenarios. Some studies have compared the AC power flow model with the DC model, highlighting that while the DC model works well for simpler cases, such as single-line failures or scenarios without cascading failures, it fails to accurately capture the dynamics of cascading failures in more complex situations. The AC power flow model provides more precise results, particularly when multiple lines fail simultaneously, but it is computationally more demanding. Thus, a trade-off exists between the accuracy of the AC model and the computational efficiency of the DC model.

Overall, the DC power flow model remains an invaluable tool in cascading failure analysis due to its simplicity and speed, particularly in scenarios where rapid evaluations and predictions are necessary. Despite its limitations in capturing transient dynamics, it continues to be an essential tool for studying the robustness of power systems.

### 2.2.3 Illustrative Example

We consider a basic cascading failure simulation process that begins with an initial disturbance in the power grid. Subsequently, network components may fail successively due to overload protection, and this iterative process continues until no further failures occur. As illustrated in Figure 2.5, the cascading failure simulation can be organized into six modules:

1. Network initialization: the initial state of the network is established in this step. Both the structural parameters (e.g., topology, component connectivity) and electrical parameters (e.g., line admittances, generation and load profiles) are defined.

2. Initial contingencies: the cascading process is triggered by an initial disturbance. This may involve a structural change, such as the outage of one or more components, or an abrupt change in electrical parameters, such as significant deviations in bus voltages or line currents.

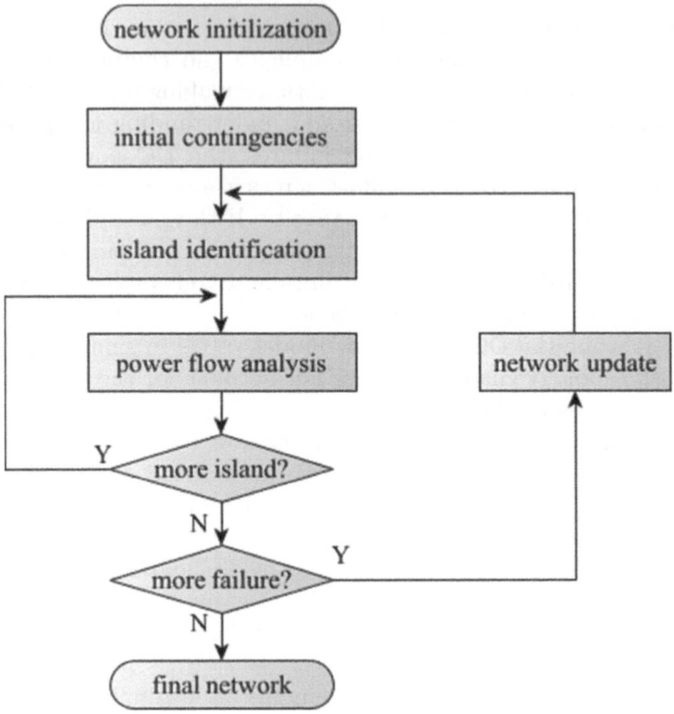


**FIGURE 2.5**
Flow chart of the cascading failure simulation framework.

3. Island identification: structural changes in the network may cause it to split into multiple disconnected subnetworks (islands). This step identifies all such islands, each of which must be analyzed separately in subsequent computations.

4. Power flow analysis: within each identified subnetwork, a power flow calculation is performed to update the power distribution across transmission lines and other components. Once system-wide power flows are updated, the network is examined for overloaded elements. Here, overload refers to a condition in which the load on a component exceeds its maximum permissible capacity. If no overloads are detected, the simulation terminates, yielding the final network state. If overloads occur, the process proceeds to the next step.

5. Network update: overloaded components are assumed to trip due to protection mechanisms. This leads to structural updates in the network topology. The updated network then returns to the island identification step, and the process iterates until no further failures occur.

To illustrate the cascading failure simulation process, we employ "MATPOWER" [309], a widely used power flow analysis toolbox, on two benchmark test cases, namely, IEEE 39 Bus system (39 nodes and 46 branches) and IEEE 300 Bus system (300 nodes and 411 branches). In both cases, the initial contingencies are modeled as the random outage of two transmission lines. The maximum load-carrying capacity of each component is set to twice its initial loading level. Both AC power flow and DC power flow models are applied for comparison. The impact of cascading failures is quantified by recording the percentage of load loss in each network.

As shown in Figure 2.6a, for the smaller IEEE 39 Bus system, the results obtained from the AC and DC power flow models exhibit only minor differences in cascading failure outcomes. In contrast, Figure 2.6b demonstrates that for the larger IEEE 300 Bus system, the choice of power flow model has a more pronounced influence. Compared with the computationally efficient DC model, the AC model more accurately captures the operational characteristics of real-world power grids, leading to more reliable results. However, the higher computational complexity of the AC model significantly reduces simulation efficiency.

Therefore, the choice between AC and DC power flow models in cascading failure analysis should be informed by practical considerations, carefully balancing the trade-off between computational efficiency and the reliability of results. This decision should take into account the scale of the system and the specific objectives of the study.

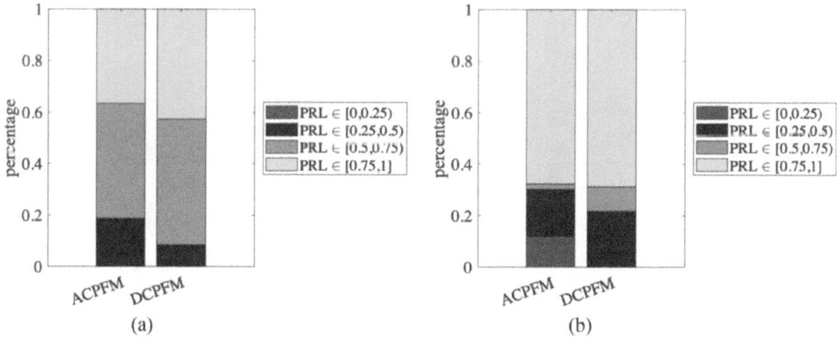


**FIGURE 2.6**
Cascading failure outcomes, expressed as the percentage of remaining load (PRL) within four predefined ranges after the cascading failure process, under AC and DC power flow models (ACCFM and DCCFM). For the IEEE 39 Bus system (a), the two models produce similar PRL distributions, suggesting limited model sensitivity in smaller networks. In contrast, for the IEEE 300 Bus system (b), substantial differences arise between the two models, underscoring the trade-off between the computational efficiency of the DC model and the higher fidelity of the AC model in large-scale networks.

## 2.3 Summary

This chapter reviews common representations of power systems, with a particular focus on the application of network models for representing complex, large-scale power grids. We discuss the use of network properties—such as degree distribution, betweenness centrality, and clustering coefficient—for characterizing the robustness of power networks. Furthermore, the mechanisms underlying cascading failures and the fundamental models for power flow calculation are examined to illustrate how cascading failure processes can be simulated and systematically analyzed.

# 3

# *Power System Protection Strategies*

The reliability and safe operation of modern power systems rest upon the effective protection of their complex, tightly interconnected infrastructures. As transmission networks expand, generation portfolios diversify, and operational dynamics grow increasingly uncertain, these systems are exposed to a wider spectrum of vulnerabilities, from routine equipment malfunctions to rare but high-impact disturbances such as natural disasters or coordinated cyber-physical attacks. Without timely and well-coordinated protective action, such events can trigger rapid fault propagation, escalate into cascading failures, and ultimately lead to large-scale blackouts. Protection strategies, therefore, play a dual role in the broader context of power system robustness. First, they must prevent minor incidents from developing into instability, ensuring that the system remains within safe operational boundaries. Second, when severe contingencies arise, they must act swiftly to contain damage, facilitate restoration, and safeguard critical infrastructure. Achieving this requires a comprehensive, structured defense framework that addresses the full life cycle of disturbances—from early detection and prevention to emergency containment and post-event recovery. One proven approach is the layered defense strategy, in which protection and control functions are organized according to both the temporal scale of system stress and the severity of its evolution. In this chapter, we examine three key layers: (i) preventive protection and control; (ii) emergency protection and control; (iii) recovery protection and control. Together, these layers form the operational backbone for ensuring not only robustness (the ability to withstand stresses without loss of stability) but also resilience (the capacity for rapid and effective recovery in the aftermath of extreme disruptions) [30]. The remainder of this chapter explores the principles, mechanisms, and coordinated deployment of each layer, emphasizing their role in safeguarding modern power systems against cascading failures and maintaining the reliability upon which society depends.

## 3.1 Preventive Protection and Control

Preventive protection and control in power systems encompass a suite of proactive measures aimed at maintaining system stability and security under

DOI: 10.1201/9781003622130-3

normal operating conditions and during minor disturbances. Its core objective is to anticipate and mitigate potential threats before they escalate into severe contingencies, thereby preventing the system from entering an emergency state.

A key focus of this protective layer is to ensure that the power network can withstand the unexpected loss of any single component, a requirement formalized by the widely adopted $N-1$ security criterion, without compromising operational integrity.

Preventive strategies involve the early detection and isolation of incipient faults, dynamic adjustment of operating parameters, and optimal utilization of system redundancy. By addressing issues at their inception, operators can effectively prevent the onset of cascading failures that often result in widespread outages.

In today's highly interconnected grids, preventive protection integrates advanced monitoring technologies, predictive analytics, and coordinated control actions. These tools enable robust system performance even in the face of common disturbances such as minor line faults, equipment outages, or transient load fluctuations.

### 3.1.1 Relaying Protection Systems

One of the most fundamental preventive measures in a power grid is the relaying protection system [11]. Its main purpose is to detect abnormal electrical conditions quickly and to isolate the faulty equipment so that the rest of the system can continue operating. This prevents local problems from spreading into larger disturbances.

- Overcurrent Protection Example

  In a healthy circuit, the current stays within the normal operating range determined by the equipment's rating. When a short circuit occurs, the current can suddenly rise to many times its normal value. A simple overcurrent relay measures the circuit current $I$ and compares it with a preset threshold $I_{\text{set}}$:

  $$\text{If } I > I_{\text{set}} \;\Rightarrow\; \text{trip the breaker} \tag{3.1}$$

  For example, if the rated line current is 200 A and the relay is set at $I_{\text{set}} = 300\,\text{A}$, any sustained current beyond that will trigger the breaker to open and isolate the fault. This action usually occurs within cycles (tens of milliseconds) to minimize damage.

- Overvoltage Protection Example

  Overvoltage can occur due to lightning strikes, sudden load rejection, or resonance phenomena. Excessive voltage can damage insulation or sensitive equipment. Overvoltage relays monitor the line voltage $V$ and compare it to a limit $V_{\text{set}}$, i.e.,

  $$\text{If } V > V_{\text{set}} \;\Rightarrow\; \text{trip or signal corrective action} \tag{3.2}$$

In practical systems, surge arresters may also be installed in parallel to divert excess energy to ground during very fast transients, protecting equipment while the relay decides on further isolation steps.

- Selective Isolation and Auto-Reclosing

  The principle of selectivity means only the faulty section is disconnected. For example, if an overcurrent occurs on one feeder of a substation, only that feeder's breaker should open, not all the feeders.

  In many transmission lines, faults are temporary, such as a lightning strike causing a momentary flashover. Auto-reclosing attempts to close the breaker after a short time delay, e.g., 0.5 s. If the fault has been cleared, the line is restored without operator intervention, reducing outage time and avoiding unnecessary network stress.

### 3.1.2 Local Automatic Controls

Local automatic controls are fast-acting devices installed directly on power system equipment, designed to keep key system quantities—like voltage and frequency—within acceptable ranges, without the need for manual operator intervention. Because they react within seconds or even fractions of a second, they help stabilize the grid against small disturbances before they develop into bigger problems.

- Automatic Voltage Regulator

  The automatic voltage regulator (AVR) is connected to the excitation system of a generator, where it controls the strength of the magnetic field in the rotor [56]. Its primary function is to maintain the generator's terminal voltage as close as possible to the specified reference value $V_{\text{ref}}$.

  When the voltage measured at the generator terminals $V_{\text{measured}}$ decreases slightly—for instance, as a consequence of a sudden increase in load—the AVR responds by increasing the excitation current $E_{\text{field}}$. This action strengthens the rotor's magnetic field and raises the terminal voltage toward the reference value. A simplified proportional control relationship can be written as

$$E_{\text{field}} \propto (V_{\text{ref}} - V_{\text{measured}}) \tag{3.3}$$

  Consider the following operating scenario. The generator's voltage reference is set to 13.8 kV under normal conditions. If an abrupt increase in load causes the terminal voltage to drop to 13.5 kV, the AVR will detect this deviation and automatically increase the excitation current. As a result, the terminal voltage returns to 13.8 kV within a matter of seconds. Such automatic and rapid voltage correction prevents damage to customer equipment and inhibits the onset of broader voltage instability in the power system.

- Governor and Primary Frequency Control

  The governor is a control device connected to the turbine—whether steam, water, or gas—that drives the generator [27]. Its role is to regulate the mechanical input power $P_m$ in response to small deviations in system frequency $f$.

  A drop in system frequency typically indicates that the electrical demand has become slightly greater than the available supply. For example, when the frequency decreases from 50 Hz to 49.95 Hz (assuming a nominal system frequency of 50 Hz), the governor detects this deviation and increases the flow of fuel, steam, or water to the turbine. This adjustment raises the mechanical power input, thereby helping to restore the frequency toward its nominal value.

  The fundamental *droop control law* governing this response can be expressed as:

$$\Delta P_m = -\frac{1}{R} \cdot \Delta f \tag{3.4}$$

  where $\Delta P_m$ is the change in mechanical power, $\Delta f$ is the frequency deviation from the nominal value, and $R$ is the droop constant (for instance, 5%).

  To illustrate, consider a power system with a nominal frequency of 50 Hz and an observed frequency of 49.95 Hz, corresponding to a deviation of $-0.05$ Hz. With a droop setting of $R = 0.05$ (5%), the governor's response would be calculated as

$$\Delta P_m = -\frac{1}{0.05} \times (-0.05) = 1.0 \tag{3.5}$$

  This result represents an increase of one per unit of rated mechanical power, which is promptly delivered to the generator to counteract the frequency drop. Such action forms the first line of defense in primary frequency control, mitigating frequency deviations before slower control layers, such as secondary regulation, are engaged.

If AVRs and governors were absent, even small disturbances could cause continuous voltage decay or frequency drift, eventually forcing operators to take larger corrective measures—or even shutting down parts of the grid. By having these actions occur automatically and locally, the system can maintain stability and comfort, often without customers even noticing that a disturbance occurred.

### 3.1.3 Operational Security Criteria

Operational security criteria are rules and limits that ensure the power system can keep operating safely, even when something goes wrong. They are like safety margins in engineering design.

- $N$-1 Security Criterion

  The $N$-1 security criterion is a fundamental standard in power system planning and operation. It requires that the system remain within all defined safety limits—covering parameters such as conductor temperature, bus voltages, and transient stability—following the sudden loss of any single critical component [218]. Such a component may include a transmission line, transformer, or generating unit. The engineering intent of this criterion is to ensure that the loss of one element does not initiate cascading failures or lead to unacceptable service interruptions. In practical terms, the remaining network elements must be capable of accommodating the redistributed load without exceeding their operational limits.

  To illustrate, consider a scenario in which power is supplied from Plant A to City B through two parallel transmission corridors. Under normal operation, the first corridor carries 90 MW and the second carries 80 MW. Each transmission line has a thermal loading limit of 100 MW. If the first line is suddenly disconnected from service, the second line must carry the total demand of 170 MW. Because 170 MW exceeds the thermal limit, this configuration fails to meet the $N$-1 security requirement.

  To comply with the criterion, the initial loading must be arranged such that, after the loss of any one line, the surviving line's loading does not exceed its rating:

  $$P_{\text{post-fault}} \leq P_{\text{limit}} \tag{3.6}$$

  Adhering to this principle allows system operators to maintain reliable service during single-component contingencies, thereby reducing the risk of widespread outages.

- Spinning Reserve

  Spinning reserve refers to the additional active power capacity that is already synchronized to the system and ready to be dispatched immediately, but is not currently being utilized [103]. This reserve is provided by generators that are online, operating below their maximum output, and capable of increasing their generation within seconds to minutes. Its primary role is to respond promptly to sudden generation losses or rapid increases in demand, thereby supporting system frequency and preventing stability deterioration.

  Consider an operating condition in which the total system demand is 900 MW, while the combined maximum capacity of the online generators is 1,000 MW. The available spinning reserve can be calculated as:

  $$P_{\text{reserve}} = P_{\text{max, online}} - P_{\text{demand}} \tag{3.7}$$

  Substituting the given values:

  $$P_{\text{reserve}} = 1{,}000 \text{ MW} - 900 \text{ MW} = 100 \text{ MW} \tag{3.8}$$

This reserve corresponds to approximately 11% of the current load. If, for example, an online generator rated at 80 MW unexpectedly trips, the remaining units can immediately increase their output by drawing from the spinning reserve. This rapid response helps maintain system frequency close to its nominal value (commonly 50 Hz or 60 Hz) and prevents frequency instability during the contingency event.

- Reactive Power Reserve

  Reactive power ($Q$) plays a critical role in sustaining acceptable voltage profiles throughout the power network. The term reactive power reserve refers to the unused capacity of generators, capacitor banks, synchronous condensers, or other reactive compensation devices that can supply or absorb $Q$ as system conditions change [254]. This reserve can be mobilized quickly to counteract voltage deviations during disturbances or varying load conditions.

  Ensuring adequate reactive power reserve is essential because voltage stability depends not only on active power balance but also on the system's ability to provide sufficient reactive support. Even when active power demand is fully satisfied, insufficient $Q$ reserve can lead to progressive voltage decline and, in severe cases, voltage collapse. Such risks are particularly pronounced during periods of heavy loading or on long transmission corridors with significant reactive losses.

  Consider, for example, a generating plant with the capability to supply up to +50 MVAr of reactive power (capacitive) or absorb down to −30 MVAr (inductive). Under normal conditions, the plant supplies +30 MVAr, leaving a remaining capacitive reserve of +20 MVAr. If the system load increases and bus voltage begins to decline, this reserve can be dispatched to enhance voltage support. The voltage at the relevant bus must be maintained within the prescribed operational limits:

$$V_{\min} \leq V_{\text{bus}} \leq V_{\max} \tag{3.9}$$

  By ensuring that such margins exist, system operators can maintain voltage stability and compliance with operational standards even under changing conditions.

Operators and energy management systems constantly monitor $N$-1 compliance, spinning reserve levels, and reactive power margins in real time. If margins get too small, they may redispatch generators, start fast-start units, or switch reactive devices to restore security margins before any fault actually happens. These criteria are therefore a preventive shield, ensuring that common faults don't escalate into emergencies.

### 3.1.4 Preventive Control Actions

Preventive control actions are proactive adjustments made to the power system's operating point before any actual fault occurs, based on security analysis. The main idea is that we change them in advance to avoid entering an emergency state if we know certain conditions might cause trouble after a fault. Often, these actions are guided by results of $N$-1 studies, contingency analysis, or other "what-if" simulations that predict how the system would behave if something failed.

- Security-Constrained Economic Dispatch

  Security-constrained economic dispatch (SCED) is an optimization framework employed in modern power system control centers to determine the most cost-effective dispatch of generating units while simultaneously ensuring compliance with all relevant security limits. Unlike purely economic dispatch [24], SCED explicitly incorporates operational constraints such as the $N$-1 security criterion, bus voltage limits, and reserve requirements into the dispatch problem.

  The primary engineering objective of SCED is to achieve economic efficiency without compromising system security. By embedding preventive security checks directly into the dispatch algorithm, operators can ensure that economic decisions do not inadvertently push the system toward unsafe operating points. This integration is particularly important under high loading conditions or during network reconfigurations, when the margin to security violations can be small. To illustrate, consider a simplified system in which three generators supply a total demand of 500 MW. The dispatch must minimize the total operating cost $C_{\text{total}}$ subject to the following constraints:

  1. Power balance constraint:

     $$\sum_{i=1}^{3} P_{G_i} = P_{\text{demand}} \tag{3.10}$$

     where $P_{G_i}$ is the power generated by the $i$th generator and $P_{\text{demand}}$ is the power demanded.
  2. Security constraint:

     $$P_{\text{line},k} \leq P_{\text{limit},k} \quad \forall k \tag{3.11}$$

     $$V_{\text{min}} \leq V_{\text{bus},m} \leq V_{\text{max}} \quad \forall m \tag{3.12}$$

     where $P_{\text{line},k}$ is the power flow in line $k$, $P_{\text{limit},k}$ is the maximum power flow permitted in line $k$, $V_{\text{bus},m}$ is the voltage of bus $m$, $V_{\text{min}}$ and $V_{\text{max}}$ are the minimum and maximum bus voltage, respectively.

3. Reserve requirement:

$$P_{\text{reserve}} \geq P_{\text{req}} \tag{3.13}$$

where $P_{\text{reserve}}$ and $R_{\text{req}}$ are the reserve and requested power, respectively.

If real-time measurements or forecasts indicate that any of these limits are approaching violation, the SCED algorithm automatically adjusts generator outputs or recommends network reconfiguration to re-establish adequate margins, thereby avoiding unsafe states before they occur. By embedding such security criteria directly into economic dispatch, SCED ensures that cost minimization is always aligned with operational reliability, thus strengthening the preventive control layer of overall system robustness.

- Reducing Power Flow on a Critical Line

  Operating transmission elements near their thermal or stability limits increases the risk of overloads if network topology changes unexpectedly. A common preventive measure is to reduce loading on critical lines in advance, thus ensuring that potential post-contingency flows remain within limits.

  Consider a situation where a particular transmission element, Line X, carries 480 MW under normal operation, with a thermal limit of 500 MW. Contingency analysis shows that if a parallel line (Line Y) were to trip, the flow on Line X would rise to 560 MW, exceeding its limit. To mitigate this risk, the operator redispatches generation so that the normal flow on Line X is reduced to 400 MW, which satisfies the limit condition:

$$P_{\text{post-fault}} \leq P_{\text{limit}} \tag{3.14}$$

  This proactive adjustment avoids an overload, preserves network security, and eliminates the need for emergency interventions during the contingency event.

- Starting Additional Generating Units to Restore Reserve

  Adequate spinning reserve is necessary to maintain frequency stability when unexpected generation losses occur. If forecasts indicate that reserve margins will fall below the specified threshold, starting additional units in advance ensures the system can respond to contingencies without frequency violation.

  In one scenario, the forecast shows that load will peak at 18:00, causing available spinning reserve to drop to 4% of load, below the required 5% margin. To address this, at 17:30, the operator brings online a 100 MW fast-start gas turbine. When the peak load arrives, the reserve margin is restored above the minimum, providing sufficient capacity to compensate instantly if a large generator trips. This action preserves frequency stability during the most stressed period of the day.

- Adjusting Voltage Support Devices

Voltage stability limits can become critical when reactive power margins are low, especially in parts of the network that are electrically distant from major generation sources or are served by long transmission lines. Preventive use of reactive compensation devices increases voltage support and reduces the risk of voltage collapse under rising load conditions [254].

For example, studies predict that during the evening peak, the eastern part of the network will experience low reactive power reserves, making it vulnerable to voltage decline. To counter this, operators switch on capacitor banks and increase generator excitation in that area during the afternoon. This preventive adjustment provides additional reactive power ($Q$) support, increases the local voltage stability margin, and ensures that bus voltages remain within operational limits given by equation (3.9). By securing reactive margins before the evening peak, the system avoids emergency voltage control actions and maintains secure operation even under heavy loading.

## 3.2 Emergency Protection and Control

Emergency protection and control refers to the set of fast-acting measures deployed after a severe disturbance has already occurred, and the power system is at imminent risk of instability or collapse. Unlike preventive strategies that act before faults happen, emergency control comes into play when the impact of the disturbance exceeds the design security margin, such as multiple simultaneous outages, the tripping of a large generating unit, or cascading line failures. In these situations, the system may experience rapidly deteriorating conditions, including frequency collapse, voltage collapse, or loss of synchronism (angle instability), within seconds.

The objective at this stage is to stop the instability from spreading and to pull the grid back into a dynamically stable or at least controllable state. Actions can be automatic, triggered by protective relays or special protection schemes (SPS), or semi-automatic, initiated by operators aided by real-time decision support tools. In extreme cases, the priority may shift from preserving the entire interconnected system to segmenting it into smaller stable islands ("controlled islanding") to prevent a total blackout.

In essence, emergency protection and control form the last solid defensive wall before system collapse, aiming to contain the disturbance, protect critical infrastructure, and preserve as much supply as possible for the remaining customers.

### 3.2.1 Fast Fault Clearance and Special Protection Schemes

In the realm of emergency control, fast fault clearance (FFC) functions as an immediate "hemorrhage control" measure — swiftly isolating the fault to remove the disturbance source and constrain the initial impact. In contrast, SPS or Remedial Action Schemes (RAS) serve as "patient stabilization" mechanisms — acting after fault isolation to prevent rapid deterioration of system stability. When implemented in a coordinated manner, these two measures establish a multilayer, real-time defense framework, which is critical for preventing cascading failures in heavily loaded, highly interconnected power systems.

- Fast Fault Clearance in High-Voltage Transmission Systems

  The duration for which a fault remains on a power system has a direct and critical influence on transient stability and the overall stress experienced by network components [174]. When fault currents persist for more than a few hundred milliseconds, they may induce severe voltage depressions, cause a loss of synchronism among generators, and lead to a sequence of protective disconnections elsewhere in the network. Such conditions accelerate the progression toward widespread system collapse.

  Fast fault clearance refers to the rapid detection and isolation of the faulted network section by operating circuit breakers without delay. In extra-high-voltage transmission systems, this typically means that the entire fault clearance process—from detection to circuit isolation—should be completed within approximately 50–100 ms, while distribution networks generally require clearance within a range of about 100–200 ms. Reducing the clearance time limits the amount of energy that the fault injects into the system, thereby preserving a higher transient stability margin. This relationship may be qualitatively expressed as

$$\text{Short } t_{\text{clear}} \implies \text{lower fault stress and higher system stability margin} \tag{3.15}$$

  where $t_{\text{clear}}$ is the clearance time.

  A representative operating scenario illustrates this effect. A 500 kV transmission line experiences a three-phase short circuit near its midpoint. In the first case, the protection system detects and clears the fault in 80 ms, and dynamic simulations show that all generators in the affected area remain in synchronism. In the second case, the clearance is delayed to 200 ms, leading to a divergence of rotor angles and eventual loss of stability. The difference between the two outcomes underscores the importance of reducing clearing times to the shortest technically feasible duration.

  Achieving such performance requires the use of high-speed protective relays, such as distance, differential, and instantaneous overcurrent relays, operating in coordination with circuit breakers designed for rapid tripping, often supported by dedicated communication channels to accelerate fault detection in multi-terminal configurations. When applied effectively, fast fault clearance serves as the first layer of emergency defense, greatly reducing the likelihood that severe faults will evolve into system-wide instability.

- Special Protection Schemes / Remedial Action Schemes

  While fast fault clearance removes the initial fault from the system, the network may still be at risk of instability due to sudden power imbalances, overloads, or abnormal power flows created by the contingency. SPS or RAS are pre-planned, automatic, wide-area control actions designed to detect such conditions and apply corrective measures in under a few seconds—often in fractions of a second [303].

  These schemes rely on wide-area measurements gathered from SCADA systems or phasor measurement units (PMUs), and employ logic based on extensive offline contingency studies. Actions may include any combination of the following:

  1. Generation rejection: tripping specific generators to prevent overspeed or overfrequency after a large load loss.
  2. Network reconfiguration: opening or closing lines/switchgear to reroute power flows and prevent overload cascades.
  3. Controlled braking resistors: absorbing excess mechanical energy to damp large power swings.

  Imagine an interconnected system where a 1,000 MW generator complex supplies two regions via a pair of 400 kV lines. A fault trips one of these lines, forcing all 1,000 MW through the remaining line. Although the current does not exceed the thermal limit, stability simulations show that the sudden transfer could cause large-angle swings between the sending and receiving areas, risking loss of synchronism within seconds. To prevent this, the SPS immediately sends a trip signal to one generating unit producing 300 MW at the complex. This generation rejection reduces transmitted power to a stable level, damping the swing and allowing the remaining units to stay in synchronism with the rest of the grid. As a result, the disturbance is contained without the need for load shedding.

  By acting on a wide-area scale and executing coordinated countermeasures almost instantaneously, SPS/RAS prevent temporary overloads and imbalances from escalating into system-wide instability or blackouts.

### 3.2.2 Emergency Load Shedding

Emergency load shedding is an emergency control measure that becomes necessary when the equilibrium between generation and demand is abruptly disrupted and other rapid control mechanisms, such as governor response or automatic voltage regulator support, are insufficient to restore stability [27]. The underlying strategy is to deliberately reduce a portion of the system's electrical demand in order to protect the remainder of the grid from complete collapse. Depending on the severity and nature of the disturbance, this action can be initiated automatically by specially designed protection relays or under semi-automatic operator supervision. Owing to its direct and often customer-impacting nature, emergency load shedding is generally regarded as the final line of defense before an uncontrolled cascading failure leads to a total blackout.

Two principal variants are commonly employed in emergency control practice: under-frequency load shedding (UFLS), which safeguards against severe active power deficits, and under-voltage load shedding (UVLS), which addresses dangerous declines in voltage stability margins arising from reactive power shortages.

- Under-Frequency Load Shedding (UFLS)

  A sudden loss of major generation capacity creates an immediate imbalance in active power. The mechanical power provided by turbine-driven generators, denoted $P_m$, instantaneously falls below the electrical load $P_e$, initiating a system-wide frequency decline. The rate of this decline can be described by the swing equation:

$$2H\frac{df}{dt} = \frac{P_m - P_e}{S_{\text{base}}} \tag{3.16}$$

  where $H$ is the normalized inertia, $S_{\text{base}}$ is the machine rating, and $f$ is the frequency. If the frequency falls below critical thresholds, such as 48.5 Hz in a 50 Hz system, generators may stall or trip on under-frequency protection, precipitating a rapid system collapse.

  Consider a system carrying a load of 10,000 MW at a nominal frequency of 50 Hz. The sudden tripping of a 1,500 MW generating unit represents an active power deficit of 15 percent. Primary governor action within the first two seconds can provide only about 600 MW, leaving a remaining deficit of approximately 900 MW. Predictive stability analysis indicates that without corrective actions, the frequency would fall to around 48.2 Hz within five seconds—a level that risks widespread generation disconnection.

  To avert collapse, the UFLS scheme automatically disconnects roughly 900 MW of demand through a series of rapid, pre-programmed stages. For instance, one block of load may be shed when the frequency reaches 49.2 Hz and another when it drops to 48.8 Hz. Staging the disconnection in this way prevents excessive load loss if the system frequency recovers after the initial stage. Although a segment of consumers experiences outages, the system frequency stabilizes near 49.3 Hz, maintaining synchronism among generating units and buying operators crucial time to start reserve generation and restore supply.

- Under-Voltage Load Shedding (UVLS)

Large deficits in reactive power support or extreme loading of transmission corridors can initiate a progressive decline in system voltage, a phenomenon known as voltage instability, which may culminate in voltage collapse. By rapidly disconnecting selected loads, typically those of lower priority or with contractual interruptibility, the network's reactive power demand is reduced, supporting voltage recovery and preventing mass tripping of equipment.

A representative situation arises in a coastal city supplied via a long-distance 220 kV transmission corridor delivering 500 MW, with limited local generation. During the summer peak, a large share of the city's reactive power is imported over these lines. Following the outage of a major 220 kV circuit into the area, the remaining path becomes heavily loaded, and the voltage at the receiving-end bus declines from 1.00 pu to 0.88 pu within seconds. Voltage stability studies of the area indicate that if the bus voltage falls below 0.85 pu, a large number of induction motors will begin stalling, abruptly increasing reactive demand and driving the system toward collapse.

The UVLS protection logic detects the voltage dipping past 0.88 pu and immediately sheds around 150 MW of interruptible industrial and commercial load. The resulting reduction in reactive consumption allows the voltage to recover to approximately 0.92 pu within a few seconds, averting further voltage decay and maintaining supply to the majority of the city. In this way, UVLS acts as a targeted and decisive intervention to sustain voltage stability during extreme reactive power stress conditions.

While under-frequency and under-voltage schemes focus on removing load to balance the system, there are also situations where the opposite approach is needed — removing generation or reducing electrical stress. One example is over-frequency generator tripping control. This condition can arise when large blocks of load are suddenly lost, for instance, due to a major network section being isolated. The generators feeding that section experience a sudden surplus of mechanical power $P_m$ over electrical load $P_e$, causing the system frequency to rise above safe limits (e.g., 50.5 Hz in a 50 Hz system). If governors cannot reduce output fast enough, automatic tripping of selected generating units can be used to quickly correct the imbalance, stabilizing frequency before protective relays in other parts of the system operate.

Another case is over-voltage protection in emergencies. Severe over-voltages can occur when lightly loaded long transmission lines remain energized (Ferranti effect) or after large load rejections. If bus voltage rises above a set threshold (e.g., 1.10 pu), protective devices such as over-voltage relays may trip selected lines, shunt reactors may be switched in to absorb reactive power, or in extreme cases, generators may be instructed to increase reactive power absorption to bring voltages back within limits.

These measures, while less common than load shedding, are vital complements in the emergency control toolbox, ensuring that both extremes—too little load or too much voltage—are managed to keep the system within safe operating boundaries.

### 3.2.3 Dynamic Stability Control

Dynamic stability control in emergency situations refers to fast-acting measures designed to arrest or damp large rotor angle swings and other electromechanical oscillations before generators lose synchronism with the grid. Unlike preventive damping systems such as Power System Stabilizers (PSS), which operate continuously, emergency dynamic stability control is triggered only when a severe disturbance has already caused large deviations in generator speeds, angles, or bus voltages [109].

When a major fault or loss of a key transmission link occurs, power transfer patterns can change abruptly. This sudden change affects the electrical torque in generators, causing them to accelerate or decelerate relative to the rest of the system. If the difference in rotor angles between two areas exceeds a critical limit (often around 180°), synchronism will be lost, splitting the system frequency and leading to widespread outages. The goal of dynamic stability control is therefore to add damping or synchronizing torque quickly enough to keep rotor angles within stable bounds.

- Braking Resistor for Transient Stability

  Consider a large hydro generator exporting power over a long 400 kV line to a distant load center. A short circuit occurs near the receiving end and is cleared within 100 ms. Although the fault is removed quickly, the sudden drop and recovery of electrical power $P_e$ during the fault period causes the generator rotor to accelerate relative to the system, increasing its rotor angle $\delta$ rapidly.

  To counter this, a dynamic braking resistor connected near the generator is switched in immediately after fault clearance. By temporarily absorbing 200 MW of excess electrical energy, the resistor increases $P_e$ sharply during the first swing, reducing the net accelerating power $P_m - P_e$. This mitigates the "accelerating area" and reduces the amplitude of the rotor angle swing, preventing it from crossing the stability boundary. After about one second, the resistor is switched off, and the system returns to normal operation without any loss of load. The key principle (Equal-Area Criterion) is given by

  $$\text{Accelerating Area} \downarrow \Rightarrow \delta_{\max} \downarrow \tag{3.17}$$

  where $\delta_{\max}$ is the maximum rotor angle of the generator relative to the infinite bus that can be reached during the first swing, determining stability or loss of synchronism; $P_m$ and $P_e$ refer to mechanical power input from the turbine (MW) and electrical power output to the grid (MW), respectively. Accelerating area is the area on the $P$–$\delta$ curve where $P_m > P_e$, representing the net accelerating energy (MW·rad) during instability analysis.

- Emergency Excitation Boost to Prevent Loss of Synchronization

  In a multi-machine system, loss of a major tie-line between two regions suddenly reduces the synchronizing torque, making the rotor angle difference between the areas grow quickly. If not corrected, the areas will drift apart in frequency. An emergency excitation boost [2] is a control function that temporarily drives the AVR into a high-ceiling voltage mode. By rapidly increasing the generator's field current, the internal generated voltage $E$ rises, allowing the machine to supply more electrical power $P_e$ for the same rotor angle $\delta$. This strengthens the synchronizing force and improves transient stability margins.

For example, after a tie-line trip, excitation boost on two large thermal units increases their terminal voltages by 10% for a few seconds. As a result, the maximum power transfer capability temporarily increases:

$$P_{\max} = \frac{E \cdot V}{X} \tag{3.18}$$

where $P_{\max}$ represents the maximum transmittable electrical power (MW). $E$, $V$, and $X$ refer to generator internal voltage magnitude (in per unit or kV), bus voltage magnitude at the receiving end (in per unit or kV), and the total synchronous reactance between generator and system (in per unit or ohms), respectively. By increasing $E$ or $V$, the system can transfer more power for a given rotor angle, reducing the risk of loss of synchronism.

### 3.2.4 Controlled Islanding

Controlled islanding is an emergency control strategy in which the power system is deliberately split into two or more electrically independent "islands" to prevent a complete system-wide blackout when instability is unavoidable. Unlike equipment tripping that occurs automatically in response to local faults, controlled islanding is a planned and coordinated separation, based on pre-studied scenarios, that seeks to keep each island in a dynamically stable and self-sufficient condition after separation [5].

The key idea is that if synchronism between different parts of the grid is going to be lost, due to severe angle instability, cascading outages, or widespread frequency dips, it is better to separate the system in a controlled fashion, making sure each island has balanced generation and load, rather than allowing uncontrolled separation with no regard for balance. Typical objectives of controlled islanding include (i) preventing large-scale frequency collapse in the whole system, (ii) preserving critical infrastructure in at least some regions, and (ii) allowing faster service restoration by keeping portions of the grid energized.

Controlled islanding is truly an "emergency of last resort" measure — it is only executed when simulations or real-time indicators show that synchronous operation of the entire system cannot be maintained. It requires careful pre-planning, including identifying optimal cutsets (breaker locations), balancing generation and load in each island, and ensuring critical services remain online. When executed successfully, it can mean the difference between a total regional blackout and a partial, quickly recoverable outage.

- Separation to Protect Against Angle Instability

Consider a large interconnected system where Region A (generation-rich) supplies 3,000 MW to Region B (load-rich) via three 500 kV tie-lines. A double-circuit fault trips two of the three tie-lines. The heavily loaded remaining line now carries nearly its thermal limit, and stability analysis shows that rotor angle difference $\delta$ between Regions A and B will exceed 180° within 5 seconds — a sure sign of loss of synchronism.

A controlled islanding scheme trips the last tie-line at a calculated moment, intentionally separating Region A and Region B. Because studies have pre-matched load and generation in each island, Region A is left exporting only within itself at a slightly higher frequency (e.g., 50.1 Hz), while Region B operates at a slightly lower frequency (e.g., 49.7 Hz) until corrective actions such as load shedding or generator dispatch restore balance. The key stability indicator is given below,

$$\Delta\delta \to 180^\circ \quad \Rightarrow \quad \text{trigger islanding} \tag{3.19}$$

where $\Delta\delta$ refers to the rotor angle difference between generation areas (degrees), and 180° is the approximate point of loss of synchronism in a two-machine equivalent.

- Islanding to Contain a Frequency Collapse

In a coastal power network, most generation is local, but 800 MW of additional power is imported from an inland grid over high-voltage lines. A severe storm causes cascading line trips inland and results in the sudden loss of 500 MW of import. Frequency in the coastal network begins to drop rapidly from 50 Hz toward the under-frequency load shedding threshold of 49.2 Hz.

The islanding controller detects that if the coastal network remains connected to the unstable inland grid, it will continue to experience power swings and frequency oscillations. At 49.5 Hz, the system executes a separation at a pre-planned set of breakers, isolating the coastal network from the inland instability. The coastal island now contains just enough spinning reserve to arrest the frequency decline at 49.3 Hz without triggering large-scale load shedding, while the inland network proceeds into further protective actions separately. For the frequency balance principle after islanding, it can be written as

$$\Delta f \propto \frac{P_m - P_e}{2H} \tag{3.20}$$

where $\Delta f$ refers to the rate of change of frequency (Hz/s). $P_m$ and $P_e$ are the total mechanical input power from generation in the island (MW) and the total electrical load demand in the island (MW). $H$ is the inertia constant of the island (seconds), and a higher $H$ slows frequency change.

## 3.3 Recovery Protection and Control

Recovery protection and control refers to the set of coordinated technical and operational measures applied after a major power system blackout or widespread outage has already occurred. At this stage, preventive and emergency defenses have either been exhausted or bypassed, and large sections of the network have been de-energized. The immediate goals are to contain and stabilize the blackout area, prevent further spread into unaffected regions, and then restore the electricity supply in a safe, orderly, and timely manner.

Recovery control differs fundamentally from preventive and emergency actions. Preventive protection operates before faults occur, emergency controls act during the fault or instability, whereas recovery protection starts after the collapse has already taken place [167]. In this phase, the system's operating state is no longer intact; parts of the grid may be isolated with no generation, loads are disconnected, and voltage and frequency references are lost in de-energized areas. The restoration process involves several critical challenges:

1. Establishing initial energized "islands" using black start units — generator units capable of starting without any external power supply;
2. Gradually re-energizing transmission lines and substations while maintaining a balance between generation and load in each energized island;
3. Synchronizing islands and reconnecting them into a single, stable interconnected system when conditions allow.

Successful recovery protection requires strict procedure planning and real-time coordination between control centers, field crews, and generation operators. Poorly sequenced or overly rapid restoration can cause secondary blackouts due to overloads, frequency excursions, or voltage instabilities. When properly planned and executed, recovery protection not only shortens outage duration but also minimizes the economic, social, and safety impacts of major power failures.

### 3.3.1 Black Start and Initial Energization

Black start is the process of restoring power to the grid, starting from generating units capable of self-starting without an external electricity supply. In a system-wide blackout, most conventional generators cannot start because their auxiliaries, such as lubrication pumps, excitation systems, and boiler fans, require power from the grid [94]. Black start units circumvent this limitation by using alternative energy sources or stored energy to supply their own startup power. Typical black start-capable generation includes (i) small diesel generators (often a few MW) used to energize the auxiliaries of larger plants, (ii) small hydropower units located near load centers, and (iii) Certain gas turbines with battery or compressed-air starters. Once online, these units form the initial energized islands of the restoration process. They supply the auxiliary power to bring larger generating units into operation and then start energizing nearby transmission lines and substations. The early stages of energization focus on building a minimal but stable "skeleton" network capable of carrying essential control power and supplying critical loads. This phase lays the foundation for the entire restoration process. A well-executed black start strategy ensures that generation is available where it is most needed, that voltage profiles are kept within safe ranges during line energization, and that the restoration sequence proceeds without causing further trips or equipment stress. Poor planning at this stage can delay restoration by hours or even trigger secondary collapses.

- Hydropower Black Start

  A 30 MW hydropower station located upstream of a major city is designated as a black start unit. When a blackout occurs, the operator opens the intake gates, allowing water to flow to spin the turbine. The station's excitation system and control devices are powered from an onsite battery bank, enabling it to reach rated speed and voltage within 8 min. The station first energizes a 110 kV substation that supplies auxiliary systems of a 500 MW thermal power plant. With power to its pumps and fans restored, the thermal unit can begin its startup sequence, ready to contribute to grid restoration within several hours.

- Gas Turbine Black Start and Line Energization

  A 50 MW aero-derivative gas turbine, equipped with a battery-assisted starter, begins its black start sequence after detecting a total loss of grid voltage. Within 10 min, it reaches rated load and energizes a nearby 220 kV transmission line to a critical substation. The no-load energization current $I_{\text{mag}}$ for the line is carefully monitored to avoid overvoltage conditions caused by the Ferranti effect:

$$V_{\text{receiving}} \approx V_{\text{sending}} \left[ 1 + \frac{(X_C - X_L)}{Z_0} \right] \tag{3.21}$$

  where $V_{\text{sending}}$ and $V_{\text{receiving}}$ are the voltage at the sending end of the line (kV) and Voltage at the receiving end (kV), respectively. $X_C$, $X_L$, and $Z_0$ are the line capacitive reactance, the line inductive reactance, and the line surge impedance, respectively. By keeping the receiving-end voltage within $\pm 5\%$ of nominal, the operator safely energizes the line and allows the remote substation to power additional black start units.

### 3.3.2 Sequential Load Restoration and Network Reconfiguration

Sequential load restoration is the process of bringing customer demand back online step-by-step after a blackout, while maintaining real-time balance between generation and demand and avoiding overloading, frequency dips, or voltage collapse. This process runs in parallel with network reconfiguration, which involves gradually reconnecting substations, lines, and transformers to re-establish the desired grid topology. After black start units and initial energized islands have been established, operators follow pre-planned restoration schedules that specify: (1) which loads to energize first (critical loads such as hospitals, water pumps, communication systems), (2) the size and timing of each load block, and (3) which transmission paths to re-energize for stability and supply needs.

Because the system is still fragile and inertia is low at this stage, large, sudden load steps must be avoided. The golden rule is

$$P_{\text{load-step}} \leq P_{\text{available reserve}} \tag{3.22}$$

where $P_{\text{load-step}}$ and $P_{\text{available reserve}}$ represent the size of the load block to be restored (MW) and the spinning or fast-start capacity margin available in the energized island (MW).

The post-blackout system has low inertia, small stability margins, and limited reserves. Sequential load restoration ensures that generation can keep up with added demand in real time, while network reconfiguration carefully rebuilds the grid topology without causing overloads or dangerous transients. If these actions are rushed or uncoordinated, they can cause secondary collapses, undoing hours of restoration work. Proper staging not only speeds up recovery but also maximizes the amount of load restored within each safe operational step.

- Staged Load Addition to Maintain Frequency Stability

  A black-started island has 200 MW of generation online, with 20 MW of spinning reserve. The first loads to be restored are critical services. The restoration plan calls for the process, as shown in Figure 3.1.

  If a load step larger than the available reserve had been added (e.g., 30MW), frequency could have dropped below 49.5 Hz, triggering under-frequency load shedding (UFLS) and undoing part of the restoration progress.

- Network Reconfiguration to Avoid Overloads

  In the second hour of restoration, two energized islands — North and Central — are ready to be connected via a 220 kV tie-line. Power flow studies indicate that if the link is closed too early, the line could carry 150 MW over its 120 MW thermal rating due to mismatched frequencies and voltages between the islands.

  To prevent this, operators first reconfigure the North island by routing an extra 40 MW of generation to local loads, reducing the power surplus. They also use transformer tap adjustments to match voltages within ±1% between the islands. Once these conditions are met, the tie-line breaker is closed, synchronizing the two islands without overload or transient shock.

### 3.3.3 Synchronization of Islands and Final System Integration

Following a major blackout, once black start units have energized initial islands and sequential load restoration has stabilized those islands, the final stage is to synchronize and integrate these islands into a single, interconnected power system [194]. Before closing the tie-lines between energized sections, three parameters must be within acceptable limits at the synchronization points:

$$\begin{cases} |V_1 - V_2| \leq 5\% \text{ (voltage magnitude difference)} \\ |f_1 - f_2| \leq 0.1 \text{ Hz (frequency difference)} \\ |\theta_1 - \theta_2| \text{ close to zero, typically within } 10^\circ \end{cases} \tag{3.23}$$

where $V_1$ and $V_2$ represent the voltage magnitude at each side of the tie point (kV); $f_1$ and $f_2$ represent the frequency in each island (Hz); $\theta_1$ and $\theta_2$ represent the voltage phase angles in each island (degrees). These limits ensure that the moment of closing will not produce destructive inrush currents or sudden mechanical shocks to generator rotors.

This stage is the culmination of the entire recovery process. Until synchronization is completed, the grid operates as smaller, less resilient islands with limited mutual support. Properly executed integration restores the grid's full transmission capacity, enables economic load sharing, re-establishes normal

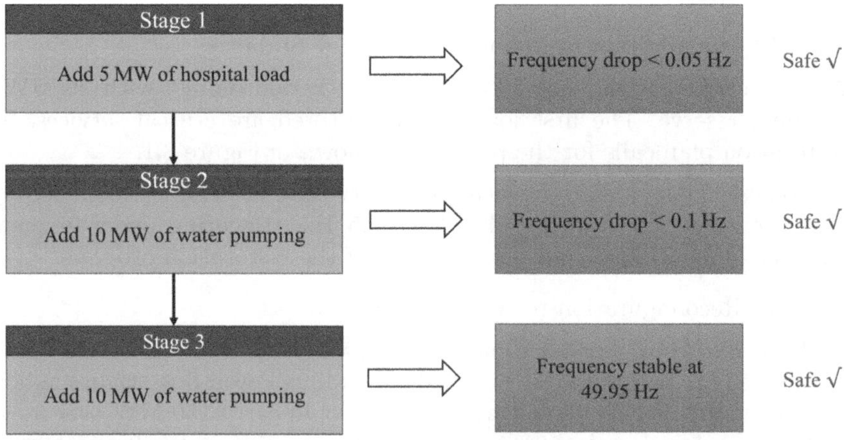


**FIGURE 3.1**
Example of a three-stage load restoration strategy in a black-started island to ensure frequency remains above protection thresholds.

reserve margins, and returns protection and control systems to their designed configurations.

Poor synchronization planning or execution can instantly reverse restoration progress, potentially causing generator trips, severe power swings, and even re-triggering a wider blackout. For this reason, real-time measurements, strict operational criteria, and sometimes automated synchronization systems are indispensable tools for system operators.

- Manual Synchronization Between Two Regions

  During the system restoration process following a major disturbance, it is common for different parts of the network to operate temporarily as electrically isolated islands. Before these islands can be reconnected, their electrical parameters must be closely matched to avoid damaging transients. Synchronization is the process of ensuring that the voltages, frequencies, and phase angles at the interconnection point are within acceptable tolerances prior to closing the tie-line circuit breaker.

  In one instance, the North and Central regions had been operating independently for several hours while restoration procedures progressed. At the designated 400 kV interconnection bus, control room operators monitored a synchroscope, which continuously compared the conditions on either side of the open breaker. The voltage magnitudes in the two regions were 398 kV and 402 kV, corresponding to a difference of approximately one percent. The system frequencies were measured at 50.02 Hz and 50.07 Hz, indicating a difference of only 0.05 Hz. The phase angle between the regions was four degrees and steadily decreasing as operators adjusted generation to bring the systems into closer alignment.

  When the synchroscope display indicated that the phase difference was approaching zero—commonly represented by the pointer nearing the "12 o'clock" position—the operator issued the close command for the interconnecting breaker. The closure occurred smoothly, without any noticeable transient or disturbance. Immediately upon energization of the tie-line, the two formerly separate islands began to operate as a single synchronized system, sharing generation resources and reserve capacity. This careful process of manual synchronization ensured a safe and stable reconnection, a critical step in accelerating the overall restoration of the power system.

- Automatic Synchronizer with Pre-Balancing

  In a larger restoration, East and West islands must be integrated via two 220 kV tie-lines. An automatic synchronizer device measures voltage, frequency, and angle on both sides, and sends fine control signals to generator governors in the East island to match the West island's frequency at exactly 50 Hz.

  To avoid high instantaneous transfers at the moment of connection, both control centers pre-dispatch generation so that the net expected power flow is less than 15% of each tie-line rating. Once the automatic device detects that the voltage difference is less than 2%, frequency difference is less than 0.02 Hz, and angle difference is less than 3°, it issues the simultaneous close signal to both tie-line breakers. The islands merge smoothly, with post-closing power flows matching planned values and no oscillations observed.

### 3.3.4 Remarks on Recovery Protection and Control

Recovery protection and control form the final line of defense in the resilience framework of a power system. After a large-scale blackout, it enables operators to re-establish supply through a carefully orchestrated sequence: initiating black start units, energizing critical transmission paths, restoring load in manageable stages, reconfiguring the network, and finally synchronizing restored islands into a unified and stable grid. When executed according to well-designed plans, these measures can significantly reduce outage duration and limit the economic and social impact of major power failures.

However, recovery protection operates after a collapse has already occurred. The central mission of this book lies upstream in the disturbance chain: to understand, assess, and enhance the robustness of power systems so that widespread instability and cascading failures can be avoided in the first place. For this reason, while we have outlined the essential principles and practices of recovery protection in this section, no further in-depth treatment of these third-layer strategies will be pursued in the remainder of this text. The chapters that follow will focus on preventive and mitigation-oriented approaches, i.e., the methods, models, and operational tools that keep the grid from ever entering a full blackout scenario.

## 3.4 Summary

This chapter examines the existing frameworks designed to safeguard power systems against failures and ensure the reliable operation of the critical and complex processes involved in electricity supply and distribution. It explores key operational challenges, vulnerabilities, technical limitations, and the ongoing advancements in protection systems currently implemented within our power infrastructure. Alongside Chapters 1 and 2, this chapter provides essential background knowledge needed to support the detailed analysis and model development presented in subsequent chapters. These models will address cascading failures and outline crucial strategies for enhancing the robustness and integrity of modern power systems.

# 4

# *Cascading Failure Models of Evolving Power Grids*

Cascading failures are one of the most serious threats to the secure operation of a power system. A seemingly minor disturbance can trigger a sequence of dependent outages that, if unchecked, may escalate into a large-scale blackout. Modeling these processes is essential for understanding how failures propagate, estimating blackout risk, and identifying effective measures to strengthen system robustness.

In the past, most modeling approaches focused solely on the physical power network. However, as power systems evolve into more complex, interconnected, and dynamic infrastructures, new factors must be considered. Modern grids increasingly feature tight coupling with cyber communication networks, increased integration of renewable energy sources that reduce system inertia, and growing shares of inverter-based resources with distinct protection and control behaviors. These changes significantly alter the dynamics of cascading failures, making traditional models inadequate for accurate risk assessment.

This chapter presents a progression of cascading failure models, starting from fundamental deterministic and stochastic approaches and advancing to frameworks that incorporate cyber-physical interactions, dynamic frequency control in high-renewable systems, and the specific disconnection mechanisms of inverter-based generators. The final section compares the strengths, limitations, and application scopes of these models, providing the foundation for the robustness assessment and enhancement strategies explored in the following chapters.

## 4.1 Failure Propagation Mechanism

With the power network representations and power flow calculation methods introduced in Chapter 2, we can derive the mechanisms by which overload-induced failures propagate in a power system. When a line or node in the network exceeds its operational limits — whether thermal capacity, voltage tolerance, or stability margin — protection mechanisms are triggered to remove the component from service. This removal alters the network topology and

DOI: 10.1201/9781003622130-4

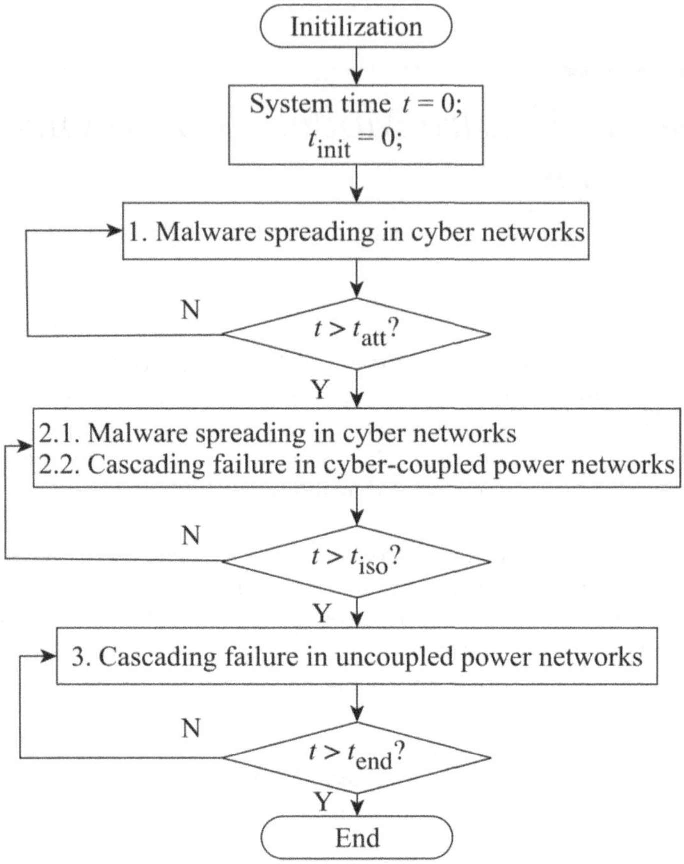


**FIGURE 4.1**
Flow chart of cascading failure.

redistributes power flows, which may in turn overload other components and cause further tripping. Repetition of this process can lead to cascading failures.

Two general modeling approaches are used to simulate this overload-propagation chain. The first is deterministic modeling, in which any component exceeding a predefined threshold is immediately considered non-functional and is disconnected from the network. The second is stochastic modeling, in which the underlying voltage and current stresses of individual components follow deterministic circuit equations, but the timing between successive failure events is governed by probabilistic processes. Deterministic models emphasize reproducibility and simple worst-case analysis, while stochastic models capture uncertainty and variability in the cascade evolution. Together, they form the foundation on which more advanced cascading failure frameworks are built in later sections.

### 4.1.1 Deterministic Modeling Approach

In a deterministic cascading failure model, once a link or node in the network fails, the topology of the power system changes, resulting in a redistribution of power flows according to the network equations introduced in Section 2.2.2. If this redistribution causes any remaining components to operate beyond their rated capacity, those components are immediately taken out of service. The process repeats until all remaining components can operate within their limits.

The deterministic framework is highly intuitive: exceeding the limit causes disconnection immediately. It is widely used for worst-case scenario analysis due to its simplicity and reproducibility. In this subsection, the cascading process is described in three sequential stages, as illustrated in Figure 4.1.

1. Initialization settings: At the start of the simulation, basic operating parameters of the power network are defined. For the voltage setting, generator buses are initialized at 1.0 p.u. voltage; all non-generator (load) buses sink 1.0 p.u. of current. For the transmission line settings, each line has an initial admittance. Regarding the component ratings, the current loading of a transmission line is the magnitude of the line current. Its capacity is defined as

   $$\mathrm{C}_{\mathrm{line}} = (1+\alpha) \times I_{ij}(0) \tag{4.1}$$

   where $I_{ij}(0)$ is the initial current under the starting topology, and $\alpha$ is the line safety margin.

   The power loading of a node is defined as

   $$\mathrm{L}_{\mathrm{node}} = V_i(0) \times I_i(0) \tag{4.2}$$

   where $I_i(0)$ is the total current flowing out of node $i$, and the node's capacity is

   $$\mathrm{C}_{\mathrm{node}} = (1+\beta) \times V_i(0) \times I_{0i}(0) \tag{4.3}$$

   where $\beta$ is the node safety margin.

   In practice, economic constraints limit safety margins; In simulations, we normally set $\alpha = 0.2$ for lines and $\beta = 0.5$ for nodes. With these settings, power flow equations are solved for the initial state, yielding the nodal voltages, line currents, and initial loadings. Any component whose initial loading exceeds its capacity is immediately removed.

2. Placement of initial failure: To initiate a cascading sequence, one or multiple components (line or node) are randomly or intentionally selected as the initial failure. This component is removed from the network, triggering the first topology change.

3. Cascading iteration: Following the removal of a component, the network is re-evaluated:

(a) The updated network is decomposed into connected subgraphs.

(b) Subgraphs without generators contain only unserved nodes, which are considered out of service.

(c) In subgraphs with at least one generator, the power flow equations are solved to determine the new power flow distribution.

(d) Any remaining component whose loading exceeds its capacity is immediately removed.

These steps are repeated iteratively, with the network topology, flows, and loadings updated at each step, until no component exceeds its limit. The system at this stage is considered to have reached its final balanced condition, which may correspond to a fully or partially served network.

The deterministic approach offers clear rules and fast simulation, making it suitable for identifying the structural weaknesses of a network under stress. However, it ignores random variations in timing, operator interventions, or protective device behavior, which are addressed in the stochastic models of the following section.

### 4.1.2 Stochastic Modeling Approach

The stochastic modeling approach extends the deterministic overload-tripping framework by incorporating randomness in failure timing. While the sequence of failure locations is still governed by deterministic circuit-based power flow equations (Chapter 2), the time intervals between consecutive failures are represented by stochastic processes. This enables the model to capture the variability introduced by diverse and uncertain physical mechanisms, such as manufacturing tolerances, environmental influences, and protection system delays, that are difficult to fully describe deterministically.

A distinguishing feature of this approach is that it allows the entire dynamic profile of a cascading failure to be simulated in continuous time, rather than only as a sequence of discrete "immediate" outages. The model consists of two key layers: (1) Component-level failure mechanism: The tripping behavior of individual nodes or links is described probabilistically, based on their overload extent and protection settings; (2) Network-level collective dynamics: The overall cascade evolution is treated as a system of interacting elements, analyzed using an extended Chemical Master Equation (CME)/Gillespie simulation method to determine both the time to the next failure and which component fails next.

#### 4.1.2.1 Component-Level Transition Dynamics

The power network is represented as a graph $G$, where nodes correspond to electrical stations (generators or substations), and links represent transmission lines. Nodes and links are collectively referred to as elements, each with two

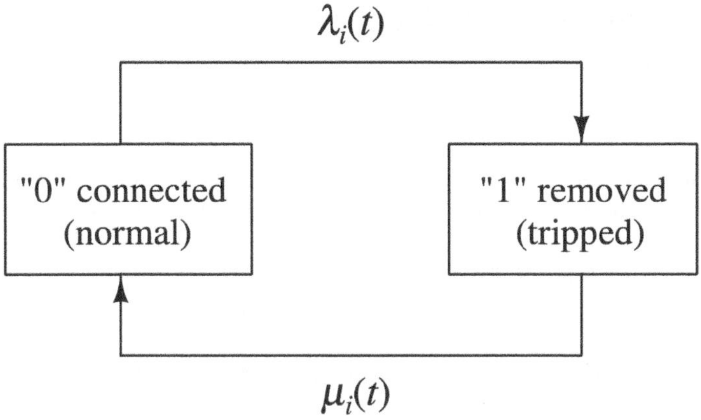


**FIGURE 4.2**
Dynamic description of failure in terms of state transitions. State "0" is the normal connected state; state "1" is the removed or tripped state. Arrows represent transitions between different states, while self-loop arrows are not displayed in the figure.

possible states: $s_i(t) = 0$ means element $i$ is connected and operating; $s_i(t) = 1$ means element $i$ is tripped or disconnected.

A component's state transition from 0 to 1 is called an Element State Transition (EST), which is shown in Figure 4.2. The transition rate $\lambda_i(t)$, i.e., probability per unit time of moving from state 0 to 1, consists of two additive parts:

$$\lambda_i(t) = \lambda_i^0(t) + \lambda_i^1(t) \tag{4.4}$$

where $\lambda_i^0$ is the baseline failure rate (independent of load), estimated from historical data, and $\lambda_i^1$ is the tripping rate due to overload protection.

When $L_i(t) > C_i$ (load exceeds capacity), $\lambda_i^1(t)$ increases linearly with the overload extent, i.e.,

$$\lambda_i^1(t) = \begin{cases} a_i \dfrac{L_i(t) - C_i}{C_i}, & \text{if } L_i(t) > C_i \\ 0, & \text{otherwise} \end{cases} \tag{4.5}$$

where $L_i(t)$ is the load on element $i$ from the latest power flow calculation, $C_i$ is the rated capacity, and $a_i$ is a unit scaling factor (trips/sec).

In fast cascading failure simulations, overload-induced trips dominate, so $\lambda_i^1(t) \gg \lambda_i^0(t)$ and $\lambda_i(t) \approx \lambda_i^1(t)$. Repair (transition $1 \to 0$) is possible in the general framework, with recovery rate $\mu_i(t)$, but is neglected ($\mu_i = 0$) in high-speed cascade scenarios due to the short time scale.

#### 4.1.2.2 Network-Level Failure Propagation

The state of the network at time $t$ is

$$S(t) = \{s_1(t), s_2(t), \ldots, s_m(t)\} \tag{4.6}$$

with $m$ total elements. A cascade evolves as a sequence of such states $S(t_1) \rightarrow S(t_2) \rightarrow \ldots$, where each step corresponds to exactly one overloaded element tripping.

The extended Gillespie method is applied to determine: (1) the time to the next trip ($\tau$), and (2) the next failed component, given the set of overloaded elements.

At time $t_1$, with set $\Omega_0$ of operating elements, the total network transition rate is

$$\lambda^*(t_1) = \sum_{i \in \Omega_0} \lambda_i(t_1) \tag{4.7}$$

which acts as a measure of the system-wide overloading stress. The probability that no trip occurs for a duration $\tau$ is

$$Q(\tau) = e^{-\lambda^*(t_1)\tau} \tag{4.8}$$

From this, the time to the next trip is

$$\tau = \frac{\ln(1 - z_1)}{-\lambda^*(t_1)}, \quad z_1 \sim U(0,1) \tag{4.9}$$

#### 4.1.2.3 Selecting the Next Failed Component

While multiple elements may be overloaded, only one trip per step is used for modeling clarity. The probability that element $i$ is the next to trip is proportional to its own rate, i.e.,

$$\text{Relative probability:} \quad rf_i = \frac{\lambda_i(t_1)}{\lambda^*(t_1)} \tag{4.10}$$

With a second uniform random number $z_2$, the next element is selected according to the cumulative probabilities of $rf_i$. The detailed derivation process is provided in Appendix A.

#### 4.1.2.4 Simulation Procedure

The stochastic cascading failure simulation proceeds as follows, as illustrated in Figure 4.3.

1. Initialization: Assign generator voltages, load currents, line admittances, and element capacities. Compute baseline flows via power flow equations.
2. Plant initial failure: Randomly remove one or multiple elements; set $S(t_0)$ accordingly.

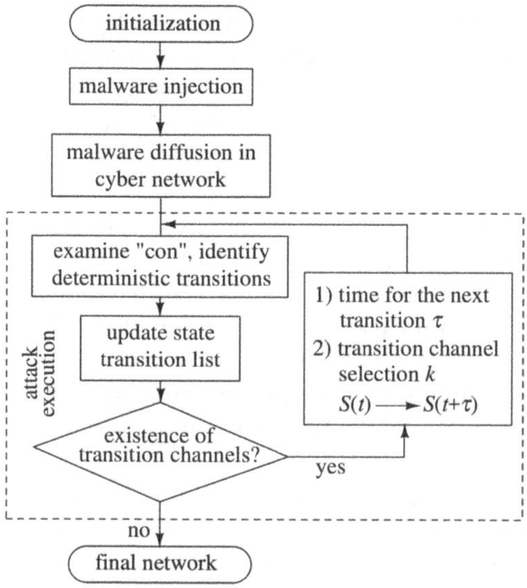


**FIGURE 4.3**
Flow chart for simulating the dynamic propagation of cascading failure.

3. Iterate:
    (a) Identify network islands. Elements in islands without generators remain unpowered ($s_i = 0$ but load $= 0$).
    (b) For each active island (with generators), compute new power flows and determine $\lambda_i$ for all connected elements.
    (c) If overloaded ($\lambda_i > 0$), sample $\tau$ and select next element to trip.
    (d) Advance time $t \leftarrow t+\tau$, update $S(t)$, and repeat until all $\lambda_i = 0$.

The result is a time-stamped sequence of failures showing both the order and timing of events, providing a much richer profile than purely deterministic models.

## 4.2 Cascading Failure Models in Cyber-Coupled Power Systems

This section presents a network-based stochastic model for analyzing cascading failures in cyber-physical power systems (smart grids). The model explicitly represents the interdependence between the physical power grid and a coupled cyber network, where malfunction or removal of a cyber node can disrupt the associated power node, and vice versa. By incorporating three key mechanisms, namely, power overloading, malware contagion in the cyber layer, and cross-layer interdependencies, the framework captures the complex dynamics of blackout propagation under cyberattack scenarios.

Cyber–physical systems (CPS) have emerged as essential infrastructures, combining physical processes with computational and communication capabilities to enable intelligent monitoring, control, and optimization across residential, commercial, industrial, and military domains [126, 217]. Among their diverse applications, the smart grid is one of the most critical examples: an electric power network augmented with advanced cyber infrastructure for sensing, communication, and control [292]. While this integration enhances efficiency and reliability, it also creates new cybersecurity vulnerabilities. The 2015 Ukrainian power grid outage, caused by a targeted malware attack, underscored that cyber intrusions can trigger real, large-scale physical disruptions [241].

Research on cascading failures in such systems has evolved along two main lines. Network science approaches model power grids as graphs, where nodes represent generators, substations, or loads, and edges represent transmission lines [265, 278]. These models are useful for large-scale vulnerability analysis but often treat grid elements as abstract entities. Power systems engineering approaches, in contrast, focus on detailed electrical stability phenomena, such as voltage collapse and frequency instability [144]. More recently, hybrid approaches have emerged, combining network-based formalisms with embedded physical models to ensure that cascading processes reflect actual operational constraints [55].

In modern smart grids, cyber assets such as smart meters, intelligent sensors, and automated controllers are pervasive, and supervisory control and data acquisition systems (SCADA) are themselves susceptible to cyberattacks [13]. Attackers can compromise SCADA to execute disruptive actions such as system desynchronization [255] or false data injection [165]. This has led to growing interest in modeling interdependent networks—or coupled networks, where the power grid and cyber network are linked via functional dependencies. Early studies by Buldyrev et al. [42] proposed iterative models of cascading failure across such interdependent systems, showing how coupling can amplify vulnerability.

However, purely topological or statistical physics–based coupled-network models omit crucial engineering realities. In real systems, Kirchhoff's laws dictate the distribution of power flows in the grid, while communication protocols govern cyber data exchanges. Ignoring these mechanisms can yield unrealistic or inconsistent results. To address this, recent works [52, 158] have integrated AC or DC power flow models into interdependent network studies, enabling more accurate assessments of operational robustness.

A central challenge is defining realistic interdependence between the cyber and power layers. For example, in the interactive model by Cai et al. [43], tripping of power components can disable data centers in the cyber layer, while cyber component failures can overload transmission lines. Other studies have examined the effects of multiple attack types—including denial-of-service (DoS), replay, and false data injection—on cascading failure in cyber–physical grids [266], highlighting the growing complexity of the problem.

Existing modeling efforts often suffer from three key limitations:

1. Omission of ongoing malware propagation: Many models assume malware fully compromises the cyber network instantaneously and ignore continued infection during the physical cascade [242].
2. Oversimplified cyber–power interdependence modeling: Percolation-based dependency assumptions do not map cleanly onto engineering realities.
3. Omission of active attack–defense interactions: Few models address how attackers and defenders strategically allocate their resources during the cascade.

The model developed in this section addresses these gaps. Specifically, the model (1) integrates a network-based framework with a physical power flow model to account for realistic electrical operation; (2) represents bidirectional interdependence between cyber and power layers, with simultaneous malware contagion in the cyber network and overload-induced tripping in the power network; and (3) incorporates strategic attack–defense dynamics, where both sides allocate finite resources to attack or protect targeted elements.

This study examines multiple resource allocation strategies (including uniform, degree-based, capacity-based, and random strategies) and evaluates their performance using four key temporal markers of the cascade: start time, attack time, isolation time, and end time. By combining these features, the model offers a practical, engineering-oriented approach to assessing the vulnerability of cyber-coupled power systems and identifying effective defense strategies, including a tit-for-tat response, where defenders mirror attackers' strategies, which is found to be effective under most conditions. Furthermore, we introduce the relative coupling correlation coefficient to characterize coupling patterns between the power and cyber layers. This parameter serves as both a functionality indicator and a means to assess the impact of coupling patterns on robustness. Four classes of coupling patterns are defined, based on two cyber-node criticality metrics (node degree and node betweenness) and two power-node criticality metrics (node degree and node capacity). Using a network-based model that explicitly includes physical processes, simulations show that a higher relative coupling correlation coefficient correlates with a lower robustness index, indicating reduced resistance to cascading failures.

### 4.2.1 Model Description

This section describes the model used for simulating cascading failures in cyber-coupled power systems. The model aims to explicitly incorporate malware attack processes occurring in the cyber layer. The integrated framework

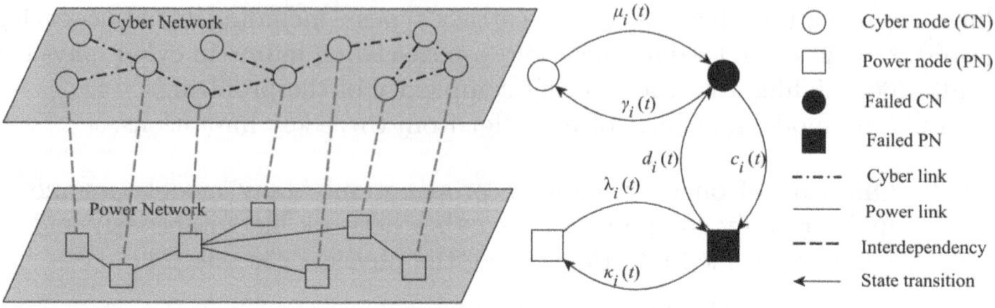


**FIGURE 4.4**
Coupled network consisting of a cyber network and a power network, with state transitions showing infection of a cyber node, overload tripping of a power node, and attack to a power node from a cyber node.

enables the study of different attack–defense strategy combinations and their effect on failure propagation patterns.

#### 4.2.1.1 Cyber–Coupled Power Systems

We model the cyber–physical system as a pair of interdependent networks. Power network $A$ consists of $N_A$ nodes (buses, substations) and links (transmission lines). Cyber network $B$ consists of $N_B$ nodes (control devices, communication nodes). The cyber network monitors and controls the power network, while the power network provides power to the cyber network. Interdependency is represented by vertical dashed links between nodes in $A$ and $B$, as shown in Figure 4.4.

#### 4.2.1.2 Coupling Structure

We define the coupling strength using the formulation of Parshani et al. [204]. First, a fraction $q_A$ of nodes in $A$ are connected to nodes in $B$. Second, a fraction $q_B$ of nodes in $B$ are connected to nodes in $A$. For simplicity, we assume one-to-one coupling, although the model can be extended to other coupling patterns, which can be denoted by

$$N_{A-B} = q_A N_A = q_B N_B \tag{4.11}$$

#### 4.2.1.3 Coupling Pattern

In the network-based model, the number of cyber nodes is typically larger than the number of power nodes. The first step is to select a subset of cyber nodes for one-to-one connection with the power nodes. We assume that the selected cyber nodes can directly control their coupled power nodes. The deployment of cyber facilities introduces intelligent control and monitoring functions into modern power grids. To implement automatic control, operational data such as bus voltages and currents can be collected by sensors, e.g., power measurement units (PMUs), and processed to make control decisions such as load shedding or optimal transmission switching.

According to the specific control and monitoring functions [126], cyber nodes coupled to the power system can be broadly categorised as: (1) *data-collecting* nodes, (2) *decision-making* nodes, and (3) other functional nodes (e.g., data processing, data exchange). Appropriately distributing data-collecting and decision-making cyber nodes within the cyber network facilitates effective control of the coupled system [283].

The first step in coupling a cyber network to a power network is to determine which decision-making cyber nodes will be connected to the power nodes. Using the controller placement strategy used in software-defined networks [115], we select decision-making cyber nodes to minimize the packet propagation latency between them and other cyber nodes [115]. The *average propagation latency* $\epsilon$ is defined as

$$\epsilon = \frac{1}{N-K} \sum_{a \in A \setminus A_c} d(a, A_c) \tag{4.12}$$

where $A_c$ is the set of decision-making cyber nodes, $A \setminus A_c$ is the set of all other cyber nodes, $K$ is the number of decision-making cyber nodes, and $d(a, A_c)$ is the shortest path length between a cyber node $a \in A \setminus A_c$ and its nearest decision-making cyber node.

A diagrammatic explanation of $\epsilon$ is given in Figure 4.5, where two basic communication topologies—star and tree—are illustrated. For the star topology, $\epsilon = (1/6)\times(1+1+1+1+1+1) = 1$ (left), $\epsilon = (1/6)\times(1+2+2+2+2+2) = 11/6$ (middle), and $\epsilon = (1/6) \times (1+2+2+2+2+2) = 11/6$ (right). For the tree topology, $\epsilon = (1/6) \times (1+1+2+2+2+2) = 10/6$ (left), $\epsilon = (1/6) \times (1+1+1+2+3+3) = 11/6$ (middle), and $\epsilon = (1/6) \times (1+2+2+3+4+4) = 16/6$ (right). In both topologies, the lowest latency occurs in the left-hand scenario.

For a given cyber network, the placement of decision-making nodes greatly influences $\epsilon$. Central placement can yield lower latency and hence higher control efficiency—an important requirement for communication systems in modern smart grids [263]. In practice, decentralised modular communication architectures, such as those used in Japanese power grids, have demonstrated low latency and delay variation in system monitoring, enabling adaptive protection and emergency control to localise disruptions and prevent large-scale blackouts [230].

The *coupling pattern* refers to how cyber and power nodes are interconnected. The choice of coupling pattern has a profound impact on the robustness of the coupled network. Prior work has shown that higher *inter-similarity* between networks tends to increase robustness. One such inter-similarity metric is the *inter degree-degree correlation* (IDDC), which is given in equation (2.1). It is noted that:

1. Large positive $r$: *assortative coupling*—high-degree nodes in one network connect to high-degree nodes in the other.

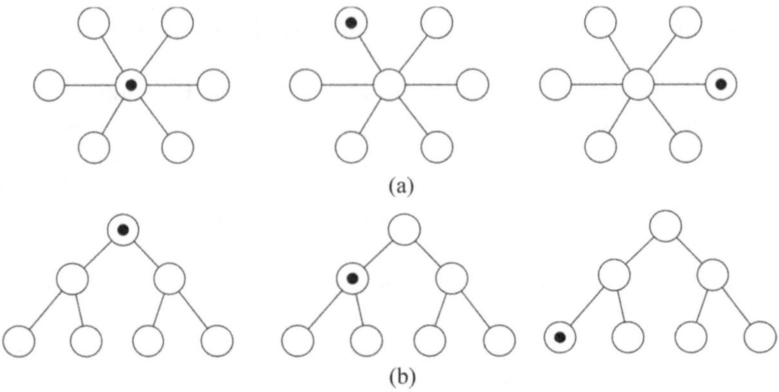


**FIGURE 4.5**
Average propagation latency of (a) star network when one decision-making cyber node (with a dot inside) is placed in three different locations. (b) Tree network when one decision-making cyber node (with a dot inside) is placed in three different locations.

2. Large negative $r$: *disassortative coupling*—high-degree nodes in one network connect to low-degree nodes in the other.

Figure 4.6 shows examples of coupling patterns: left-assortative ($\rho > 0$), middle-intermediate, right-disassortative ($\rho < 0$).

In our study of cyber-coupled power systems, apart from using node degrees, we consider a few other network-based metrics for describing node criticality. Here, we characterize the coupling pattern by extending the inter-degree-degree correlation coefficient to an inter-criticality-criticality correlation coefficient. This new metric, called *relative coupling correlation coefficient*, is defined as

$$\rho = \frac{\mathfrak{m}^{-1}\sum_{(i,j)\in M} I'_{A,(i,j)} I'_{B(i,j)} - [\mathfrak{m}^{-1}\sum_{(i,j)\in M} \frac{1}{2}(I'_{A,(i,j)} + I'_{B,(i,j)})]^2}{\mathfrak{m}^{-1}\sum_{(i,j)\in M} \frac{1}{2}(I'^2_{A,(i,j)} + I'^2_{B,(i,j)}) - [\mathfrak{m}^{-1}\sum_{(i,j)\in M} \frac{1}{2}(I'_{A,(i,j)} + I'_{B,(i,j)})]^2} \tag{4.13}$$

where $I'_{A,(i,j)}$ and $I'_{B,(i,j)}$ are the *normalized node criticality metrics* of two nodes in the two interconnected networks, respectively, when they are connected by edge $m$. As explained before, the existence of an interconnected edge $m$ connecting cyber and power nodes implies that the cyber node can take a malicious action against its coupled power node.

The normalized node criticality for the power network $A$ is

$$I'_A = \frac{I_A - \min(I_A)}{\max(I_A) - \min(I_A)} \tag{4.14}$$

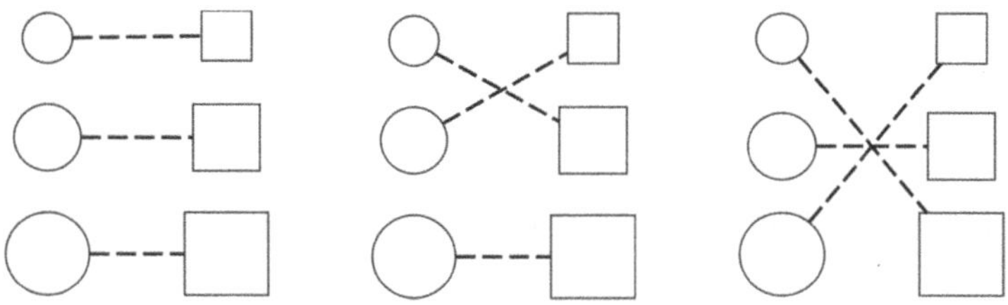

**FIGURE 4.6**
Illustration of relative coupling correlation coefficient $\rho$.

and for the cyber network $B$, it is

$$I'_B = \frac{I_B - \min(I_B)}{\max(I_B) - \min(I_B)} \tag{4.15}$$

Here, $I_A$ and $I_B$ are the (unnormalized) node criticality metrics in the power and cyber networks, respectively. For cyber networks, we focus on two centrality-based metrics: node degree and node betweenness [191], as these capture how well a node is connected and its influence on data transmission [48]. In power networks, we use node degree and node capacity as criticality measures, consistent with prior robustness studies [38, 186].

The coupling pattern is determined by sorting power and cyber nodes by their criticality and linking them in a prescribed order.

1. Ascending-to-ascending order yields assortative patterns ($\rho > 0$).
2. Ascending-to-descending order yields disassortative patterns ($\rho < 0$).

Besides these extremes, intermediate patterns can be described and quantified using (4.13). In this study, we consider four classes of coupling patterns according to the chosen node criticality metrics in each layer:

1. degree-to-degree (d2d)
2. degree-to-betweenness (d2b)
3. capacity-to-degree (c2d)
4. capacity-to-betweenness (c2b)

#### 4.2.1.4 Dynamics of Cascading Failure

Two processes govern cascading failures in a coupled system. The first one describes the malware contagion in the cyber network. Cyber nodes can be infected by malware, and once infected, can launch malicious actions on their coupled power nodes (interdiction attacks [113]). Denote a normal cyber node by a white circle, and an infected node by a solid black circle. Following the results from reference [136], the infection rate for cyber node $i$ is

$$\mu_i(t) = \sum_{j \in \Omega} \beta_{ij} \tag{4.16}$$

where $\Omega$ is the set of infected neighbors of node $i$ and $\beta_{ij}$ is the infection rate from node $j$ to node $i$.

For fast cascades, recovery is ignored ($\gamma_i(t) = 0$). In an infinitesimal time interval $dt$, the probability that node $i$ becomes infected is

$$T_1 : P[s_{B_i}(t + dt) = 1 \mid s_{B_i}(t) = 0] = \mu_i(t)\, dt \tag{4.17}$$

The second process relates to the overload tripping and unserved disconnection in the power network. A power node (square in Figure 4.4) can fail in two ways: (1) overload trip, and (2) immediate failure if isolated from all generators. For a normally operating power node,

$$T_2 : P[s_{A_i}(t + dt) = 1 \mid s_{A_i}(t) = 0] = \lambda_i(t)\, dt \tag{4.18}$$

where $\lambda_i(t)$ is the stochastic overload trip rate. When the coupled cyber node is infected, malware increases the trip rate by $c_i(t)$, giving

$$T_3 : P[s_{A_i}(t + dt) = 1 \mid s_{A_i}(t) = 0] = \big(\lambda_i(t) + c_i(t)\big)\, dt \tag{4.19}$$

Here, $c_i(t)$ quantifies the added failure risk due to cyber compromise. We assume power-to-cyber coupling is negligible during cascades ($d_i(t) \equiv 0$); thus, failure is unidirectional from cyber to power.

#### 4.2.1.5 Attack–Defense Interaction

We now incorporate strategic interactions into $c_i(t)$, i.e.,

$$c_i(t) = \begin{cases} \chi_i(t) - \psi_i(t), & A_i \in A' \\ 0, & A_i \notin A' \end{cases} \tag{4.20}$$

where $A'$ is the set of power nodes coupled with cyber nodes, $\chi_i(t)$ and $\psi_i(t)$ represent attack rate (malware-induced risk) and defense rate (protection effect), respectively.

Both rates have the same physical units as $\lambda_i(t)$ and can be positive or negative (net effect limited to $\geq 0$ failure probability), i.e.,

$$T_3' : P[s_{A_i}(t + dt) = 1 \mid s_{A_i}(t) = 0] = \max\big\{\, 0, \lambda_i(t) + c_i(t) \,\big\}\, dt \tag{4.21}$$

From the attacker's perspective, a vulnerable power node is one whose coupled cyber node is infected. Let $V$ be the set of vulnerable nodes. The attack budget $X$ is constant, allocated via four strategies:

1. Even distribution ($S_{A1}$):

$$\chi_i(t) = X, \ \forall A_i \in V \tag{4.22}$$

2. Degree-based ($S_{A2}$): proportional to node degree $k_{A_i}$, which is given by

$$\chi_i(t) = \frac{k_{A_i} X}{\frac{1}{N_A} \sum_{A_i} k_{A_i}} \tag{4.23}$$

3. Capacity-based ($S_{A3}$): proportional to node capacity $C_{A_i}$, which is given by

$$\chi_i(t) = \frac{C_{A_i} X}{\frac{1}{N_A} \sum_{A_i} C_{A_i}} \tag{4.24}$$

4. Random ($S_{A4}$): random allocation subject to $\sum \chi_i(t) = N_V X$.

From the defender's perspective, $\psi_i(t)$ is applied to all cyber-coupled power nodes $A'$ with budget $\Psi$, allocated via four analogous strategies:

1. Even distribution ($S_{D1}$): $\psi_i(t) = \Psi$, $\forall A_i \in A'$;
2. Degree-based ($S_{D2}$): proportional to $k_{A_i}$;
3. Capacity-based ($S_{D3}$): proportional to $C_{A_i}$;
4. Random ($S_{D4}$): random allocation with $\sum \psi_i(t) = N_{A'} \Psi$.

#### 4.2.1.6 Analysis of Failure Propagation

Here, we introduce four critical time points to permit detailed time series analysis. The aim of the time series analysis is to investigate the effect of the launch time determined by the attacker on the severity of the cascading failure in coupled systems. The Ukrainian blackout report has highlighted two important points [15,145]. First, the attacker started shutting down the power grid by action of a malware which has intruded and diffused over the cyber network for a certain period of time. Second, the operator of the power system switched from automatic mode to manual mode of the control operation. This operation was executed to stop the spreading of failure in the power network caused by cyber coupling. In other words, switching to the manual control isolates the infected cyber network from the power system. When the cyber network carrying malware is disconnected, the power network cannot be controlled by the attacker. Correspondingly, we define four critical time points to construct a failure propagation profile.

1. Intrusion time ($t_{\text{init}}$): the time when the malware begins to infect the cyber network.
2. Attack launch time or launch time ($t_{\text{att}}$): the time when the attacker launches the attack. Before $t = t_{\text{att}}$, the malware can diffuse to more cyber nodes, and the attacker may obtain more information to effectively impair the power grid. It is noted that $t_{\text{att}} \geq t_{\text{int}}$ and $t_{\text{att}}$ will serve as a variable for examining cascading failure in coupled systems.

3. Isolation time ($t_{\text{iso}}$): the time when the cyber network is isolated from the power system, making the power grid unaffected by the malware. Thus, $c_i(t)$ in equation (4.21) is equal to zero after $t = t_{\text{iso}}$. The value of $t_{\text{iso}}$ is determined after confirming that the failure is caused by cyberattacks. There are a number of approaches in the detection of cyberattacks [175,181]. Here, we implement one feasible solution, and specifically, we assume that the cybersecurity scheme starts to determine whether an attack has originated from the cyber network when the first failure of a power network occurs at $t = t_{\text{fail}}(m)$ where $m$ indicates the number of failed power nodes. Here, according to SCADA-specific intrusion detection applied in power grids [306], if more cyber nodes are infected when the detection is initialized, the probability that the data containing abnormal activity can be captured is higher. Thus, it is appropriate to assume that the detection time is inversely proportional to the number of infected nodes in the cyber network at the time when an action of detection is taken. Here, the $t_{\text{iso}}$ is given in the basic form of

$$t_{\text{iso}} = \frac{T_{\text{ISO}}}{P_{\text{ICN}}} \tag{4.25}$$

where $T_{\text{ISO}}$ is a constant determined by the amount of resource available to the defender for detecting the intrusion, and $P_{\text{ICN}}$ is the percentage of the infected cyber nodes at the time when the first power node fails.

4. End time ($t_{\text{end}}$): the time at which the cascading failure ends. In practice, the power system has its own protection scheme to stop cascading failure within a period of time. Moreover, if there is no protection scheme terminating a cascading failure, then the propagation failure should end at $t_{\text{end}}$, which is the time when there is no more power overloading either in nodes or links.

It is noted that this model explicitly couples the physical power flow process with cyber malware propagation, while embedding game-like attack–defense allocation mechanics.

### 4.2.2 Case Study

Simulations of failure propagation in a coupled system can be performed according to the flow chart shown in Figure 4.7, which comprises three main processes, namely, malware spreading on the cyber network, cascading failure in the cyber-coupled power system, and cascading failure in the uncoupled power grid (after isolation is effected).

As a preliminary study, in this subsection, we specifically examine the effect of launch time on the possible scenarios of cascading failure in the coupled system. We consider the coupled system comprising the UIUC-150 Bus power

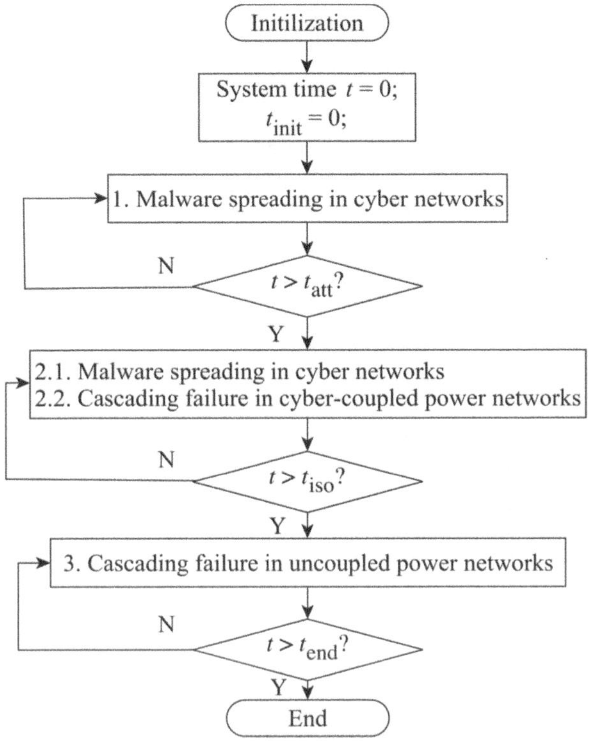


**FIGURE 4.7**
Flow chart of simulation of failure propagation in a coupled system.

system [127] and a cyber system realized by a Gnutella peer-to-peer network containing 6301 nodes and 20777 edges [146]. We randomly select $q_A N_A$, where $q_A = 1$ power nodes to connect with the same number of cyber nodes in a one-to-one connection style. In particular, we consider the entire power network being fully coupled with the cyber network. Furthermore, in the simulations, we set the current limits as $C_i = (1 + \alpha) \times I_i(\text{normal})$, where $I_i(\text{normal})$ is the total current flowing out of a power node under the normal load demand condition. In particular, the tolerance index $\alpha$ is set as 0.2, which implies that the capacity of each power node is 1.2 times larger than its load. To launch the simulation study, the default parameters are set as follows:

1. The failure rate in power system $a_i$ is 0.21 min$^{-1}$ and the infection rate $\beta_{ij}$ is one-tenth of $a_i$.
2. The attack and defense strengths are assumed to be the same, with the assumption that the attacker and defender have the same amount of resources to attack and defend one power node. Thus, the ratio between two strengths is one, i.e., $X/\Psi = 1$.

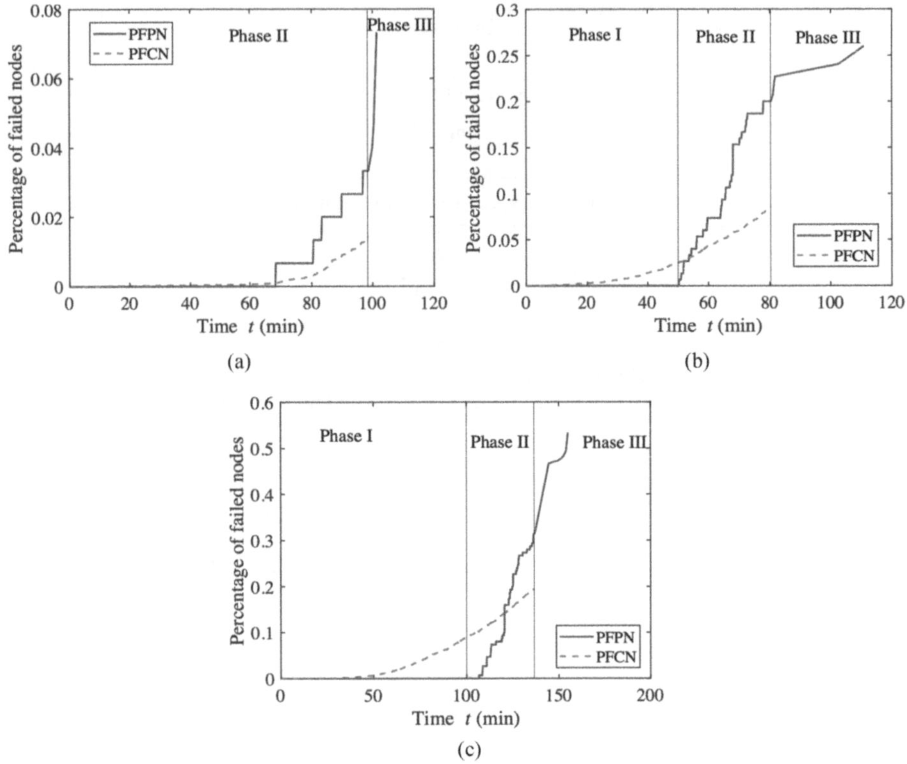


**FIGURE 4.8**
Failure propagation in coupled systems under the condition of random coupling with (a) $t_{\text{att}} = 0$; (b) $t_{\text{att}} = 50$ min; (c) $t_{\text{att}} = 100$ min.

3. According to prior reports on malware analysis [117,128], if a cyber network is fully infected within several tens of minutes, the malware can be detected, and then the potential cascading failure originated from cyberattacks can be confirmed. Here, $T_{\text{ISO}}$ is set as 10 min.

We introduce two essential metrics for characterizing the extent of the failure, namely, *percentage of failed power nodes* (PFPN) and *percentage of failed cyber nodes* (PFCN), which are defined as follows:

$$\text{PFPN}(t) = \frac{n_{\text{unserved}}(t) + n_{\text{removed}}(t)}{p} \tag{4.26}$$

$$\text{PFCN}(t) = \frac{n_{\text{infected}}(t) + n_{\text{shutdown}}(t)}{q} \tag{4.27}$$

where $n_{\text{unserved}}(t)$ and $n_{\text{removed}}(t)$ represent the number of power nodes in state 1 and 2 at time $t$, respectively. Similarly, $n_{\text{infected}}(t)$ and $n_{\text{shutdown}}(t)$

are the number of cyber nodes in state 1 and 2 at time $t$, respectively. Note that a large PFPN($t$) (PFCN($t$)) means that a large total area of disconnected fragments of the power grid (cyber network) is out of operation.

#### 4.2.2.1 Phases of Failure Propagation

This case study illustrates the dynamic process of cascading failure in a coupled cyber–physical power system under a malware-driven attack. Three distinct phases are observed in the full propagation process, as shown in Figure 4.8, which plots the percentage of failed nodes in both the power network and the cyber network over time for three representative simulation scenarios. The initial malware injection occurs at $t = 0$ ($t_{\text{init}} = 0$), and the curves reveal how the interaction between malware spread, cyber–power coupling, and overload-induced failures shapes the progression of the cascade.

In Phase I ($t_{\text{init}} \leq t < t_{\text{att}}$), the malware spreads through the cyber network, progressively compromising more control devices and gaining intelligence on the power grid topology and operation. Simultaneously, the attacker's access to operational assets increases, but no direct attack is yet launched on the power grid. This is reflected in Figure 4.8c by the steadily rising dashed curve prior to $t_{\text{att}} \approx 100$ min, showing the growing percentage of infected cyber nodes. In scenarios where $t_{\text{att}} = 0$, such as in Figure 4.8a, Phase I is absent, and the attack on the power grid begins immediately.

In Phase II ($t_{\text{att}} \leq t < t_{\text{iso}}$), the attacker actively uses the malware-compromised cyber nodes to launch control-based attacks on their coupled power nodes. Two processes occur concurrently, including the continued malware spread in the cyber network and the onset of cascading failure in the power network caused by the direct cyber-induced outages and the overload tripping from redistributed power flows. The solid curve in Figure 4.8n (between 50–80 min) shows the rapid increase in failed power nodes. Notably, a "staircase-like" propagation pattern is visible: repeated bursts of failures occur as successive cyberattacks destabilize different parts of the grid, each burst followed by topology adjustments. As only malware-triggered attacks affect the power network in this phase, it is referred to as the active attack phase.

In Phase III ($t_{\text{iso}} \leq t \leq t_{\text{end}}$), once the cyber network becomes isolated from the power grid control layer, further failures occur solely due to overload-induced tripping in the power network (corresponding to transition $T_1$ in our model). Malware-induced actions no longer play a role. In Figure 4.8b and c, the percentage of failed power nodes continues to rise until no component exceeds its capacity, marking the cascade endpoint at $t_{\text{end}}$. This overload-driven tail somewhat resembles post-disturbance decay patterns observed in earlier single-layer studies [302].

Two main observations emerge from varying the attack initiation time $t_{\text{att}}$:

1. Active phase duration matters: A larger $t_{\text{att}}$ shortens the active attack phase, reducing the window for coordinated cyber-induced damage.
2. Impact of cyber infection level at $t_{\text{att}}$: Severe outages occur when many cyber nodes are already infected at the start of the active attack. For example, when 10% of cyber nodes are infected at $t_{\text{att}}$ (Figure 4.8c), more than 50% of power nodes fail, whereas lower infection levels (Figure 4.8a and b) lead to much milder impacts.

By using the four critical time markers, namely, $t_{\text{init}}$ (start), $t_{\text{att}}$ (attack launch), $t_{\text{iso}}$ (isolation), and $t_{\text{end}}$ (cascade end), this analysis provides a detailed temporal profile of cyber–power cascading failures across multiple failure mechanisms. It is noted that this model offers a practically consistent and operationally relevant view of cascading robustness in cyber–physical power systems.

#### 4.2.2.2 Attack–Defense Strategy Interactions

In this case study, the model developed in Section 4.2.1 is applied to a large number of numerical simulations in order to examine in depth how different combinations of attack and defense strategies influence the propagation of cascading failures in a coupled cyber–physical power system. The results provide a more comprehensive understanding of the interplay between adversarial actions launched from the cyber layer and the corresponding countermeasures.

To assess the extent and speed of blackout progression in different scenarios, we make use of the Average Failure Rate (AFR), defined as

$$\text{AFR}(t_{\min}, t_{\max}) = \frac{N_{\text{FPN}}(t_{\max}) - N_{\text{FPN}}(t_{\min})}{t_{\max} - t_{\min}} \tag{4.28}$$

where $N_{\text{FPN}}(t)$ denotes the number of failed power nodes at time $t$, while $t_{\min}$ and $t_{\max}$ define the evaluation time window. Here $t_{\max}$ is always set to be $t_{\text{end}}$, the time when cascading failure concludes, which can vary across simulation runs. The AFR measures the average rate of power node failures over the specified time span. If $t_{\min} = 0$, the AFR evaluates the average failure rate from the start of the malware launch to the end of the cascade, capturing the entire event. Alternatively, if $t_{\min} = t_{\text{att}}$, it measures the average failure rate in the active attack phase, namely from the moment the attacker begins targeting the power system directly. Naturally, $\text{AFR}(0, t_{\text{end}})$ is generally smaller than $\text{AFR}(t_{\text{att}}, t_{\text{end}})$, since the latter excludes the typically slower malware propagation period before the active attack. In this analysis, both forms are used to gain insight: the first to characterize the overall rate and severity of the failure process, and the second to focus on the rapidity of the power system collapse once cyber intrusion translates into physical damage. All reported AFR values are obtained by averaging over multiple simulation runs.

To investigate the influence of attacker–defender interactions, we design a set of "games" in which each player adopts one of the four strategies described in Section 4.2.1. With four attack strategies and four defense strategies, this yields sixteen distinct combinations. Each combination corresponds to a specific pairing of an attack resource allocation scheme and a defense resource allocation scheme. In each simulation, the malware launches from the cyber network into the power network at a certain attack start time $t_{\text{att}}$, which is varied systematically from 0 to 2,000 min. For every sampled value of $t_{\text{att}}$, the cascading process is simulated and the values of $\text{AFR}(0, t_{\text{end}})$ and $\text{AFR}(t_{\text{att}}, t_{\text{end}})$ are computed.

The full set of results is presented in Figures 4.9 and 4.10. In Figure 4.9, the attack strategy is held fixed while the defense strategy varies, covering all sixteen possible combinations. For example, Figure 4.9a and e show $\text{AFR}(0, t_{\text{end}})$ and $\text{AFR}(t_{\text{att}}, t_{\text{end}})$, respectively, for the case where the attacker uses Strategy I, facing each of the four defense strategies. Equivalent groupings are provided for the other attack strategies. Figure 4.10 offers the complementary perspective in which the defense strategy is held constant while the attack strategy is varied.

From the analysis, several observations emerge. First, for a given pair of strategies, there exists an optimal attack launch time $t_{\text{att}}$ at which the impact on the power network is maximized. Contrary to what might be expected, attacking later does not necessarily improve the attack's effectiveness, even though more reconnaissance and information gathering are possible when malware propagates longer before activation. The reason is that launching the attack too late shortens the active attack phase, providing an opportunity for defensive measures, such as isolating the power network from the cyber network, to take effect before the malware-induced failures can become widespread. This finding is consistent with real-world events such as the 2015 Ukrainian power grid incident, in which timely intervention helped limit the escalation of the attack.

Second, when the attacker follows a structured, non-random targeting strategy—such as an even, degree-based, or capacity-based allocation of attack resources—the most effective defensive response, in terms of reducing AFR, is to adopt the corresponding strategy on the defense side. In other words, a "tit-for-tat" approach, where the defender mirrors the attacker's resource allocation scheme, yields the best mitigation results for these structured scenarios. This alignment between attacker and defender strategies appears consistently across the combinations presented in Figure 4.9a–c and e–h. An exception arises when the attacker uses a random allocation strategy: in that case, no clear benefit is obtained from mirroring, because the defender's random allocation will almost never match the attacker's by chance, as seen in Figure 4.9d and h.

Finally, the complementary results in Figure 4.10, in which the defense strategy is fixed, show that tit-for-tat symmetry does not necessarily yield the most effective attack. In these runs, Attack Strategy III (capacity-based allocation) often appears to be highly effective against a range of defensive schemes, highlighting asymmetry in optimal choices between attacker and defender roles. Although the numerical values of $\mathrm{AFR}(0, t_{\mathrm{end}})$ and $\mathrm{AFR}(t_{\mathrm{att}}, t_{\mathrm{end}})$ differ, their qualitative behavior is similar, reinforcing the consistency of these conclusions.

Overall, this case study demonstrates that the temporal parameters of an attack and the strategic alignment between attackers and defenders play pivotal roles in determining the scale and speed of a cascading failure in a cyber–coupled power system. The results highlight the practical importance of modeling targeted attack and defense strategies jointly, as their interaction can significantly alter both the vulnerability and the resilience of the coupled infrastructure.

### 4.2.2.3 Tradeoff between Robustness and Functionality

Before examining the effect of the average propagation latency $\epsilon$ and the relative coupling correlation coefficient $\rho$ on the robustness of cyber-coupled power systems, we define a new robustness index to evaluate the robustness of such systems. Here, we adopt the definition of vulnerability introduced by Schneidera et al. [228], which considers the size of the largest affected component during all possible malicious attacks. Similarly, to assess the *vulnerability* of a cyber-coupled power system against cascading failures due to cyberattacks, we examine the extent of failures in the power network under various scales of malware infection in the cyber network. This can be expressed as:

$$V = \frac{1}{N_S} \sum_{s \in S} S_{\mathrm{PN}} \mid_{S_{\mathrm{CN}} = n_s} \tag{4.29}$$

where $S_{\mathrm{PN}}$ is the extent of cascading failure in the power network (number of failed power components), $S_{\mathrm{CN}}$ is the extent of infection in the cyber network (number of infected cyber nodes), $S$ is the set of $N_S$ infected cyber subnetworks, and in each cyber subnetwork $s$, $n_s$ denotes the number of infected cyber nodes. For example, $n_s = 10$ indicates that 10 cyber nodes have been infected. When $n_s$ cyber nodes are infected, the resulting size of cascading failure in the power network is recorded as $S_{\mathrm{PN}} \mid_{S_{\mathrm{CN}} = n_s}$. By calculating the average according to equation (4.29), the vulnerability of a cyber-coupled power system can be quantified. A larger value of $V$ indicates that a cyberattack leads to more severe cascading failures in the power grid.

For instance, consider two cyber-coupled power systems with distinct coupling patterns, denoted as CCPS1 and CCPS2. Suppose the sequence of $S_{\mathrm{CN}}$ (number of infected cyber nodes) is $10, 20, 30$, and the corresponding $S_{\mathrm{PN}}$ (number of failed power nodes) values are $10, 20, 30$ for CCPS1 and $15, 25, 35$ for CCPS2. In this case, CCPS1 yields $V = 20$, while CCPS2 yields $V = 25$, indicating that CCPS1 is less vulnerable.

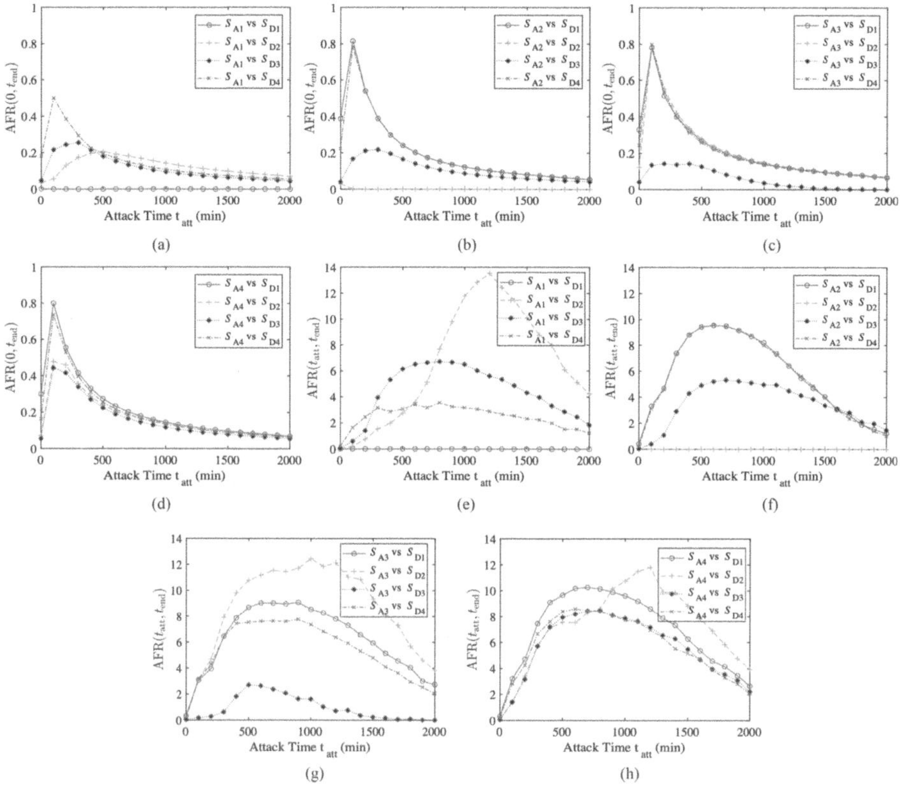


**FIGURE 4.9**

Comparison of attack-defense interaction for two average failure rates, i.e., $\text{AFR}(0, t_{\text{end}})$ and $\text{AFR}(t_{\text{att}}, t_{\text{end}})$. (a) & (e) Attack strategy I versus defense strategy I, II, III or IV, (b) & (f) attack strategy II versus defense strategy I, II, III or IV, (c) & (g) attack strategy III versus defense strategy I, II, III or IV, (d) & (h) attack strategy IV versus defense strategy I, II, III or IV. All graphs plot the mean value of AFR over 500 simulation runs.

Furthermore, a *robustness* index $R$ can be derived from the vulnerability measure as:

$$R = 1 - \frac{V}{N_A} \tag{4.30}$$

where $N_A$ is the size of the power network. The normalisation ensures that $R \in [0, 1]$. A larger $R$ value indicates that the power system can survive with a smaller fraction of failed nodes under cyberattacks.

The experiments are carried out as follows:

- Step I: For a given cyber network, identify the decision-making nodes, and calculate several values of $\epsilon$ by considering different possible selections of decision-making nodes.

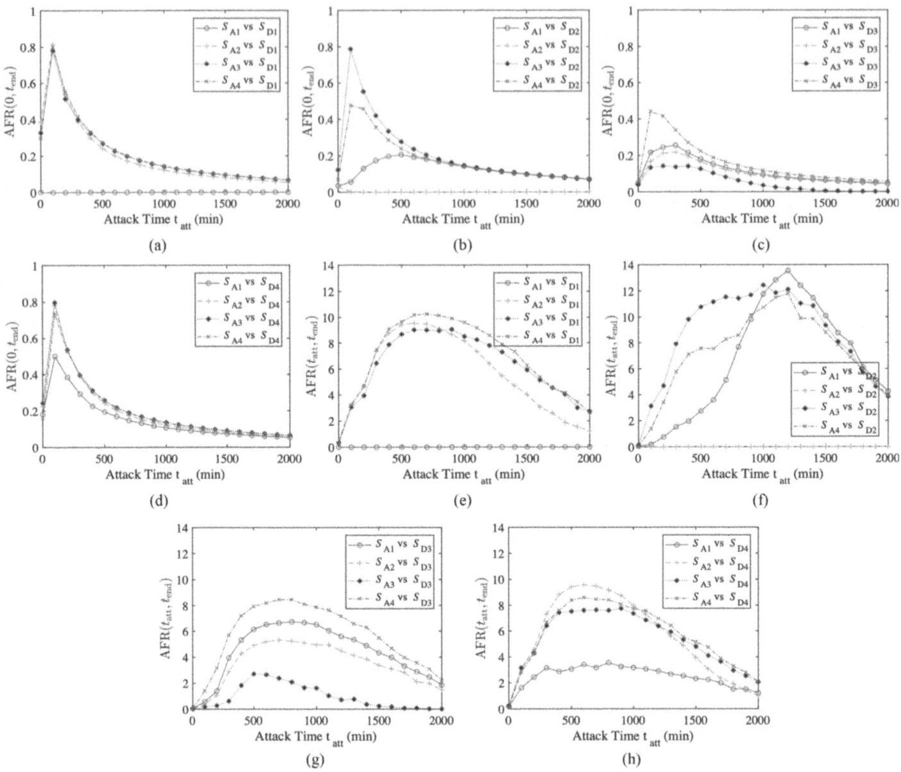


**FIGURE 4.10**
Comparison of attack-defense interaction for two average failure rates, i.e., AFR$(0, t_{\text{end}})$ and AFR$(t_{\text{att}}, t_{\text{end}})$. (a) & (e) Attack strategy I, II III or IV versus defense strategy I, (b) & (f) attack strategy I, II III, or IV versus defense strategy II, (c) & (g) attack strategy I, II, III or IV versus defense strategy III, (d) & (h) attack strategy I, II, III or IV versus defense strategy IV. All graphs plot the mean value of AFR over 500 simulation runs.

- Step II: Based on the chosen decision-making cyber nodes, generate various coupling patterns corresponding to different $\rho$ values to couple the cyber network with the power network.

- Step III: For each coupling pattern, perform cascading failure simulations and compute the robustness index $R$ from the simulation results. In particular, record the number of failed power components when $10\%, 20\%, \ldots, 100\%$ of the cyber nodes are infected. Calculate $V$ according to equation (4.29) and determine $R$ from equation (4.30). In all simulations, a cyber-to-power attack is triggered when malware begins to spread in the cyber network.

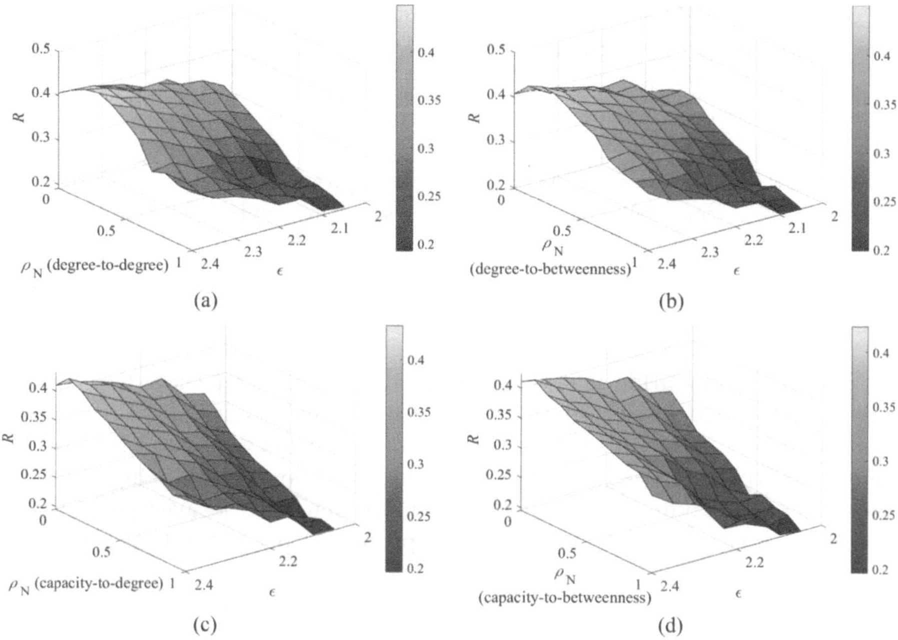


**FIGURE 4.11**
Effect of relative coupling correlation coefficient and average propagation latency on the robustness of cyber-coupled power systems. Four combinations of adoption of criticality indices are considered, namely, (a) degree-to-degree, (b) degree-to-betweenness, (c) capacity-to-degree, and (d) capacity-to-betweenness.

Figure 4.11 shows the effect of coupling patterns on the robustness of cyber-coupled power systems against cascading failure. For ease of comparison, a normalized $\rho$, defined as $\rho_{\mathrm{N}} = (\rho - \rho_{\min})/(\rho_{\max} - \rho_{\min})$, is used. According to Figure 4.11a, with the same value of $\rho_{\mathrm{N}}$, e.g., $\rho_{\mathrm{N}} = 1$, $r$ decreases from 0.2 to 0.15. Moreover, given the same value of $\epsilon$, namely, $\epsilon = 2.4$, the value of $r$ is also decreasing from 0.4 to 0.2. Furthermore, we consider four different combinations of adoption of criticality metrics in power and cyber networks, i.e., degree-to-degree, degree-to-betweenness, capacity-to-degree, and capacity-to-betweenness, which are depicted in Figure 4.11a–d. In general, the surfaces shown in Figure 4.11c and d are flatter than those shown in Figure 4.11a and b. The difference implies that taking node capacity as the node criticality in power networks leads to less variation for the choice of the decision-making cyber nodes and lower sensitivity of the coupling pattern on the robustness of cyber-coupled power systems. Referring to these diagrams, where a lighter color refers to a higher robustness index, two key observations can be made:

1. A smaller average propagation latency leads to a less robust cyber-coupled power system.

2. For a given average propagation latency, a larger relative coupling correlation coefficient leads to a less robust cyber-coupled power system.

The first observation is clearly consistent with the fact that the decision-making cyber nodes selected based on a smaller average propagation latency are more likely to be infected by malware in cyber networks. In other words, under the condition of the same number of cyber nodes being infected, more decision-making cyber nodes are infected when the average propagation latency is smaller. A small average propagation latency thus intensifies the cascading failure in power grids. The second observation, however, does not agree with some previously reported work [205] which has demonstrated that assortative coupling patterns contribute to improving the robustness of coupled networks. In our simulation, we take into consideration the effects of power overloading based on a physical power flow model, contagion, and interdependence between power grids and cyber networks on failure propagation in coupled systems. The results indicate that the assortative coupling pattern degrades the robustness of coupled systems.

Adopting assortative coupling patterns is generally highly desirable in cyber-coupled power systems. One of the main reasons for their widespread adoption is to maintain high reliability of controlling the power network while keeping the control cost low. In particular, a higher relative coupling correlation coefficient $\rho$ indicates that high-criticality power nodes tend to be linked to high-criticality cyber nodes. This pairing improves reliability and operational efficiency because high-criticality cyber nodes can be regarded as more capable of controlling high-criticality power nodes. On the one hand, an assortative coupling pattern enables more control functions to be realized within coupled networks under the constraint of fixed control resources. On the other hand, if low-criticality cyber nodes are coupled to high-criticality power nodes (a dissortative coupling pattern), the control cost increases because more cyber nodes would be needed to support control functions in those power nodes. Without incurring the extra cost, the system's functionality—including intelligent monitoring and control—will degrade.

In summary, a high relative coupling correlation coefficient is beneficial for enhancing control and monitoring capability, while a high robustness index $R$ is also essential. Therefore, a trade-off exists in coupling-pattern design: it is necessary to balance functionality and robustness to achieve an optimal configuration for cyber-coupled power systems.

## 4.3 High Renewable Energy Integration and Frequency Dynamics

The increasing use of renewable energy sources, together with the widespread deployment of power electronics–based generation, presents critical challenges to the secure operation of modern power systems and significantly elevates the risk of large-scale blackout events. This section presents a modeling framework for analyzing cascading failure processes in power systems with integrated renewable generation. The framework couples a network-based representation of physical power flow redistribution with realistic frequency control operations, encompassing primary, secondary, and emergency control schemes. Special attention is devoted to the fundamental differences between synchronous machine–based generators and power electronics-based generators, particularly in their frequency control responses during failure propagation. Incorporating practical control actions into the model is essential for reproducing dynamic behaviors observed in historical blackout events. Simulation studies further demonstrate that higher levels of renewable integration markedly increase the probability of large-scale blackouts. The complexity of modern grids, arising from the accelerating integration of renewable and power electronics technologies, necessitates modeling approaches that faithfully represent essential physical processes while leveraging advanced network analysis techniques to produce results consistent with real-world systems for both assessment and prediction.

The high level of integration of renewable energy sources (RES) has emerged as a defining trend in the transformation of modern power systems. Driven by government policies promoting the transition to carbon-neutral energy systems, many countries are setting targets for 100% renewable based grids as part of the global response to energy shortages, climate change, and environmental pollution [143]. Statistical data confirm that mature renewable technologies, most notably wind and solar generation, have experienced rapid growth in their share of global electricity production over the past decade [129]. While the shift towards clean energy offers undeniable environmental and sustainability benefits, power systems with a high level of RES integration may face new vulnerabilities that threaten their operational robustness [251].

In recent years, unexpected shutdowns of renewable resources have been implicated in severe blackouts, raising concerns among power system operators and engineers. Two notable examples are the 2016 South Australia blackout and the 2019 United Kingdom blackout. On 28 September 2016, extreme weather caused extensive damage to transmission and distribution assets. This triggered a chain of events, including the sudden shutdown of multiple online generators following a rapid system frequency drop, which ultimately led to a state-wide outage lasting 85 hours [19]. On 9 August 2019, heavy rain and lightning storms in the UK caused the loss of 150 MW of generation, followed closely by the sudden disconnection of another 1,131 MW. The resulting frequency decline activated the under-frequency load shedding scheme, disconnecting approximately 1 GW of load and depriving over one million customers of electricity for more than 30 min [83]. In both cases, the high integration of renewable generation, especially inverter-based resources, was identified as a contributing factor in hastening frequency decline and exacerbating the blackout, underscoring the need to better understand the interaction between RES integration and cascading failures.

Cascading failure modeling has been an active research area for over two decades [162]. Approaches grounded in complex network theory analyze the topological robustness of power grids, treating buses as nodes and transmission lines as links. While such topological models are valuable for abstract, large-scale analyses, they often neglect electrical properties, operational constraints, and stability phenomena. This omission can lead to unrealistic conclusions and even misleading recommendations for blackout prevention. To address these shortcomings, hybrid approaches have been developed that combine bottom-up methods (grounded in detailed electrical models) with top-down network science concepts. Our previous work has followed this combined methodology, introducing models for both stand-alone power systems [153, 301, 302] and cyber–physical coupled systems [159, 161, 299].

A growing body of literature has examined the impact of high RES integration on the vulnerability of power systems to cascading failures. Athari and Wang [16, 17] used DC and AC power flow–based models to assess the influence of wind generation integration and uncertainty. Liu et al. [168] proposed a framework and accelerated algorithm to trace cascading failure paths under high wind integration. From a top-down probabilistic perspective, Monte Carlo–based [29] and graph-based [287] studies have examined reliability in renewable-rich power systems. At the same time, bottom-up approaches have addressed stability mechanisms specific to inverter-based technologies, including harmonic stability [264], nonlinear and transient stability [305], and other issues surveyed in recent reviews on power systems with high levels of renewable integration and power electronic devices [231].

The consensus from these studies is that without effective protection and management, higher renewable integration, accompanied by the widespread deployment of power electronics, tends to exacerbate the risk of cascading failures. However, many existing works do not explicitly embed the frequency control process into their cascading failure frameworks [10]. As a result, they cannot fully reproduce the failure propagation sequences that occur in high-RES systems, where frequency response dynamics play a critical role. Since inverter-based RES resources differ fundamentally from synchronous machines, especially in their low inertia, limited active power contribution to frequency control, and tendency to trip under large frequency deviations, ignoring frequency control leads to incomplete risk assessments.

### 4.3.1 Model Description

A power system, comprising generators, substations, and loads interconnected by transmission lines, can be represented as a network. In this representation, nodes correspond to functional components of the system, while links correspond to transmission lines. Two principal classes of nodes are identified: power generation nodes, which supply electrical power, and power load nodes, which consume it. The interconnections between these nodes, combined with their electrical characteristics, determine how power is transmitted and distributed across the network. Given the specified generation outputs, load demands, and the complete line admittance matrix, the real power flowing through each transmission link can be determined by performing a power flow analysis. A fundamental condition for a valid power flow solution is that the system must maintain a balance between the total generated power and the total consumed power (including losses).

In the method described here, the frequency control function is explicitly included in the cascading failure model. This function is responsible for maintaining the equilibrium between generation and load by regulating system frequency through primary, secondary, and emergency frequency control actions. Integrating frequency control into the model allows us to capture the operational mechanisms that are critical for realistic and practical simulations of cascading failures in power systems with a high integration of renewable energy sources.

In view of the rapid progression of failure events during a cascade, the short-term transient dynamics of voltages, currents, and generator rotor angles can be neglected when computing post-failure operating conditions. As in prior work on cascading failure simulation, the instantaneous post-event power flows are determined using the DC power flow model, which provides a computationally efficient and sufficiently accurate approximation for overload analysis. The DC model assumes constant voltage magnitudes at all buses and neglects reactive power flows, enabling the rapid recalculation of power distributions following each topological change in the network. This simplification is well suited for studying the step-by-step evolution of cascades where the focus is on active power balance, frequency excursions, and overload-induced tripping.

#### 4.3.1.1 Frequency Control Function

A power system, composed of generators, substations, and loads interconnected by transmission lines, can be represented as a network in which nodes correspond to functional components and links to transmission lines. Two principal node classes are recognized: power generation nodes, which supply electrical power, and load nodes, which consume it. Given specified generation outputs, load demands, and a complete line admittance matrix, the active power flowing through each link is determined by a power flow analysis. A prerequisite for a valid calculation is that total generated power and total consumed power (including losses) are balanced.

In the method described here, the cascading failure model explicitly incorporates the system frequency control function, whose role is to maintain this balance between generation and load during disturbances. Frequency stability is directly linked to this balance: any mismatch manifests as a deviation from nominal frequency (50 Hz in this study), triggering a control sequence to restore it. Three frequency control stages are included:

1. Primary frequency control: rapid local adjustment of generator output via droop response.
2. Secondary frequency control: slower, centralized automatic generation control (AGC) to restore nominal frequency and scheduled tie-line flows.
3. Emergency control: immediate UFLS (under-frequency load shedding) or OFGS (over-frequency generator shedding) when deviations are large.

For a power system with $K$ control areas, the relationship between system frequency deviation $\Delta f(t)$ and net power imbalance $\Delta P_{\mathrm{U}}(t)$ in control area $k$ is given by

$$\Delta P_{\mathrm{U}}(t) = 2H\frac{d\Delta f(t)}{dt} + D\Delta f(t) \tag{4.31}$$

where $H$ is the aggregated inertia and $D$ is the aggregated load damping, computed as

$$H = \frac{\sum_{i=1}^{n_g} H_i P_{gi}^{\text{rated}}}{\sum_{i=1}^{n_g} P_{gi}^{\text{rated}}} \qquad D = \sum_{i=1}^{n_g} D_i \tag{4.32}$$

with $n_g$ the number of generators in the area, $H_i$ the inertia of generator $i$, $D_i$ its load damping, and $P_{gi}^{\text{rated}}$ its rated output.

The magnitude of the total generation–load imbalance in each control area is

$$\Delta P_{\mathrm{U}}(t) = \sum_{i=1}^{n_g} \Delta P_{\mathrm{m}i}(t) - \sum_{i=1}^{n_l} \Delta P_{\mathrm{L}i}(t) - \Delta P_{\mathrm{tie}}(t) \tag{4.33}$$

where $\Delta P_{\mathrm{m}i}$ is the mechanical power change of generator $i$, $\Delta P_{\mathrm{L}i}$ is the load change at bus $i$, and $\Delta P_{\mathrm{tie}}(t)$ is the deviation of actual from scheduled tie-line flows, i.e.,

$$\Delta P_{\mathrm{tie}}(t) = \sum_{i=1}^{T} \left(P_{\mathrm{tie,actual,i}}(t) - P_{\mathrm{tie,scheduled,i}}(t)\right) \tag{4.34}$$

with $T$ tie-lines for area $k$.

The primary control provides a fast local response to frequency deviations, automatically adjusting generator mechanical power output based on droop characteristics. Using a simplified governor–turbine model [277], the dynamics for generator $i$ are given by

$$\Delta P_{\mathrm{m}i}(t) + T_{\mathrm{GT}i}\frac{d\Delta P_{mi}(t)}{dt} = -\frac{\Delta f(t)}{R_i} \tag{4.35}$$

where $T_{\mathrm{GT}i}$ is the turbine–governor time constant and $R_i$ is the droop coefficient. For example, $R_i = 0.1$ Hz/MW means a 0.1 Hz deviation changes output by 1 MW.

Secondary control, or AGC, aims to restore system frequency to nominal and re-establish scheduled inter-area power flows. It acts via an Area Control Error (ACE), i.e.,

$$\mathrm{ACE}_k(t) = \Delta P_{\mathrm{tie,k}}(t) + \beta\,\Delta f(t) \tag{4.36}$$

with $\beta = 1/R_k + D_k$.

A PI (proportional–integral) controller generates a total adjustment signal $\Delta P_C(t)$, allocated among generators [152]. The AGC loop for generator $i$ is

$$\Delta P_{mi}(t) + T_{mi}\frac{d\Delta P_{mi}(t)}{dt} = \Delta P_{Ci}(t) - \frac{\Delta f(t)}{R_i} \tag{4.37}$$

When deviations exceed safe thresholds, emergency measures are applied. UFLS rapidly disconnects the load to arrest a frequency drop, while OFGS reduces generation to curb over-frequency. The magnitude of power shed in each step, $\Delta P_{\mathrm{UFLS}}(t)$ or $\Delta P_{\mathrm{OFGS}}(t)$, is proportional to the deviation magnitude $|\Delta f(t)|$, following common industry practice [27].

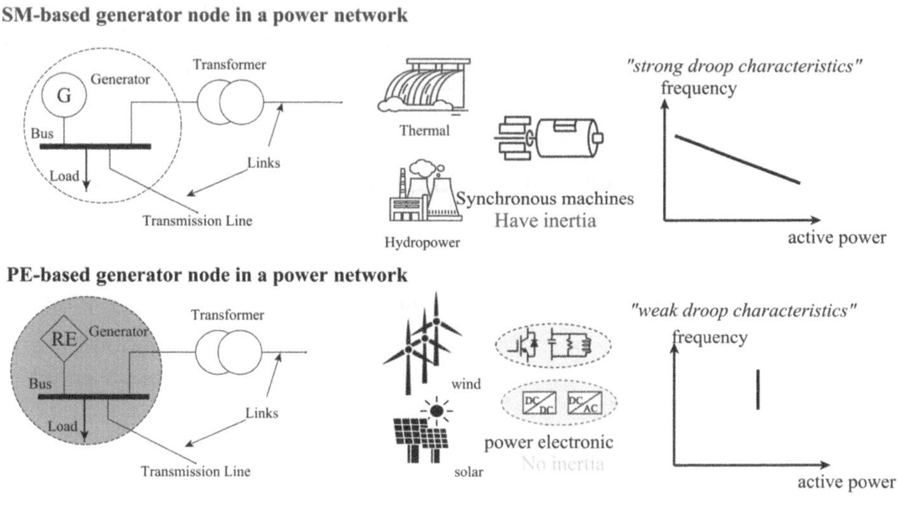


**FIGURE 4.12**
Comparison between a synchronous machine-based generator node and a power electronic-based generator node.

#### 4.3.1.2 Synchronous Machine-Based versus Power Electronics-Based Generator Nodes

Synchronous machine-based (SM-based) power generators play an essential role in a transitional power system where the frequency should be synchronized during normal system operation. Synchronous machine-based power generators having inertia and droop characteristics participate in the process of frequency control for reducing frequency deviation. However, the increased use of renewable energy sources is expected to affect the frequency control function in power systems. To assess the effects of renewable source integration on cascading failure of a power system, we identify three distinct features of a power electronics-based (PE-based) generator, namely a renewable source, when comparing to a synchronous machine-based power generator, as shown in Figure 4.12. In fact, PE-based generator nodes have

1. virtually no inertia, namely $H_{\text{PEG}} \to 0$, leading to smaller overall system inertia;
2. insignificant droop characteristics, namely $R_{\text{PEG}} \to \infty$, hence not contributing to frequency control; and
3. a high risk of being interrupted when the system is under large disturbances, such as a large frequency deviation.

To highlight the vulnerability of renewable energy sources to a large frequency deviation during a failure propagation process, the power to be cut in a PE-based generator node is assumed to be proportional to the magnitude and the duration of frequency deviation. In particular, we set a weighting factor, denoted by $w_{\text{PEG}}$, to represent the vulnerability level of a PE-based generator node to a frequency deviation. In other words, a larger $w_{\text{PEG}}$ implies a larger magnitude of the generated power being cut in the PE-based generator nodes.

#### 4.3.1.3 Simulation Process

The cascading failure simulation with frequency control follows the process in Figure 4.13.

1. Initialization: The full network topology and electrical parameters are loaded, including frequency control settings for all subsystems.
2. Initial failure: One or more components (such as transmission lines or substations) are removed to mimic disturbance events observed in practice [19, 83].
3. Iteration: After each failure:
    (a) The network topology is updated, and connected subnetworks are identified.
    (b) For each subnetwork, the frequency control action (primary, secondary, emergency, or forced trip) is selected based on the

calculated $|\Delta f|$. Multi-area subnetworks execute independent control loops.

(c) With the updated generation and load from control actions, the DC power flow is calculated.

(d) Any element exceeding its revised thermal limit $F_{ij} > cF_{ij}^{\text{initial}}$ $(c > 1)$ is tripped.

(e) The process repeats until no overloads remain.

This integrated simulation captures the interplay between dynamic frequency stability and sequential component outages, enabling a realistic assessment of cascading failure behavior in systems with high levels of renewable, inverter-based generation.

### 4.3.2 Case Study

The method described in Section 4.3.1 is now applied to evaluate the effect of increasing use of renewable energy sources (RES) on the cascading failure behavior and robustness of large-scale power systems. The simulation framework integrates both overload-induced tripping and frequency control processes, enabling assessment of how reduced inertia and altered control dynamics in high-RES systems influence the progression and severity of cascades.

To initiate failure propagation, two transmission lines are included in the initial failure set for each test system and are tripped during the initialisation step. Branch overload protection is modelled by setting the thermal capacity of each transmission line to $c = 1.2$ times its initial steady-state loading. After each DC power flow calculation, performed using the MATLAB® toolbox MATPOWER [309], the overloaded line to be removed is determined according to the mechanism described in earlier work [299].

The frequency control module aims to maintain balance between generation and load and thereby restore the system frequency to its nominal value of 50 Hz. Parameters such as inertia, load damping, and generator droop coefficients are set following established guidelines [27], ensuring a realistic representation of primary, secondary, and emergency control actions.

Two standard IEEE test systems are used to examine the impact of high RES integration on cascading failures. The IEEE 39 Bus system, containing three control areas connected by tie-lines, is used first to analyze in detail how RES integration affects frequency control responses during cascading events. The IEEE 300 Bus system, also consisting of three control areas, is then employed to evaluate the effect of increasing RES share on overall system robustness against cascading failure.

In implementing frequency control, the same settings as in [27] are adopted. Primary frequency control is activated whenever the measured frequency departs from the range 49.8–50.2 Hz. When the frequency moves outside the range 49.2–50.8 Hz, the emergency control scheme is triggered, invoking either

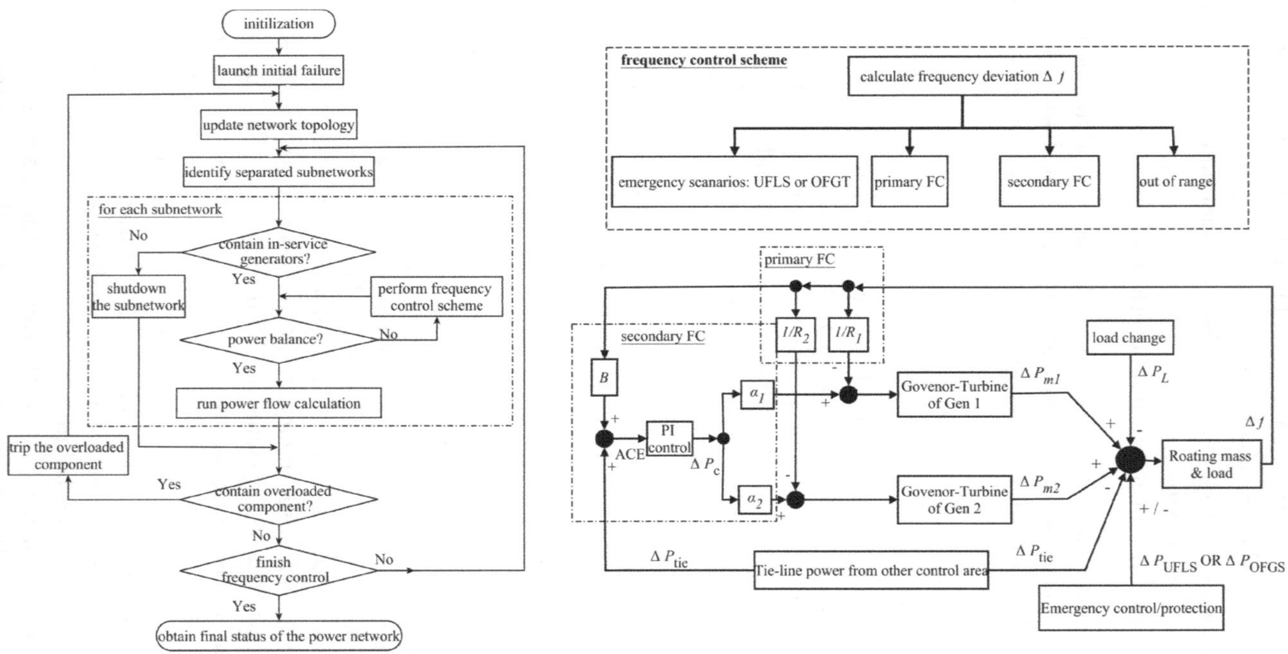


**FIGURE 4.13**
Flow chart of the cascading failure model and the involved frequency control scheme.

under-frequency load shedding or over-frequency generation shedding depending on the disturbance direction. In the AGC process, generation setpoints for participating units are updated every three seconds, with control signals computed from area control error and distributed according to the scheduling rules as described in Section 4.3.1. This experimental configuration enables direct comparison of cascade progression under different levels of inverter-based renewable generation, highlighting changes in control action timing, frequency stability margins, and ultimate blackout size in systems of varying scale.

#### 4.3.2.1 Failure Propagation

Simulations of the IEEE 39 Bus test system are presented here for all combinations of two transmission lines being initially tripped. Among these trials, two representative cases, illustrated in Figure 4.14, highlight the frequency response behavior during cascading failure for different levels of renewable energy source (RES) integration. In each simulation, the proportion of RES generation at each generator bus was increased while the remaining output was supplied by synchronous machines. Thus, a single generation node can operate with a combined portfolio of synchronous generation and inverter-based renewable (PE-based) generation.

In the first representative case (Figure 4.14a), an initial generation loss creates a substantial power imbalance between supplied and demanded power, causing a rapid frequency decline. The under-frequency load shedding (UFLS) scheme is activated as an emergency control measure to arrest this drop. Once UFLS stabilizes conditions, the system enters the primary frequency control (PFC) period, during which frequency is regulated via generator droop response. Following PFC, secondary frequency control (SFC) is applied to progressively restore frequency to its nominal value of $f_{\mathrm{n}} = 50\,\mathrm{Hz}$. At approximately $t = 36\,\mathrm{s}$, further component failures cause network separation, isolating part of the observed control area and producing a large net generation surplus. This surplus drives the frequency sharply upward, which in turn triggers the over-frequency generator shedding (OFGS) action to rapidly reduce excess generation. As frequency begins to return towards nominal, PFC and SFC again take over to restore stability, provided no further large-scale loss of generation or load occurs.

The second representative case (Figure 4.14b) exhibits a similar control sequence. Initial disturbance induces a frequency drop, UFLS is deployed to halt the decline, and the progression of PFC-SFC-OFGS is observed as further imbalances emerge during the cascading sequence. These time-domain simulations capture the cyclical nature of control activation during a complex cascade.

The simulated frequency traces demonstrate that the modeling framework reproduces the complete sequence of frequency deviation and control actions in a cascading event. This is further supported by the close correspondence

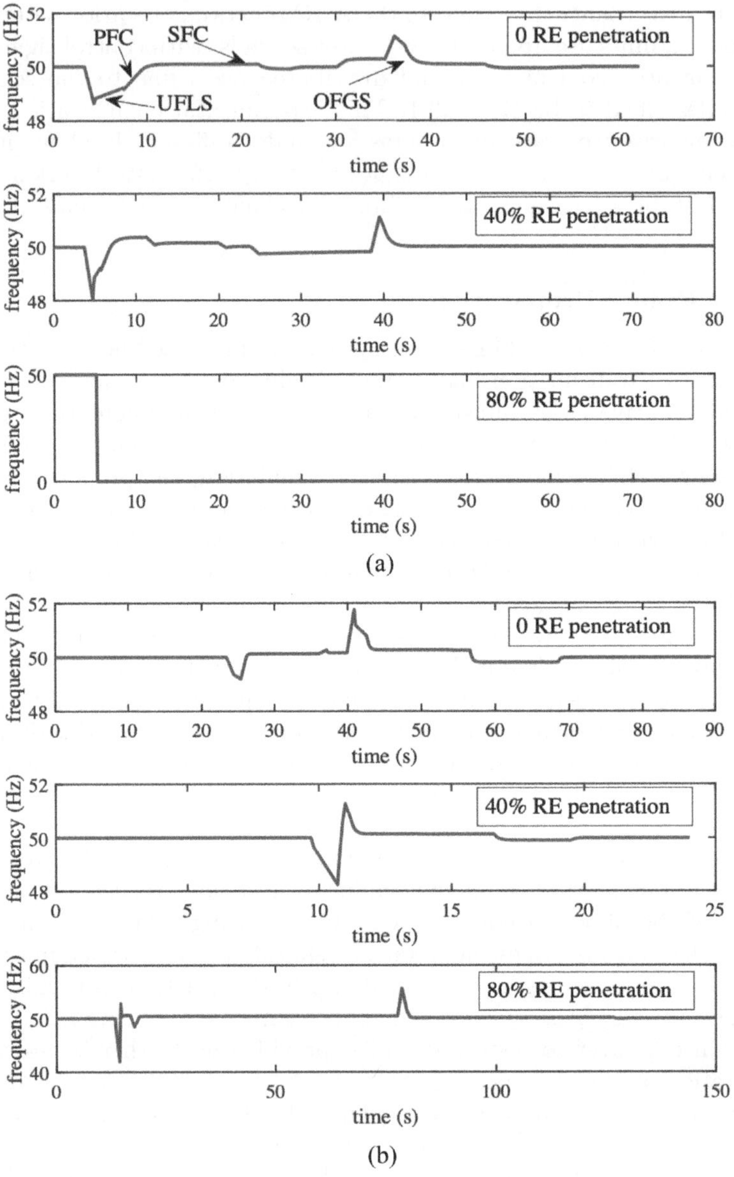


**FIGURE 4.14**
Frequency response in the cascading failure process of the IEEE 39 Bus power test case where two transmission lines are tripped initially. In two significant cases shown in (a) and (b), 0%, 40%, and 80% renewable energy sources are deployed in the power grid.

between the simulated profiles and the annotated frequency record from the August 9, 2019 UK blackout (Figure 4.15), in which generator loss, frequency decline, UFLS, and subsequent control actions are clearly observed. The high degree of similarity between simulation and recorded behavior confirms the capability of the integrated frequency–cascading failure model to reproduce real-world dynamics.

To examine how increased RES integration influences control behavior, two additional scenarios were simulated for the same disturbance patterns, with RES shares increased to 40 % and 80 % of total generation capacity. For each integration level, the frequency deviation is defined as

$$\Delta f(t) = f(t) - f_n \tag{4.38}$$

where $f_n$ is the nominal system frequency.

The results reveal a clear trend: as RES integration increases, the magnitude of frequency deviations grows during the cascade. This reflects the reduction in system inertia and the diminished participation of inverter-based units in active power–frequency control. These changes reduce the system's ability to contain rapid frequency excursions when protective measures are insufficient.

Two quantitative metrics summarized in Table 4.1 characterize this effect. The first, the peak-to-peak frequency deviation $\Delta f_{\text{pp}}$, measures the maximum excursion range, i.e.,

$$\Delta f_{\text{pp}} = \Delta f_{\max} - \Delta f_{\min} \tag{4.39}$$

The second, the power outage rate $r_{\text{outage}}$, expresses the proportion of generation lost during the cascade

$$r_{\text{outage}} = \frac{\sum_{g_i \in V_{\text{G}}} \left(P_{g_i}^{\text{init}} - P_{g_i}^{\text{end}}\right)}{\sum_{g_i \in V_{\text{G}}} P_{g_i}^{\text{init}}} \tag{4.40}$$

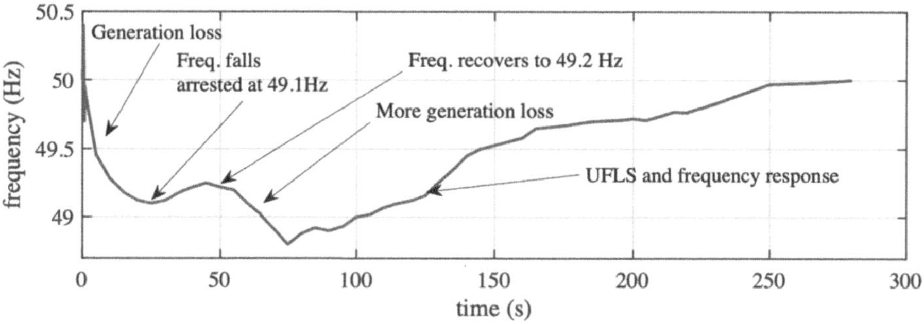


**FIGURE 4.15**
Annotated frequency trace in the UK blackout event on August 9, 2019, redrawn according to the technical report [83], showing resemblance with results generated by the proposed model.

**TABLE 4.1**
Comparison of the Peak-to-Peak Frequency Deviation ($\Delta f_{\text{pp}}$) and the Power Outage Rate ($r_{\text{outage}}$) of the IEEE 39-Bus Power Test System under Cascading Failure, When Renewable Energy Sources are Integrated into the Power System at Various Percentage ($r_{\text{RE}}$), namely, 0%, 40%, 80%

| $r_{RE}$ | 0% | 40% | 80% |
|---|---|---|---|
| $\Delta f_{\text{pp}}$ (Hz) | $0.8535 \pm 0.0121$ | $1.2821 \pm 0.0192$ | $2.9519 \pm 0.0495$ |
| $r_{\text{outage}}$ | $0.3771 \pm 0.0058$ | $0.3803 \pm 0.0056$ | $0.6486 \pm 0.0078$ |

where $V_{\text{G}}$ denotes the set of generator nodes, $P_{g_i}^{\text{init}}$ the initial generated power, and $P_{g_i}^{\text{end}}$ the generated power after the cascade ends.

Averaging over multiple simulation runs, both $\Delta f_{\text{pp}}$ and $r_{\text{outage}}$ increase with RES integration. Higher shares of PE-based generation lead to deeper frequency deviations and more extensive generation loss, indicating greater vulnerability of the system to cascading failures when inertia and frequency control participation are reduced.

#### 4.3.2.2 Robustness

To assess the impact of high renewable energy source (RES) integration on the robustness of large-scale power systems, over 1,000 simulations were conducted on the IEEE 300 Bus test system. In each run, two transmission lines were selected as the initial failure set. Simulations were carried out for five integration levels of RES generation, denoted by $r_{\text{RE}}$, corresponding to 0%, 20%, 40%, 60%, and 80% of total installed generation capacity.

Rather than reporting a single value of the power outage rate $r_{\text{outage}}$ for each run, robustness is characterized using the cumulative probability distribution of outage severity. The cumulative probability is defined as

$$p\left(r_{\text{outage}}\right) = \frac{N_{\text{sim}}\big|_{r_{\text{outage}} \geq r'_{\text{outage}}}}{N_{\text{sim}}} \tag{4.41}$$

where $N_{\text{sim}}$ is the total number of simulation runs and $r'_{\text{outage}}$ is a sampled threshold in the range 0 to 1. The metric $p(r_{\text{outage}})$ thus represents the probability that the outage in a given simulation is at least as large as $r'_{\text{outage}}$.

Figure 4.16a and b present the cumulative probability curves for different renewable integration levels, both without ($w_{\text{PEG}} = 0$) and with ($w_{\text{PEG}} = 1$) the explicit consideration of inverter-based generator vulnerability to system disturbances. In all cases, for a fixed $r_{\text{RE}}$, $p(r_{\text{outage}})$ decreases as $r'_{\text{outage}}$ increases, indicating that large-scale blackouts are less probable than small-scale ones. However, as $r_{\text{RE}}$ increases, $p(r_{\text{outage}})$ at any given $r'_{\text{outage}}$ also increases, meaning that high renewable use makes large outages more likely. This increase is particularly pronounced at $r_{\text{RE}} = 60\%$ and $r_{\text{RE}} = 80\%$, where the probability of severe cascading failure rises sharply.

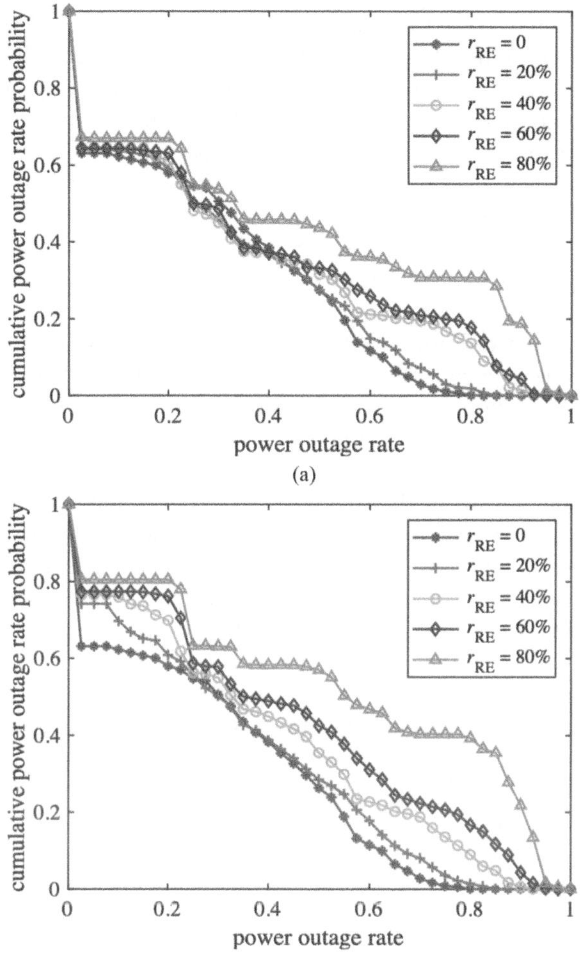


**FIGURE 4.16**
Cumulative probability of power outage rates of power systems where renewable integration (penetration) levels are set as 0, 20%, 40%, 60%, and 80%, (a) with and (b) without consideration of the vulnerability of renewable energy sources to the large frequency deviation.

The underlying cause of this increased vulnerability lies in the reduced ability of high-RES systems to respond effectively to disturbances during cascading events. Inverter-based generators contribute little or no rotational inertia and generally do not participate in active power–frequency control, weakening the system's ability to arrest frequency excursions following major disturbances. This structural fragility has motivated research into virtual inertia and synthetic droop characteristics for wind and solar plants as a way to enhance the resilience of RES-dominated grids. The results obtained here are consistent with previous studies [29,123,295] that have linked high integration levels of inverter-based generation to a heightened risk of large cascading failures. They also confirm that it is essential to explicitly model the dynamical frequency control process when assessing robustness.

Further insight is obtained by examining how the vulnerability factor $w_{\text{PEG}}$, which measures the sensitivity of a PE-based generator to frequency deviation, affects outage probability. Figure 4.17 plots the average $p(r_{\text{outage}})$ across all $r'_{\text{outage}}$ values for different integration levels and different $w_{\text{PEG}}$ settings. In all cases, the curves for nonzero $w_{\text{PEG}}$ lie above those for $w_{\text{PEG}} = 0$, indicating that greater vulnerability increases the likelihood of severe outages. The effect is especially marked when $r_{\text{RE}} > 40\%$, where a larger $w_{\text{PEG}}$ substantially raises the mean probability of widespread generation loss.

This behavior can be explained by considering that a low tolerance of inverter-based generators to frequency deviations aggravates the generation–load imbalance during disturbances. Early disconnection of such generators under large deviations forces additional load shedding and generator tripping in other parts of the system, amplifying the cascade and increasing the probability of a large-scale blackout. In effect, overly rapid protective disconnection of inverter-based resources can paradoxically increase system vulnerability rather than reduce it. Whether this quick reaction is ultimately beneficial or detrimental remains an open question and warrants detailed future study, particularly when considering the inherent variability and uncertainty of weather-dependent renewable resources. Accurate modeling of these uncertainties will be increasingly important as RES sources come to dominate future generation portfolios.

## 4.4 Network-Based Cascading Failure Model with Inverter-Based Resource Integration

The integration of inverter-based resources (IBRs) in modern power systems has intensified in recent years. While supporting the transition towards low-carbon energy systems, high integration of IBRs—implemented through power electronics–based (PE-based) interfaces—can significantly alter the dynamic behavior of the grid and, in turn, increase its vulnerability to large-scale cascading failures. This section presents a complex network–based cascading failure modeling framework that explicitly incorporates the failure mechanisms of PE-based nodes, enabling the evaluation of vulnerability in PE-integrated power systems under varying integration levels. In particular, the model accounts for the propensity of PE-based generator nodes to deviate from stable operating ranges during failure propagation and to disconnect from the network upon breaching protective thresholds. By embedding this

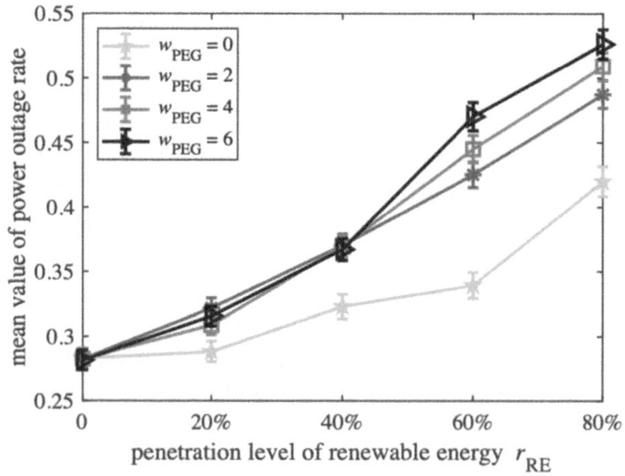


**FIGURE 4.17**
Robustness of the power system under different vulnerability levels to frequency deviation at different levels of integration (penetration) of renewable sources.

novel mechanism into the cascade simulation, the framework more accurately reflects the physical processes observed in recent blackouts. Case studies on the IEEE 39 Bus and IEEE 118 Bus systems demonstrate that neglecting the specific behavior of PE-based nodes can lead to a substantial underestimation of blackout risk. These investigations reinforce the need to incorporate IBR integration effects into cascading failure analysis for credible robustness assessment and for the development of effective strategies to maintain the resilience of evolving power grids.

The increasing complexity of critical infrastructures, such as power and transportation systems, has placed the applicability of traditional modeling and analytic methods under considerable strain. Network-based approaches have emerged as effective alternatives for studying the large-scale, interdependent behavior of such infrastructures [64]. In the context of global climate change, the power sector is undergoing a profound evolution to meet carbon neutrality targets, integrating renewable energy sources, electric vehicles, and high-voltage direct current transmission systems. This transformation is leading to the emergence of power electronics–integrated (PE-penetrated) systems [251]. Given the severe socio-economic consequences of large-scale blackouts, improving the security of these evolving PE-integrated grids and mitigating the risk of major outages have become paramount priorities [231].

Concerns about the robustness of PE-integrated power systems have been reinforced by several high-profile blackout events in which the high integration of power electronic devices was a critical factor. On 1 July 2015, subsynchronous oscillations at a wind farm in the Hami region of Xinjiang, China, activated torsional vibration protection in a thermal power plant more than

300 km away, forcing it offline [148]. On 28 September 2016, a severe storm in South Australia disabled multiple transmission lines [19], resulting in a loss of over 50% of operating wind generation due to turbine control behavior. The grid—already characterized by high renewable use and limited synchronous machines—experienced reduced system inertia, an unexpected frequency response, and ultimately a state-wide blackout. More recently, on 9 August 2019, a lightning strike on the UK transmission network caused an unbalanced voltage dip that triggered oscillatory behavior in a large wind farm [83], leading to a loss of 737 MW of wind power, a rapid rate-of-change-of-frequency event, and a cascading blackout. These incidents underscore the heightened vulnerability of modern PE-integrated grids to instability.

Extensive research has sought to address the operational and stability challenges introduced by the large-scale integration of power electronics. Kong et al. [142] provided a comprehensive review of stability issues, examining root causes related to converter controls and network characteristics. Research efforts include the design of resilient converters with current-limiting strategies [293], application of periodic small-signal analysis to multilevel converters in HVDC systems [70], and studies of synchronization stability and multi-timescale dynamics in IBR-rich systems using bifurcation and oscillation analyses [150, 172, 286]. However, the majority of these investigations have been conducted on small-scale networks with one or a few converters [157, 240], limiting the scope for understanding the complex, large-scale dynamics that occur during major blackout events.

At the system level, cascading failure modeling has been a key tool in assessing vulnerability and devising mitigation strategies for evolving grids. Previous work has captured the interplay between network topology, dynamic processes, and interdependence, with applications including cyber–physical coupling analysis [52, 73, 99, 298], the interaction of cyberattack and defence strategies [159], and the inclusion of protection mechanisms such as frequency response [163] and transient stability control [4]. Advanced methods including graph convolution networks [308] and graph-based techniques [287], have been applied to high-renewable systems. These approaches have proven beneficial for resilience enhancement, including planning [45], mitigation [149], and recovery [153]. However, a fundamental limitation persists: the failure mechanisms of PE-based nodes—generation nodes employing inverter-based resources—are generally not considered. This omission may lead to misestimation of system vulnerability and reduce the practical value of these models.

To address this gap, the method described in this section focuses on cascading failure risk analysis for large-scale power systems with increased use of power electronics. The framework integrates a PE-node failure mechanism into a complex network–based cascading failure model coupled with steady-state AC power flow. In this formulation, PE-based nodes are explicitly recognized as those equipped with inverter-based resources. These nodes, when exposed to significant disturbances such as voltage or frequency instability, may exceed their operating limits, trigger protective actions, and disconnect from the grid. The proposed failure mechanism for PE-nodes captures this protective disconnection process and integrates it into the iterative cascade simulation. In each simulation iteration, before achieving steady-state power flow redistribution, network topology and electrical parameters are updated to reflect the effects of PE-node protection alongside conventional overload tripping. Once the network state stabilizes, risk assessment metrics are derived from the final system configuration.

The modeling framework has been applied to small-scale and large-scale test systems, including the IEEE 39 Bus, IEEE 118 Bus, and Illinois 200 Bus networks. Simulation results demonstrate two key outcomes: (i) the model can reproduce salient propagation characteristics observed in real large-scale disturbances, and (ii) excluding PE-node failure effects leads to systematic underestimation of outage risk. Furthermore, the framework enables assessment of how both the percentage integration and spatial distribution of PE-based generation influence cascading vulnerability.

### 4.4.1 Model Description

To accurately capture the failure propagation process in a power system with a significant presence of power electronics devices, a cascading failure model needs to incorporate the failure mechanisms of both nodes and links within the power network. In addition to considering the conventional overload effect that can lead to node and link failure, the proposed cascading failure modeling framework integrates the failure mechanisms specific to PE-based nodes. Employing this model to simulate the failure propagation process can assess the associated power outage risks, thereby evaluating the system's reliability under various contingency scenarios.

#### 4.4.1.1 Cascading Failure Simulation

A power system can be represented as a graph $G = (V, E)$, where the set of nodes $V$ represents generators, substations, and loads, and the set of links $E$ represents transmission lines. Following an initiating contingency — such as the failure of one or more links or nodes — subsequent failures may occur as the system state evolves and predefined protection actions are triggered. This failure propagation process may ultimately result in widespread component outages or complete system collapse.

Two primary classes of cascading failure models are commonly used in research: AC and DC power flow models [47]. While DC models offer computational simplicity, AC models achieve higher accuracy in calculating post-contingency power flows and facilitate the inclusion of realistic protection mechanisms employed in real-world operation. For the analysis in this section, a steady-state AC power flow model is adopted. This choice ensures that both topological and electrical changes during the cascade are represented

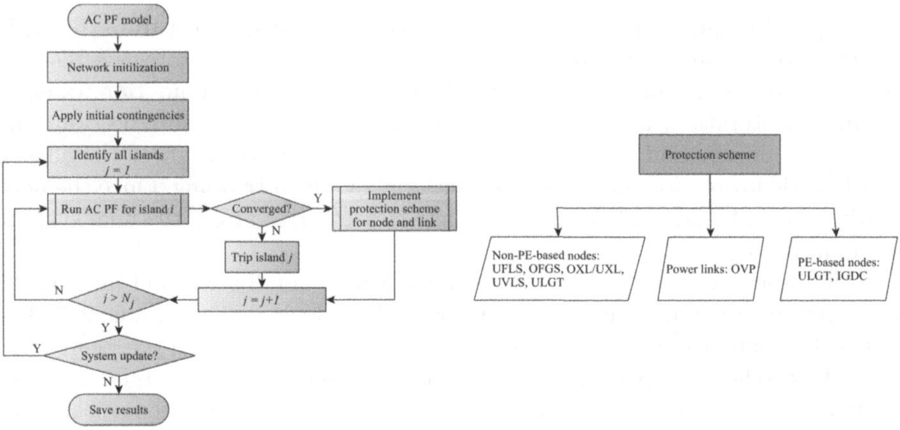


**FIGURE 4.18**
Cascading failure simulation process in PE-integrated power systems considering failure mechanisms of non-PE-based nodes, power links, and PE-based nodes. In a power network containing nodes and links, the protection scheme to be implemented: UFLS (under-frequency load shedding), OFGS (over-frequency generator shedding), OXL/UXL (over-/under-excitation limiters), UVLS (under-voltage load shedding), and ULGT (under-load generator tripping) for non-PE-based nodes, OVP (overload protection) for power links, as well as ULGT (under-load generator tripping) and IGDC (instability generation disconnected) for PE-based nodes.

accurately, enabling the simulation of operational protections associated with voltage, frequency, and overload constraints.

Figure 4.18 presents a flowchart of the main steps used in the AC cascading failure simulation, which proceed as follows:

1. Initialize network: The initialization stage sets the topological and electrical parameters of the test system, along with the operational settings for protection mechanisms. Since the analysis focuses on the impact of power electronics integration, the integration level of PE-based nodes is also defined at this stage. In this study, the PE-integrated network contains two types of generation nodes. The first ones are called PE-based nodes, which are generation nodes integrating renewable energy sources (RES) interfaced via inverter-based resources (IBRs) [183]. The second ones are called Non-PE-based nodes, which are generation nodes without renewable energy integration, typically synchronous machine-based. Rather than quantifying integration as simply the number of PE-based nodes, it is defined as the proportion of total renewable generation

in the set of PE-based nodes $v_{\mathrm{PE}k} \in V$, expressed as

$$\{P_{V_{\mathrm{PE1}}}, P_{V_{\mathrm{PE2}}}, \ldots, P_{V_{\mathrm{PE}K}}\} \tag{4.42}$$

where $K$ is the total number of PE-based nodes.

2. Apply initial contingencies: In power system terms, a contingency refers to the sudden loss or failure of a component such as a transmission line, generator, or transformer. In line with the circumstances of actual major blackout events, the initial contingency in the simulation is defined as the simultaneous outage of $m$ transmission lines, which acts as a trigger for the cascade.
3. Identify all islands: Following each contingency, the removal of components may cause the network to fragment into multiple electrically disconnected sub-networks, or islands. Each island is identified for independent analysis.
4. Run AC power flow for individual islands: For each island, AC power flow analysis is performed to determine the post-disturbance power redistribution. If the AC power flow solution for an island does not converge, the island is considered unsupplied and is tripped entirely [195].
5. Implement protection schemes for nodes and links: Once a valid power flow solution is obtained, protection mechanisms are applied according to their triggering conditions. It is assumed that most conventional protection mechanisms are not present at PE-based nodes. Firstly, inverter-based generators in RES installations typically have limited or no autonomous frequency control capability. Secondly, during cascading failures, PE-based nodes are modeled as not actively participating in frequency regulation in a timely manner. Protection schemes considered are
   (a) Non-PE-based nodes: under-frequency load shedding (UFLS), over-frequency generator shedding (OFGS), over/under-excitation limiters (OXL/UXL), under-voltage load shedding (UVLS), under-load generator tripping (ULGT).
   (b) Links: overload protection (OLP), where link $(i, j)$ is tripped if $P_{ij} > c\,P_{ij,\mathrm{initial}}$ with $c > 1$ as the overload factor.
   (c) PE-based nodes: under-load generator tripping (ULGT) and instability generation disconnection (IGDC).
6. System update: After executing all protection actions for every island, the network model is updated. If no further changes occur between iterations, the simulation terminates, and the final outage metrics are recorded.

This AC-based cascading failure simulation framework, with explicit modeling of PE-based node types and their characteristic protection behavior, enables the evaluation of blackout risk under varying levels of inverter-based generation integration. It also allows direct comparison between failure propagation in conventional and PE-integrated power systems within a consistent analytical setting.

#### 4.4.1.2 Failure Mechanism of PE-based Nodes

With the increasing integration of power electronic devices into modern electricity networks, components connected to the main grid through power electronic interfaces are more likely to activate protection mechanisms and disconnect from the grid when exposed to large system disturbances. Incorporating this failure mechanism into cascading failure modeling is essential for a realistic assessment of blackout risk and the evaluation of mitigation strategies.

Operational experience and technical reports consistently document that large disturbances, such as rapid frequency drops, voltage collapse, or oscillatory instability, can impair the stability of power electronic equipment and trigger protective disconnection. Several notable events illustrate this vulnerability. In October 2009, a grid outage in Texas, USA, involving the integration of a doubly-fed induction generator (DFIG) wind farm via a transmission line with 75% series compensation, resulted in a sub-synchronous oscillation of approximately 20 Hz. System voltage oscillations exceeded 2.0 p.u., leading to the shutdown of multiple wind turbine units and damage to crowbar circuits [3]. In December 2012, sub-synchronous resonance was observed in wind farms in northern China, causing mass tripping of turbines [260]. Similarly, during the 2019 large-scale UK blackout, oscillations in wind farms led to significant wind turbine disconnection and contributed to the subsequent propagation of failures [83].

These and other events reveal that compared to conventional synchronous generators, power electronic–interfaced renewable units exhibit higher sensitivity to large disturbances, lower tolerance to overload, and reduced fault ride-through (FRT) capability. FRT describes the ability of equipment to remain connected and operational during transient faults — a parameter critical for system stability after severe disturbances. In general, the FRT capability of inverter-based resources is significantly lower than that of synchronous machines, leading to their rapid disconnection when voltage or frequency deviations exceed permissible limits. While this protects sensitive equipment, it also reduces available generation at critical moments, potentially worsening system stability and resilience.

To accurately reflect this heightened vulnerability in simulation, instability generation disconnection (IGDC) is modeled as one of the two main protection mechanisms applied to PE-based nodes (the other being under-load generator tripping, ULGT). The framework for IGDC is formulated as follows.

For each PE-based node $v_{\mathrm{PE}i}$, the connection state is denoted by

$$S_{v_{\mathrm{PE}i}} = \begin{cases} 1, & \text{connected to the grid} \\ 0, & \text{disconnected (failed)} \end{cases} \tag{4.43}$$

The failure process is described by the transition $S_{v_{\mathrm{PE}i}} : 1 \rightarrow 0$, representing the disconnection of the node from normal operation due to instability. The failure rate (transition rate) is given by

$$\lambda_{v_{\mathrm{PE}i}} = \lambda_{0,v_{\mathrm{PE}i}} \exp(\kappa_{v_{\mathrm{PE}i}} \Delta_D) \tag{4.44}$$

where $\lambda_{0,v_{\mathrm{PE}i}}$ is the baseline failure rate for node $v_{\mathrm{PE}i}$. A larger value denotes greater inherent vulnerability. This parameter is closely related to the design and operational robustness of the grid-connected PE equipment (as set by the manufacturer). Robust designs yield lower $\lambda_0$ values; sensitive designs yield higher ones. To reflect the generally stronger protection schemes of higher-capacity units, $\lambda_{0,v_{\mathrm{PE}i}}$ is scaled proportionally to the generator's rated output power

$$\lambda_{0,v_{\mathrm{PE}i}} = \lambda_0 \frac{P_{v_{\mathrm{PE}i}}}{\sum_{i=1}^{K} P_{v_{\mathrm{PE}i}}} \tag{4.45}$$

where $P_{v_{\mathrm{PE}i}}$ is the real power output of node $v_{\mathrm{PE}i}$, and $K$ is the total number of PE-based nodes.

The exponential term in equation (4.44) follows an Arrhenius-type formulation [178], where $\kappa_{v_{\mathrm{PE}i}}$ is the system stiffness parameter for the node, indicating how sensitive the system is to disturbance. A larger value means that even minor perturbations may drive the node toward failure. $\Delta_D$ represents the disturbance magnitude experienced by the node. This quantity is determined by comparing key electrical variables—such as bus voltage or system frequency—against their nominal reference values. Since both voltage and frequency are direct indicators of system power balance, in this study $\Delta_D$ is taken to represent the power imbalance within the island to which the PE-based node belongs. This formulation allows the model to reflect both equipment-specific characteristics (through $\lambda_{0,v_{\mathrm{PE}i}}$ and $\kappa_{v_{\mathrm{PE}i}}$) and system-level disturbance effects (through $\Delta_D$). By capturing the conditional probability of disconnection under realistic conditions, the cascading failure model gains the capability to evaluate how PE-based node vulnerabilities influence the pace and extent of failure propagation across the grid.

Large system disturbances, especially when the system variables such as voltages and frequencies are below the lower limit or above the upper limit, disrupt the proper control of power electronic devices or trigger corresponding protection mechanisms, which might disconnect inverter-based resources from the main grid. Transition process from $S_{v_{\mathrm{PE}i}} = 0$ to $S_{v_{\mathrm{PE}i}} = 1$ is depicted in Figure 4.19, which illustrates the practical meanings of all related parameters. The dashed-line enclosed circle indicates the stable zone for the PE node. Hence, when disturbances affect the system, the failure process of the PE-based node can be understood as an instability event due to the operating point crossing the safety boundary. In simple terms, the radius of the circle corresponds to the magnitude of $\lambda_{0,v_{\mathrm{PE}i}}$, as a smaller circle indicates a higher susceptibility of the PE-based node to failure. The deviations of the system variables from the normal range correspond to the magnitude of $\Delta_D$, while $\kappa_{v_{\mathrm{PE}i}}$ is a scaling factor of $\Delta_D$.

Time-to-failure (TTF) is the duration that a system operates before encountering a failure. In reliability engineering [95], the TTF of systems typically conforms to an exponential distribution ($f(\mathrm{TTF}) = \lambda \exp(-\lambda \cdot \mathrm{TTF})$), particularly when the failure rate $\lambda$ remains constant. Random samples for TTF can be derived from the cumulative distribution function ($F(\mathrm{TTF}) = 1 - \exp(-\lambda \cdot \mathrm{TTF})$) using the inverse transform sampling technique. Hence, the transition of PE-based node $v_{\mathrm{PE}i}$ from a normal operation state to a disconnection (failure) state is determined by the following relationship between TTF and the failure rate:

$$t_{\mathrm{fail},v_{\mathrm{PE}i}} = -\frac{1}{\lambda_{v_{\mathrm{PE}i}}} \ln(r_n) \tag{4.46}$$

where $r_n$ is a random number that is generated from the uniform distribution in the interval $(0, 1)$. Inspired by the Metropolis-Hastings algorithm [53], Algorithm 4.1 is applied to estimate the value of $t_{\mathrm{fail},v_{\mathrm{PE}i}}$ when considering random usage numbers in equation (4.46).

One more factor determining the disconnection of inverter-based resources and the main grid is the time needed for failure observations. In other words, if the time it takes for identifying the unstable PE-based node is shorter than the time it takes for the PE-based node to disconnect, then the instability should be controllable, and the disconnection event of the PE-based node will not occur. Therefore, we have the following two scenarios, as shown in Figure 4.20:

1. *SC1:* $t_{\mathrm{iden,vPEi}} \geq t_{\mathrm{fail,vPEi}} \rightarrow$ disconnection execution.
2. *SC2:* $t_{\mathrm{iden,vPEi}} < t_{\mathrm{fail,vPEi}} \rightarrow$ connection reservation.

For scenario *SC1*, the output power of the inverter-based resources will be shut down.

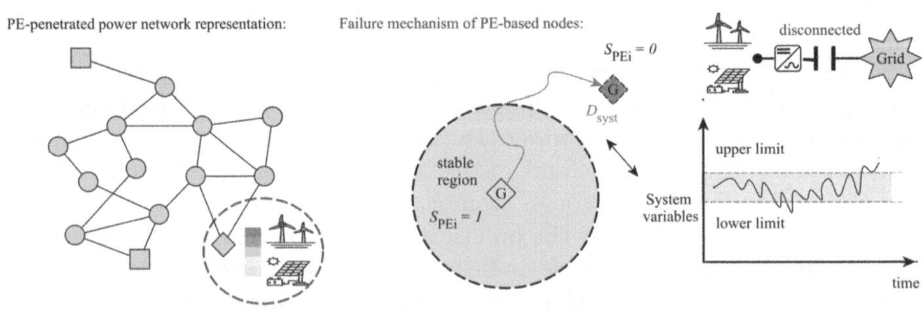


**FIGURE 4.19**
Transition process illustrating the disconnection of a PE-based node from a PE-integrated power system, from normal operation to failure state.

**Algorithm 4.1: Estimation of $t_{\text{fail},v_{\text{PE}i}}$**

**Require:** Failure rate of PE-based node $v_{\text{PE}i}$: $\lambda_{v_{\text{PE}i}}$

**Ensure:** Time-to-failure of PE-based node $v_{\text{PE}i}$: $t_{\text{fail},v_{\text{PE}i}}$

counter $\leftarrow$ 1
$f_{\text{end}} \leftarrow 1$
$list_{\text{TTF}} \leftarrow []$
$\text{TTF}_{\text{mean}} \leftarrow 0$
$\delta_{\text{TTF}} \leftarrow 10^5$
**while** $f_{\text{end}} \neq 0$ **do**
  $t_{\text{fail},v_{\text{PE}i}} \leftarrow -\ln(\text{random}(0,1))/\lambda_{v_{\text{PE}i}}$
  $list_{\text{TTF}}[\text{counter}] \leftarrow t_{\text{fail},v_{\text{PE}i}}$
  **if** counter $> 1$ **then**
    $TTF_{\text{mean}} \leftarrow \sum(list_{\text{TTF}})/\text{counter}$
    $\delta_{\text{TTF}} \leftarrow TTF_{\text{mean}} - t_{\text{fail},v_{\text{PE}i}}$
  **end if**
  **if** counter $> 10^3$ **or** $|\delta_{\text{TTF}}| < 10^{-5}$ **then**
    $f_{\text{end}} \leftarrow 0$
  **end if**
  counter $\leftarrow$ counter + 1
**end while**
**return** $t_{\text{fail},v_{\text{PE}i}}$

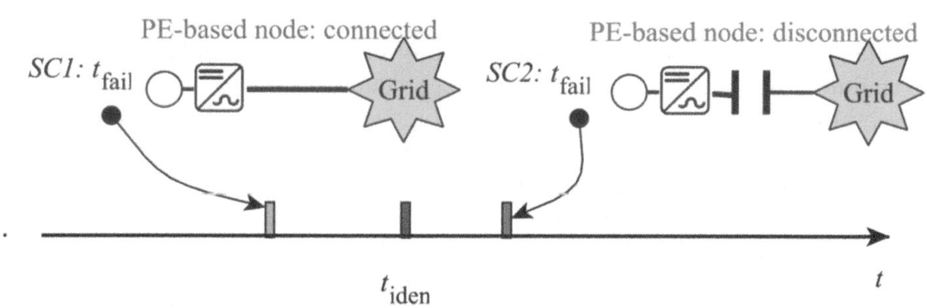


**FIGURE 4.20**
Scenarios of connection and disconnection between PE-based nodes and grid.

To further illustrate the transition from Scenario *SC1* to *SC2* for a given PE-based node, the relationships between TTF and the strength of system disturbance at three different values of system stiffness are depicted using three curves in terms of equations (4.44) and (4.46), as displayed in Figure 4.21. For the sake of simplicity, TTF and system stiffness are symbolized as $t_{\text{fail}}$ and $\kappa$, respectively. Each curve demonstrates a decrease in $t_{\text{fail}}$ as the system disturbance strength increases. Particularly, at higher values of $\kappa$ indicating greater vulnerability of the PE-based node to large disturbances, the rate of decrease in $t_{\text{fail}}$ is more pronounced. Additionally, two line segments running

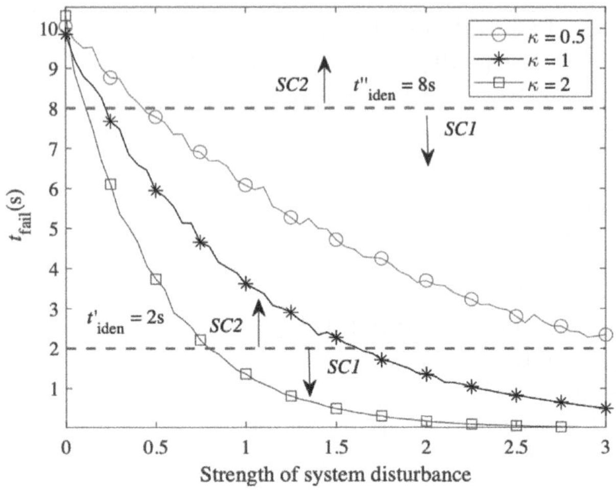


**FIGURE 4.21**
Relationship between TTF denoted by $t_{\text{fail}}$ and the strength of system disturbance for a specific PE-based node with different values of system stiffness represented by $\kappa$.

parallel to the x-axis denote PE-based nodes requiring 2 seconds and 8 seconds for failure observations. In Figure 4.21, $t_{\text{iden}}$ denotes the time needed for failure observations. In the instance of the lower $t_{\text{iden}}$ value, the PE-based node failure occurs when the ratio of active power imbalance to total system active power surpasses 80%, 160%, and 350% for $\kappa$ values of 0.5, 1, and 2, respectively.

Essentially, PE-based nodes susceptible to system disturbances are disconnected under mild disturbances, while those less vulnerable can endure more substantial disturbances and remain connected to the main grid. Conversely, for higher $t_{\text{iden}}$ values, the likelihood of disconnection for PE-based nodes remains high irrespective of their vulnerability to disturbances. These analyses can be associated with the low-voltage ride-through capability of wind turbine systems, referencing their ability to stay connected during voltage drops at the grid connection point [84, 122]. Generally, a shorter time required for failure observations and decreased vulnerability to system disturbances suggest wind turbines possess a robust low-voltage ride-through capability, ensuring uninterrupted grid operation post-system faults. Notably, traditional protective mechanisms enable conventional grid components to withstand large disturbances, with the failure mechanism in this study specific to PE-based nodes and not applicable to non-PE-based nodes. Throughout the simulation, the protective mechanisms for PE-based and non-PE-based nodes are activated based on distinct criteria, and the interplay between these mechanisms falls outside the study's scope.

### 4.4.2 Case Study

Two principal simulation tasks are carried out to investigate the influence of increasing inverter-based resource (IBR) integration on cascading failure dynamics in power grids.

In the first task, detailed simulations are performed on the IEEE 39 Bus system to reproduce the cascading failure propagation process with high fidelity, aiming to capture the key characteristics of actual large-scale power outage incidents. In addition, the IEEE 118 Bus and Illinois 200 Bus systems are employed to compare the proposed model with representative existing models, thereby assessing differences in their capability to quantify cascading failure risks.

In the second task, the IEEE 118 Bus system is used to systematically evaluate the effect of varying the integration level of PE-based nodes on the outcomes of cascading failure analysis. By varying the proportion of inverter-based generation and observing the resulting changes in failure propagation, the robustness of the network under different integration scenarios can be assessed.

#### 4.4.2.1 Simulation Settings

In the cascading failure model, we assume that the power links have an additional capacity margin to accommodate potential increases in power flow and prevent overloading. Thus, the capacity of each branch in the power network is set to 1.5 times its initial power flow ($c = 1.5$) for overload protection. The AC power flow calculation performed in the simulation iteration uses the MATLAB toolbox "MATPOWER" [309]. The settings of other non-PE-based protection mechanisms listed in Section 4.4.1 are specified according to a prior work [195]. The inverter-based resources are set to be uniformly distributed in the simulation setting previously. In other words, to determine the power generated from inverter-based resources for each PE-based node at a given integration level, the calculation involves multiplying the integration level by the total power generation of the generation node. For the PE-based node failure mechanism setting, we define a basic metric to quantify the protection strength of the system, i.e., $r_{\text{prot}} = 1/\lambda_0 t_{\text{iden}}$. This metric indicates that the decrease of either $\lambda_0$ or $t_{\text{iden}}$ enhances the protection strategy, making the PE-based nodes less vulnerable to system disturbance. In the following simulations, we fix the failure identification time, i.e., $t_{\text{iden},\text{vPEi}} = 10$ s, and tune $\lambda_0$ to assess the effect of varied protection strength on power outage risk. In addition, the value of $\kappa_{v_{\text{PE}i}}$ is set to 1 in this study.

A variety of metrics have been proposed to assess the power outage risk [108], including blackout area size, number of affected populations, blackout duration, etc. To intuitively reveal how severe the cascading failure process is, we use the *power outage rate* [163,308], which is the proportion of the total

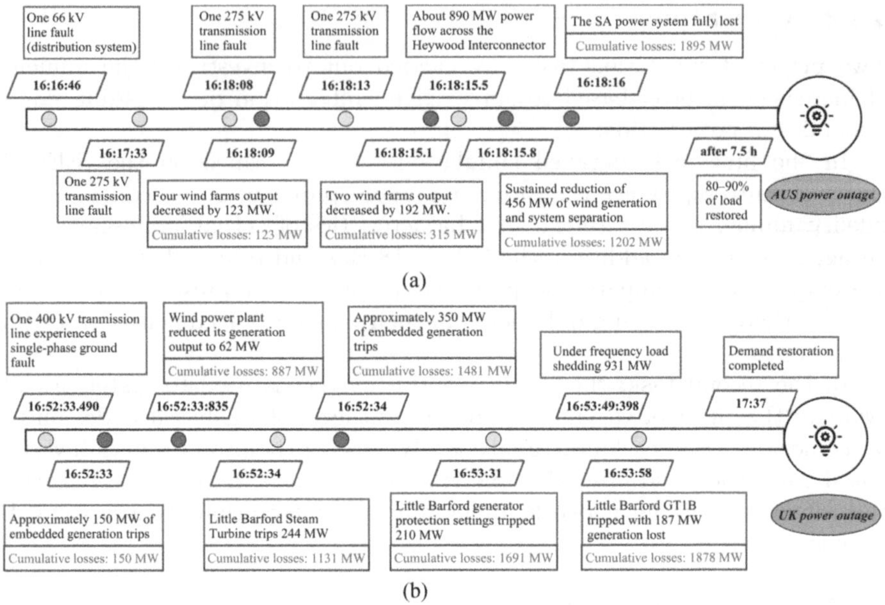


**FIGURE 4.22**
Timelines of major events in (a) 2016 South Australia blackout; and (b) 2019 UK blackout.

lost electricity demand after the cascading failure process, i.e.,

$$r_{\text{outage}} = \frac{P_d(t_0) - P_d(t_{\text{end}})}{P_d(t_0)} \tag{4.47}$$

where $P_d(t_o)$ and $P_d(t_{\text{end}})$ correspond to the total electricity demand before and after the cascading failure process, respectively.

### 4.4.2.2 Failure Propagation Process

Technical investigation reports of the 2016 South Australian blackout and the 2019 UK blackout provide detailed timelines of the major events during these cascading failure processes, as illustrated in Figure 4.22. In both incidents, wind power generators, classified as inverter-based resources (IBRs), exhibited a pronounced vulnerability to system disturbances. During the fault propagation, these units were tripped or disconnected from the grid, largely due to the strict protective logic implemented in their power electronic interfaces. Compared to conventional synchronous machine–based generators, IBRs operate within a narrower stability margin; they are therefore more prone to rapid disconnection from the grid when subjected to severe voltage or frequency deviations.

To examine this effect in simulation, two representative failure propagation processes are selected from runs using integration levels of PE-based nodes set to 30% and 50%. Here, the integration level PL is defined as the proportion of total generation capacity contributed by inverter-based resources in the PE-integrated power system, i.e.,

$$\mathrm{PL} = \frac{\sum_{k=1}^{K} P_{v_{\mathrm{PE}k}}}{\sum_{i=1}^{N_G} P_{Gi}} \tag{4.48}$$

where $P_{v_{\mathrm{PE}k}}$ denotes the generation power of PE-based node $v_{\mathrm{PE}k}$, $K$ is the number of PE-based nodes, $P_{Gi}$ is the generation of node $i$, and $N_G$ is the total number of generation nodes in the system. These two integration levels are chosen to approximate the renewable generation proportions in South Australia and in the UK at the time of the aforementioned blackouts.

The simulated failure sequences for these scenarios are shown in Figure 4.23a and b, which track total generation loss together with the number of failed transmission links and failed PE-based nodes over successive cascade iterations. Two salient observations emerge:

First, failures of transmission links due to overload protection and failures of PE-based nodes occur in alternating succession, collectively driving the cascade toward a severe outage event—approaching 5,000 MW of generation loss in these cases. This alternating fault pattern mirrors the intermediate developments described in the SA and UK blackout reports, supporting the validity of the proposed model in replicating the combined effect of overload trip events and PE-based disconnects.

Second, the disconnection of PE-based nodes markedly amplifies outage severity, leading to further generation losses even after line tripping events have subsided. For example, in Figure 4.23a and b, PE-based node failures at iteration steps 5 and 6, respectively, trigger additional system instability. Such losses of IBR capacity during disturbances can cause significant frequency or voltage deviations, in turn necessitating emergency protective measures such as load shedding. The consequent power redistribution required for rebalancing generation and demand imposes additional loading on certain lines, which may then exceed their thermal ratings and trip via overload protection. While these schemes are vital for preventing equipment damage and damming the cascade, poorly coordinated or repeated activations can escalate into extensive system-wide outages.

The findings from these cases indicate that risk assessments omitting the PE-based node failure mechanism will underestimate the true probability and scale of cascading outages. To further examine this, $N-2$ contingency simulations are performed on PE-integrated networks for the two integration levels considered. In real systems, protecting critical loads during failure propagation is a key operational objective. In the IEEE 39 Bus network, node 20 is identified (Figure 4.24a) as the most critical load, possessing the highest demand in the system. To ensure its preservation during the cascade, the power ratings of five selected transmission lines supplying node 20 are doubled relative to their initial capacities, thereby reducing the probability of their removal by overload protection.

Figure 4.24b and c summarize the simulation results, contrasting scenarios that include and exclude the PE-based node failure mechanism. Across both integration levels, omission of the PE-based mechanism produces consistently lower estimates of outage size, demonstrating that failure models lacking this behavior systematically underestimate cascading failure risk. This underestimation is particularly consequential when planning mitigation or resilience

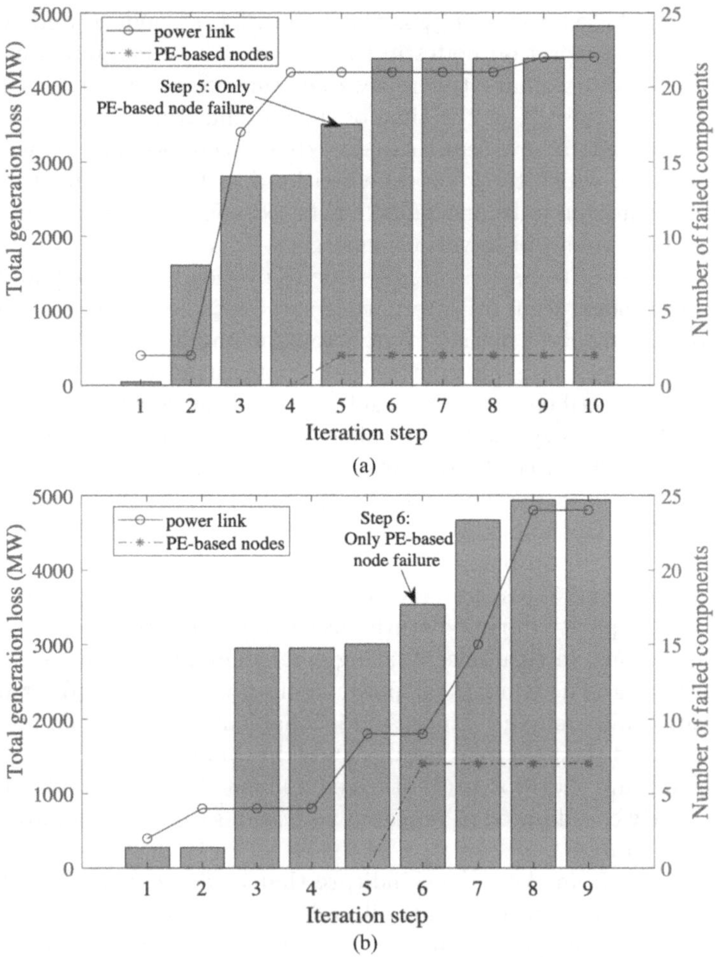


**FIGURE 4.23**
Two selected failure propagation processes in the IEEE 39 Bus system with integration levels of PE-based nodes set to (a) 30% and (b) 50%. During the process, total generation loss, number of failed power links, and number of failed PE-based nodes are recorded.

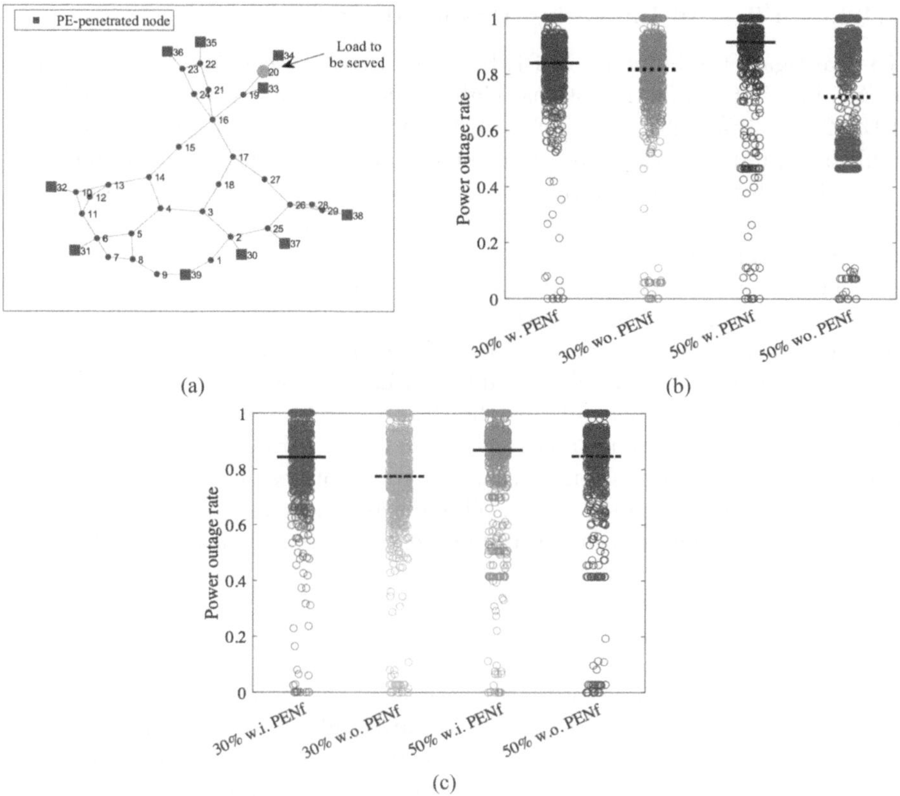


**FIGURE 4.24**
Power outage rates for all $N-2$ cascading failure simulations in the (a) IEEE 39 Bus system without (b) and with (c) considering load preservation on Bus 20 in cascading failure events. In two scenarios, both 30% and 50% integration levels of PE-based nodes with and without considering the failure mechanism of PE-based nodes (with PENf and without PENf) are adopted in cascading failure simulations. Horizontal bar for each data group represents the mean value of that group.

strategies, as it may lead to insufficient reinforcement of critical assets and a greater likelihood of unforeseen large-scale power loss.

Additionally, we compare the simulation results using our proposed models with those using two previously proposed models [66, 151]. Tables 4.2 and 4.3 illustrate that the power outage risk assessed in our proposed models is notably greater than that determined in the two existing models. This comparison shows that overlooking the failure mechanism of PE-based nodes can result in considerable underestimation of cascading failure risks in PE-integrated power systems.

#### 4.4.2.3 Effect of Increasing PE-based Nodes

To investigate the influence of PE-based nodes on cascading failure dynamics in PE-integrated power systems, eight integration levels are examined: 0%, 10%, 20%, 30%, 40%, 50%, 60%, 70%, and 80%. For each integration level, multiple simulation runs are performed, and a representative distribution is plotted to illustrate the relative probability of the power outage rate occurring within a given range. Three sets of comparative analyses are conducted to interpret the results.

The first comparison demonstrates the importance of incorporating the PE-based node failure mechanism into cascading failure models for accurate outage risk assessment. Figure 4.25a and b show that the probability of large outage events increases notably with higher integration levels. For a given integration level, simulations omitting the PE-node failure mechanism fail to fully capture the propagation patterns specific to PE-integrated systems. Averaging over all integration levels considered, such omission results in underestimation of the outage rate by up to 69%. This discrepancy widens with integration level, indicating that high-integration systems are particularly sensitive to the effects of PE-node disconnections.

The nine histograms in Figure 4.25b reveal that outage rates do not vary monotonically with integration level, but instead exhibit three distinct phases:

1. Early stage (10%–40% integration): The concentration of outage rates shifts progressively toward higher values as integration increases, indicating a rising blackout risk. Failure propagation in this range is dominated by traditional protection-driven trips, with PE-node failures acting to aggravate cascades through additional loss of generation and consequent overloads.

**TABLE 4.2**
Comparison of Power Blackout Risk (80% Interval for Power Demand Loss) on IEEE 118 Bus System under $N-6$ Contingencies

| | PE-Integration Level | | |
|---|---|---|---|
| **Cascading Failure Model** | **20%** | **40%** | **60%** |
| Model in [151] (MW) | [850, 1350] | [950, 1400] | [1050, 1450] |
| Proposed model (MW) | [2624, 2705] | [4215, 4223] | [3157, 3219] |

**TABLE 4.3**
Comparison of Power Blackout Risk (Expected Demand Not Served) on Illinois 200 Bus System under $N-2$ Contingencies

| | PE-Integration Level | | |
|---|---|---|---|
| **Cascading Failure Model** | **20%** | **40%** | **60%** |
| Model in [66] (MW) | 75 | 150 | 310 |
| Proposed model (MW) | 1452 | 1468 | 1467 |

2. Intermediate stage (40%–60% integration): A modest shift of the distribution toward lower outage values suggests a slight risk reduction. Here, cascades are jointly driven by traditional mechanisms and PE-node failures, but the latter can, in some cases, interrupt propagation by removing significant amounts of load/generation early, an effect akin to intentional load shedding for system protection.

3. Advanced stage (60%–80% integration): Outage risk increases sharply, with most outcomes showing $r_{\text{outage}} > 0.7$. In this regime, PE-node failures become the dominant driver of the cascade. The widespread and near-simultaneous disconnection of inverter-based resources following initial disturbances causes extensive generation loss, exacerbating instability and leading to high-severity blackouts.

Table 4.4 lists the 80% confidence intervals of power outage rates associated with Figure 4.25, providing an alternative view of the concentration and spread of outage rates at each integration level. These findings confirm a nonlinear relationship between integration level and outage severity, highlighting the complex role of PE-node failures in large-scale blackout dynamics.

The shifting trends in outage risk across the three phases reflect the changing interaction between PE-node disconnections and traditional protection-induced trips. At low integration, traditional protection mechanisms dominate, and increasing PE integration tends to amplify overload cascades. At intermediate integration, PE-node failures may occasionally halt propagation, acting as a mitigating factor. At very high integration, however, PE-node trips become the primary failure pathway, leading to severe widespread outages.

These observations align qualitatively with historical incidents. In the 2016 South Australian event, where renewable integration was high, blackout propagation involved frequent alternation between PE-node disconnections and traditional protection trips. In contrast, in the 2019 UK blackout, involving lower renewable integration, PE-node shedding was prominent in the initial stages (wind farm shutdown), but subsequent load loss occurred mainly through conventional protection mechanisms.

The results underscore the necessity of explicitly modeling PE-based node vulnerabilities in cascading failure analysis. Not only does omission lead to substantial underestimation of outage risk, especially in high-integration conditions, but it also obscures the nonlinear and stage-dependent effects that such vulnerabilities have on system robustness.

#### 4.4.2.4 Effect of Location of PE-Based Nodes

We further examine the effect of the spatial location of PE-based nodes in the power network on power outage risk assessment. To distinguish two specific spatial locations of PE-based nodes, two inverter-based resource allocation schemes are illustrated for a given integration level:

**TABLE 4.4**
Power Outage Risk (80% Confidence Interval of Power Outage Rate $r_{\text{outage}}$) for IEEE 118 Bus Considering Increasing Integration of Inverter-Based Resources, Obtained by Applying the Cascading Failure Model without Failure Mechanism of PE-Based Nodes (without PENf), with Failure Mechanism of PE-Based Nodes (with PENf) and $r_{\text{prot}} = 0.8$, and with Failure Mechanism of PE-Based Nodes (with PENf) and $r_{\text{prot}} = 2.5$

| | PE-Integration Level | | | | | | | |
|---|---|---|---|---|---|---|---|---|
| **Cascading** Failure Model | **10%** | **20%** | **30%** | **40%** | **50%** | **60%** | **70%** | **80%** |
| wo. PENf | [0.17, 0.18] | [0.17, 0.18] | [0.17, 0.18] | [0.17, 0.18] | [0.17, 0.18] | [0.40, 0.42] | [0.43, 0.44] | [0.46, 0.48] |
| w. PENF & $r_{\text{prot}} = 0.8$ | [0.13, 0.14] | [0.60, 0.61] | [0.68, 0.69] | [0.70, 0.71] | [0.74, 0.75] | [0.69, 0.70] | [0.76, 0.77] | [0.78, 0.79] |
| w. PENF & $r_{\text{prot}} = 2.5$ | [0.19, 0.20] | [0.23, 0.25] | [0.30, 0.31] | [0.40, 0.42] | [0.61, 0.62] | [0.68, 0.69] | [0.71, 0.72] | [0.73, 0.74] |

1. Large Capacity First Allocation (LCFA): The generation nodes are ranked in descending order based on their output power. The power generated from inverter-based resources $P_{\text{totalIBR}}$ is first allocated to the top-ranked generation node. If the output power from the top-ranked node is still below the given $P_{\text{totalIBS}}$, the second-ranked node contributes, and so on, until the power output from the generation nodes exceeds the given $P_{\text{totalIBR}}$.

2. Small Capacity First Allocation (SCFA): The allocation process follows the same steps as in the LCFA scheme, but starting with the lowest-ranked generation node and progressing to higher-ranked nodes until the power output exceeds the given total power generation $P_{\text{totalIBR}}$.

By employing two allocation schemes, when the integration level is relatively low, specifically being less than 50%, the LCFA scheme results in a centralized distribution of PE-based nodes. On the other hand, the SCFA scheme leads to a decentralized distribution of PE-based nodes. This distinction can be clearly noticed in Figure 4.26c and d, where the integration levels are set at 40% for both LCFA and SCFA schemes.

Figure 4.26a and b present the probability distribution of power outage rate under different integration levels when applying LCFA and SCFA schemes, respectively. At lower integration levels (10% to 20%), a significant shift is noted in the concentration area of power outage rates from the range $[0.2, 0.4]$ to $[0.9, 1]$ when using the LCFA and SCFA schemes. In other words, the probability of large power outage rates is higher in the case of a decentralized distribution of PE-based nodes (using the SCFA scheme) compared to a centralized distribution of PE-based nodes (using the LCFA scheme). This variation is attributed to the prevalence of traditional protection methods like overload protection, which primarily contribute to electricity demand loss during this phase.

Additionally, the decentralized distribution of PE-based nodes raises the probability of these nodes disconnecting, consequently intensifying the overall electricity demand loss. As the integration level starts to increase, the decentralized distribution of PE-based nodes appears to reduce the probability of high-value power outage rates. Based on these results, we can infer that the decentralized distribution of PE-based nodes decreases the likelihood of power nodes carrying considerable power going offline from the network, thereby avoiding more severe cascading failure. In summary, as the integration of PE-based nodes increases in the future power grid, it becomes crucial to consider their spatial distribution. Centralized distribution of PE-based nodes may effectively reduce the risk of large-scale power outage incidents when the integration rate is relatively low. Conversely, as the integration rate increases, a decentralized distribution of PE-based nodes may provide higher reliability and resilience to the power grid.

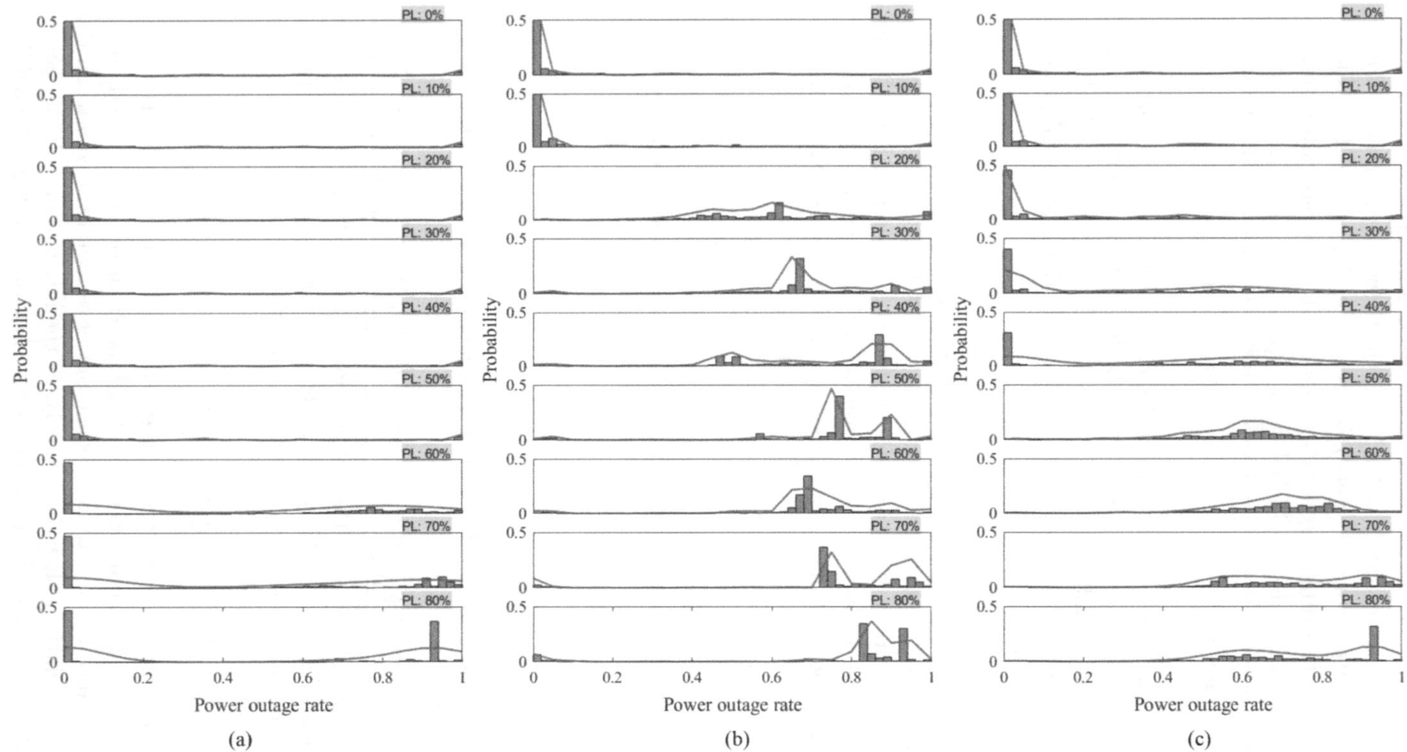


**FIGURE 4.25**
Probability distribution of power outage rates for the tested IEEE 118 Bus system with varying integration (penetration) levels (PL) of PE-based nodes under different scenarios. (a) Failure mechanism of PE-based nodes not considered; (b) and (c) failure mechanism of PE-based nodes considered, with protection ratio $r_{\text{prot}} = 0.8$ for (b) and $r_{\text{prot}} = 2.5$ for (c).

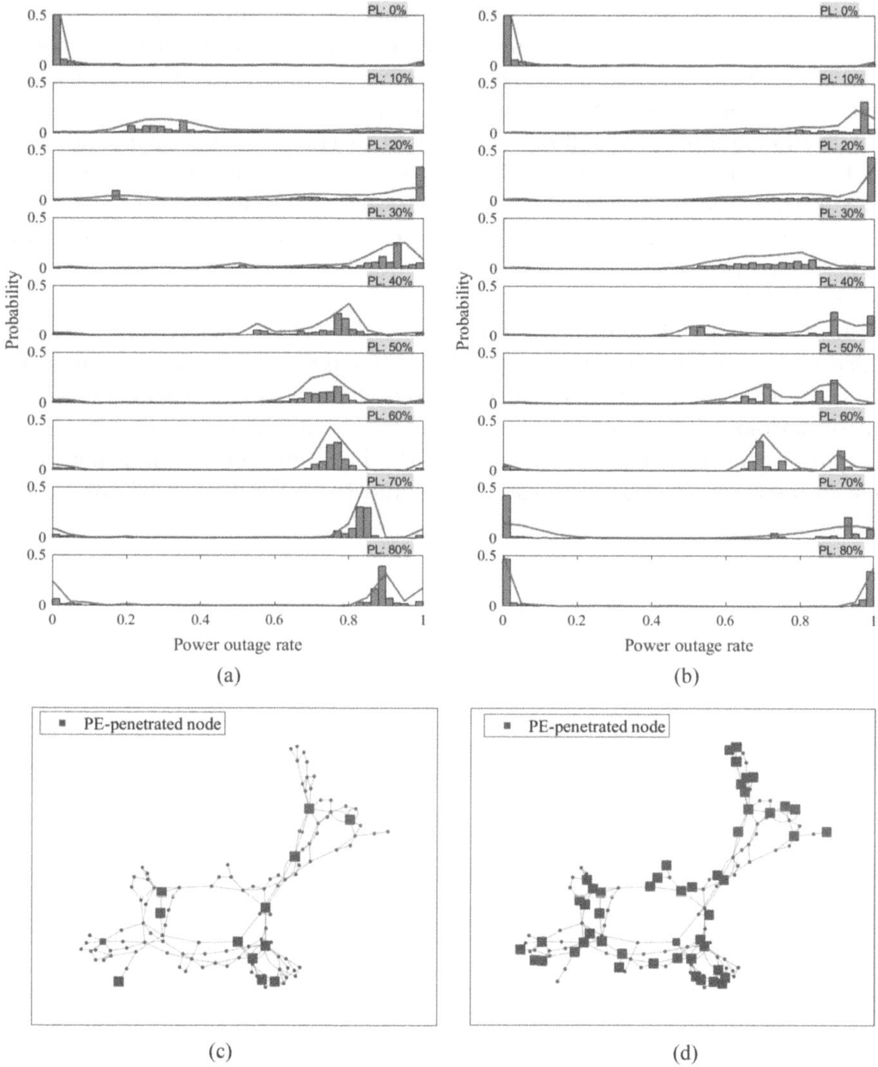


**FIGURE 4.26**
Probability distribution of power outage rates for the IEEE 118 Bus system for varying integration (penetration) levels (PL) of PE-based nodes. Two PE-based node allocation schemes are examined: (a) Large Capacity First Allocation (LCFA) and (b) Small Capacity First Allocation (SCFA). The spatial location of PE-based nodes is presented in (c) and (d) for the LCFA and SCFA schemes, respectively, both with a 40% integration level of PE-based nodes.

## 4.5 Summary

This chapter provides a comprehensive exposition on the modeling of cascading failures in modern power grids, which are increasingly characterized by the integration of power-electronics-based equipment to accommodate the growing adoption of renewable energy sources. We introduce a robust network-based framework for simulating the sequential occurrence of failure events, capturing both their deterministic and stochastic characteristics.

The modeling of power flow is grounded in the electrical properties and topological structure of power systems, governed by deterministic principles and typically requiring the solution of power flow equations. In contrast, component failures are influenced by factors such as aging, environmental conditions, and operational stress, which are effectively represented through stochastic processes.

Simulation models and methodologies are detailed in this chapter, accompanied by key insights into the propagation dynamics of cascading failures. To enhance the realism and practical relevance of the analysis, various levels of control actions—reflecting those implemented in actual power systems—are incorporated into the simulation framework. This integration enables a more accurate assessment of system robustness and the effectiveness of mitigation strategies.

# 5

# *Robustness Assessment of Evolving Power Grids*

The rapid evolution of modern power grids, driven by the increasing use of renewable energy sources, greater operational complexity, and heightened levels of uncertainty in both supply and demand, poses significant challenges to system robustness. Robustness assessment, in this context, refers to the ability to evaluate, quantify, and predict the grid's resilience to large-scale disturbances and cascading failures. Unlike robustness enhancement, which focuses on actively modifying system topology, control strategies, or operational parameters to improve performance, robustness assessment seeks to provide accurate, data-driven, and timely evaluations of system vulnerability. Such assessments form the essential foundation for informed decision-making in system operation, planning, and contingency management.

Recent advances in artificial intelligence (AI) and machine learning have opened new avenues for robustness assessment. Techniques such as graph neural networks, attention-based sequence models, and decision-system classifiers enable predictive analytics that exploit both the physical structure of the power network and its dynamic operating conditions. By integrating simulation-driven datasets with AI models capable of learning complex spatio-temporal dependencies, it is possible to produce early warnings, predictive indicators, and component criticality rankings that were previously unattainable with purely conventional analytical methods.

This chapter explores AI-enabled robustness assessment from four complementary perspectives: (1) Early risk warning using a graph attention network (GAT)-based node indispensability model; (2) Cascade path prediction through an attention-enhanced Long Short-Term Memory (LSTM) model; (3) Onset time prediction via a neural network classifier; (4) Component criticality and robustness indicator identification using decision tree analysis.

While the final study also delves into robustness enhancement by identifying network features associated with improved stability, the primary focus of this chapter remains on robustness assessment. The overarching objective is to transform complex cascade dynamics and network characteristics into actionable vulnerability metrics. These metrics empower operators and planners with predictive tools to anticipate, identify, and quantify risks, helping to mitigate catastrophic power outages before they occur.

DOI: 10.1201/9781003622130-5

## 5.1 Early Risk Warning with Partial State Observations

Fluctuations in load demand and renewable energy generation can lead to violations of the $N-1$ security criterion in power system operations. To enable timely risk alerts under such dynamic conditions, we present a Node Indispensability Estimation (NIE) model built upon a GAT. This model identifies indispensable nodes—components whose individual failure, under specific operating scenarios, would trigger overloads elsewhere in the network. Acknowledging the practical limitation that comprehensive system-wide monitoring is often infeasible, the NIE model is designed to operate using state information from only a selected subset of buses, thereby offering a scalable and efficient solution for real-time risk assessment.

The GAT-based NIE model is benchmarked against alternative neural network architectures, including graph convolutional networks (GCN) and multilayer perceptrons (MLP). Simulation studies conducted on the IEEE 30 Bus, 39 Bus, and 57 Bus test systems demonstrate that the proposed GAT-based approach significantly outperforms traditional methods in accurately identifying indispensable nodes, even when only partial state observations are available. By reliably detecting scenarios where the $N-1$ security criterion is violated, the model serves as an effective early warning tool, enabling system operators to proactively adjust operating conditions and maintain adequate security margins.

Looking ahead, future development will focus on extending the methodology to enable simultaneous evaluation of all system components and to delineate admissible operating regions—ranges of system conditions under which secure operation can be maintained. This advancement aims to enhance situational awareness and support more informed decision-making for system operators.

### 5.1.1 Background

Modern power systems are expected to comply with the $N-1$ security criterion to ensure a sufficient level of operational safety. This criterion stipulates that the system must continue to function normally following the failure of any single component [235]. However, under conditions of heavy loading, the failure of certain components can have severe consequences. A notable example occurred in 2003, when high temperatures and elevated demand placed the North Ohio grid under significant stress. The unexpected shutdown of Eastlake Unit 5 pushed the system to the brink of collapse. Subsequently, a short circuit on a critical transmission line triggered the largest blackout in North American history [307].

By predicting the indispensability of system components under current loading conditions, it becomes possible to mitigate the risk of single-component failures cascading into widespread system disruptions—particularly during periods of heavy load. In recent years, numerous researchers have focused on developing early warning mechanisms for power system operational risks. For instance, Javanbakht et al. [131] proposed a risk-aware power generation dispatch strategy that accounts for the impact of hurricanes; Wang et al. [259] introduced a cloud-based early warning model for smart grid operations; and Dong et al. [76] investigated fault early warning technologies for critical nodes in power communication networks.

However, many existing early warning approaches rely on complete status information from all components in the power system, resulting in high computational and data acquisition costs. In practice, most power utilities worldwide employ Supervisory Control and Data Acquisition (SCADA) systems to monitor and control distribution networks [79]. Yet, remote terminal units (RTUs), which are responsible for data measurement and transmission, often do not cover all busbars. To address this limitation, we present a novel early warning technique for power system operating nodes based on a GAT. The method developed in this study can accurately predict node indispensability using only partial bus information, offering a more scalable and efficient solution for real-time risk assessment.

### 5.1.2 Model Description

In this section, we outline the modelling framework for early risk warning in power systems with partial state observations. We first simulate the power flow redistribution process following load fluctuations to generate a large dataset of representative operating scenarios. The collected data is then preprocessed to ensure suitability for machine learning and to enhance model performance. Based on these processed features, we present a GAT–based NTE model, which integrates a multi-head graph attention layer with a fully connected residual network. This architecture provides strong representation-learning capability and fast convergence, enabling accurate identification of indispensable nodes even when only partial network state information is available.

#### 5.1.2.1 Data Preparation

In this section, we generate a large set of data samples for the power grid by simulating load fluctuations. Each sample represents a distinct operating condition and is labeled with corresponding node indispensability indicators. These labeled datasets serve as the foundation for training and evaluating the proposed model.

#### 5.1.2.2 Power Flow Redistribution Calculation

Suppose there are $n$ nodes in a grid. Let $P_i$ represent the external injected power of node $i$. The value of the load power is negative and the value of the generator power is positive. $\boldsymbol{P} = (P_1, P_2, \cdots, P_n)^{\mathrm{T}}$ is the injected power vector of the grid, where power balance should be achieved, i.e., $\sum_{i=1}^{n} P_i = 0$.

Interruptions such as load fluctuations and component failures can lead to power imbalance and redistribution of power flow across the grid. In response to such disturbances, frequency regulation mechanisms are activated to restore power balance. Once equilibrium is re-established, the output power of each generator node is updated accordingly and can be expressed as

$$P(g_i, t + \Delta t) = \sum_{i=1}^{n} P_i(t) \frac{P_c(g_i, t)}{\sum_{g_i \in G} P_c(g_i, t)} + P(g_i, t) \tag{5.1}$$

where $t$ is a time moment after the interruption and before the frequency regulation, $t + \Delta t$ is a time moment after the frequency regulation, $\sum_{i=1}^{n} P_i(t)$ represents the amount of the mismatched power in the grid at time $t$, and $P(g_i, t)$ and $P_c(g_i, t)$ represent the actual active power and the reserve power of generator $g_i$ at time $t$. The reserve power of the generator is defined as

$$P_c(g_i, t) = \begin{cases} P(g_i, \max) - P(g_i, t), & \text{if } \sum_{i=1}^{n} P_i(t) < 0 \\ P(g_i, t) - P(g_i, \min), & \text{otherwise} \end{cases} \tag{5.2}$$

where $P(g_i, \max)$ and $P(g_i, \min)$ are the maximum output power and minimum output power of $g_i$.

After $\boldsymbol{P}$ is updated, the DC power flow model is adopted, i.e.,

$$\boldsymbol{P} = \boldsymbol{B\theta} \tag{5.3}$$

where $\boldsymbol{\theta} = [\theta_1, \theta_2, \cdots, \theta_n]^T$ is the voltage phase vector in the power grid. $\boldsymbol{B}$ is the $n * n$ nodal admittance matrix.

By solving the above equations, the grid's $\boldsymbol{\theta}$ can be calculated. The "power flow" across each line is given as

$$P_{ij} = B_{ij} * (\theta_i - \theta_j) \tag{5.4}$$

#### 5.1.2.3 Generation of Data Samples

To evaluate node indispensability under varying conditions, we generate a diverse set of power grid operating scenarios by simulating load fluctuations. For each scenario, the indispensability of each node is assessed based on its impact on system stability. A node is labeled as indispensable if its independent removal leads to power flow redistribution that causes overloading elsewhere in the network.

Figure 5.1 illustrates the flow chart used to generate these operating conditions and estimate node indispensability. The process begins by importing

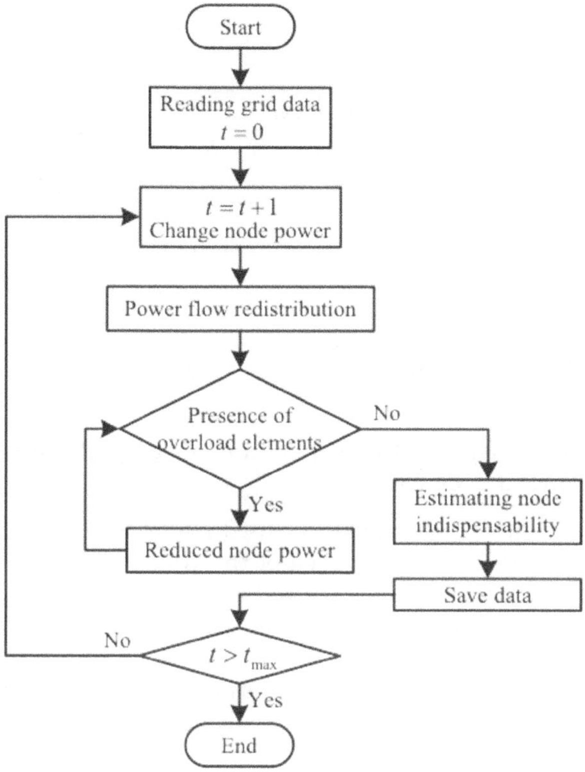


**FIGURE 5.1**
Flow chart for generating various operation conditions and estimating the node indispensability.

the original grid parameters and setting the simulation end time, denoted as $t_{\max}$. At each time step, an operating condition is generated by randomly adjusting the power at each node within $\pm 10\%$, constrained to a range of $\pm 30\%$ of the initial power. The resulting power flow distribution is then computed. If any component is found to be overloaded, the power at each node is uniformly reduced by 5% until the system returns to a non-overloaded state. Once a stable operating condition is achieved, the indispensability of each node is evaluated. This involves independently removing each node $i \in \{1, \ldots, n\}$, recalculating the power flow, and checking for overloaded nodes. If the removal of node $i$ results in overloading, it is labeled as "indispensable" (value 1); otherwise, it is labeled as "dispensable" (value 0). The electrical parameters and node indispensability labels for each operating condition are recorded. The simulation proceeds to the next time step $t = t + 1$, and continues until $t = t_{\max}$.

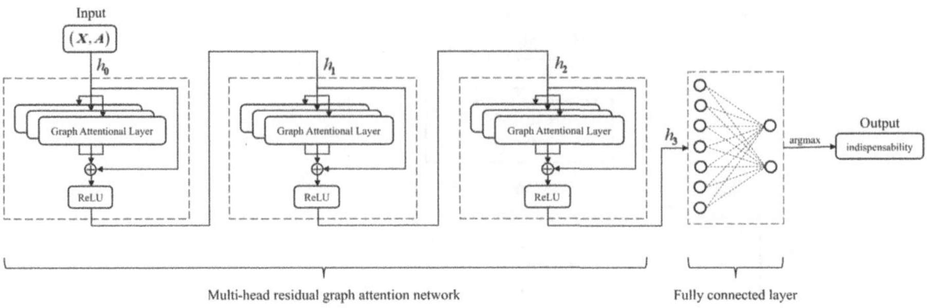


**FIGURE 5.2**
GAT-based NIE model architecture.

#### 5.1.2.4 Node Indispensability Estimation Model Based on Graph Attention Network

In this section, we introduce the GAT-based model used for estimating node indispensability and risk early warning of the grid.

For each data sample, the voltage phase vector $\boldsymbol{\theta}$ and the injected power vector $\boldsymbol{P}$, which together characterize the system's operating condition, are used as inputs to the GAT-based NIE model. The corresponding node indispensability labels serve as the model's output. Prior to model training, the input data undergoes preprocessing to enhance the neural network's ability to recognize and learn meaningful node-level features, thereby improving prediction accuracy and model robustness.

Specifically, the voltage phase vector $\boldsymbol{\theta}$ is normalized by min-max normalization, i.e.,

$$\boldsymbol{\theta}_{\text{net}}\ (t) = \frac{\boldsymbol{\theta}(t) - \min(\boldsymbol{\theta}(t))}{\max(\boldsymbol{\theta}(t)) - \min(\boldsymbol{\theta}(t))} \tag{5.5}$$

where $\boldsymbol{\theta}(t)$ represents the voltage phase vector of each node in the power grid at time $t$, and $\boldsymbol{\theta}_{\text{net}}\ (t)$ is the voltage phase vector after the min-max normalization.

For the injected power vector $\boldsymbol{P}(t)$ at time $t$, we perform the following process:

$$\boldsymbol{P}^1_{\text{net}}(t) = \frac{\boldsymbol{P}(t)}{\boldsymbol{P}(0) + \varepsilon} \tag{5.6}$$

$$\boldsymbol{P}^2_{\text{net}}\ (t) = \frac{\boldsymbol{P}(t)}{\max(\boldsymbol{P}(0))} \tag{5.7}$$

$$\boldsymbol{P}^3_{\text{net}}(t) = \frac{\boldsymbol{P}(0)}{\max(\boldsymbol{P}(0))} \tag{5.8}$$

where $\boldsymbol{P}(0)$ represents the injected power vector at the initial time of the power system, and $\varepsilon$ represents an arbitrarily small number.

We define the node feature matrix $\boldsymbol{X}$ as

$$\boldsymbol{X} = \left(\boldsymbol{\theta}_{\text{net}}\left(t\right), \boldsymbol{P}_{\text{net}}^{1}\left(t\right), \boldsymbol{P}_{\text{net}}^{2}\left(t\right), \boldsymbol{P}_{\text{net}}^{3}\left(t\right)\right) \tag{5.9}$$

In addition, we construct an adjacency matrix $\boldsymbol{A} \in \mathbb{R}^{n \times n}$ based on the connections of the power system. If there is a connection between nodes $i$ and $j$ in the power system, $A_{ij}$ and $A_{ji}$ in $\boldsymbol{A}$ are set to 1. Otherwise, they are set to 0.

Taking $(\boldsymbol{A}, \boldsymbol{X})$ as the input, we first build a GAT-based NIE model that predicts the indispensability of the node. Unlike traditional neural network architectures such as GCNs, the GAT incorporates a self-attention mechanism that adaptively assigns weights to neighboring nodes during feature aggregation. This enables the model to capture the relative importance of each neighbor, thereby enhancing its ability to learn complex topological and operational patterns in the power grid. The GAT architecture is composed of multiple graph attention layers, each contributing to a more accurate and expressive representation of node-level features. For the $l$th graph attention layer, input $\boldsymbol{h}^{(l)} = \left[h_1^{(l)}, h_2^{(l)}, \ldots, h_n^{(l)}\right]^T$, $h_i^{(l)} \in \mathbb{R}^F$, output $\boldsymbol{h}^{(l+1)} = \left[h_1^{(l+1)}, h_2^{(l+1)}, \ldots, h_n^{(l+1)}\right]^T$, $h_i^{(i+1)} \in \mathbb{R}^{F'}$. The self-attention coefficient between nodes $i$ and $j$ is

$$\alpha_{ij}^{(l)} = \text{softmax}\left(\text{LeakyReLU}\left(\vec{a}^T\left[\boldsymbol{W}^{(l)}h_i^{(l)} \| \boldsymbol{W}^{(l)}h_j^{(l)}\right]\right)\right) \tag{5.10}$$

where $\vec{a}^T$ is a map of $\mathbb{R}^{F'} \times \mathbb{R}^{F'} \to \mathbb{R}$, $\boldsymbol{W} \in \mathbb{R}^{F' \times F}$ is a learnable parameter matrix, and $h_i^{(l)}$ is the input feature of node $i$ corresponding to the attention layer of the $l$th graph. Specifically, when $l = 0$, we have

$$\boldsymbol{h}^{(0)} = \boldsymbol{X} \tag{5.11}$$

The output of the attention layer of the $l$th graph is

$$h_i^{(l+1)} = \sigma\left(\sum_{j \in \mathcal{N}(i)} \alpha_{ij}^{(l)} \boldsymbol{W}^{(l)} h_j^{(l)}\right) \tag{5.12}$$

Here, the multi-head attention mechanism and residual structure are also introduced to further improve the learning ability of the model. At this point, the output of the attention layer of the $l$th graph becomes

$$h_i^{(l+1)} = \sigma\left(h_i^{(l)} + \frac{1}{K}\sum_{k=1}^{K}\sum_{j \in N(i)} \alpha_{ij}^{(l,k)} \boldsymbol{W}^{(l,k)} h_j^{(l)}\right) \tag{5.13}$$

where $K$ is the number of heads of the multi-head attention mechanism.

**TABLE 5.1**
Training Parameters

| Test Case | Unobserved Nodes | Estimated Nodes |
|---|---|---|
| IEEE 30 Bus | 5,6,9,11,25,28 | 2 |
| IEEE 39 Bus | 2,5,6,10,11,13,14,17,19,22 | 23 |
| IEEE 57 Bus | 4,7,11,21,22,24,26,34,36,37, 39,40,45,46,48 | 17 |

### 5.1.3 Case Study

We present experimental results to validate the effectiveness of the proposed GAT-based NIE model across three IEEE test systems.

#### 5.1.3.1 Numerical Experiments

The model is implemented in Python using the Deep Graph Library (DGL) and PyTorch frameworks, while the IEEE test systems are sourced from the pypower library. Although these benchmark systems do not inherently satisfy the $n - 1$ security criterion, certain nodes still exhibit varying degrees of influence on system stability under fluctuating load conditions, allowing for meaningful indispensability estimation. In our setup, RTUs are not assigned to non-generator nodes without load, reflecting practical limitations in real-world monitoring. Consequently, the corresponding entries in the node feature matrix $\boldsymbol{X}$ for these unobservable nodes are set to zero, ensuring that the model operates under realistic data availability constraints. The dataset is partitioned into two subsets: 80% of the samples are used for training the model, while the remaining 20% are reserved for testing to evaluate its performance.

#### 5.1.3.2 Model Training and Testing

To evaluate the performance of different architectures, we trained NIE models based on GAT, Graph Convolution Network (GCN), and Multi-Layer Perceptron (MLP) using identical hyperparameters, including learning rate and batch size. The GCN-based NIE model is constructed by replacing the graph attention layers in Figure 5.2 with graph convolution layers, while retaining the same input structure as the GAT-based model. The MLP-based NIE model consists of two hidden layers and takes the node feature matrix $\boldsymbol{X}$ as input. For all three models, the output is the predicted indispensability of each node.

As shown in Table 5.1, different IEEE test systems are used with varying selections of unobserved and estimated nodes, tailored to the specific characteristics of each system. We employ a cross-entropy loss function, with class weights adjusted according to the data distribution to address class imbalance. The learning rate is initially set to 0.001 and decays by 20% every 30 training epochs to ensure stable convergence.

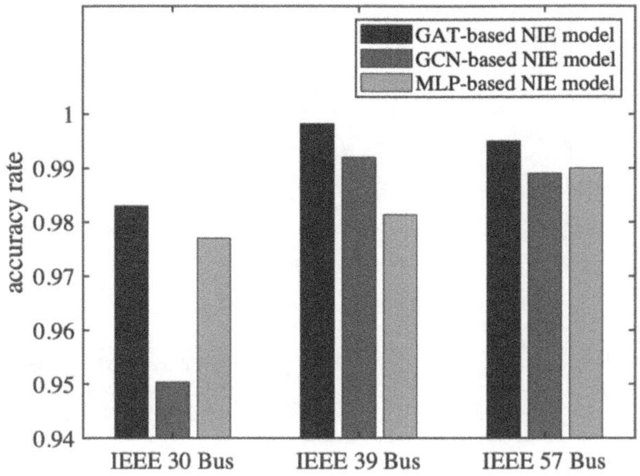


**FIGURE 5.3**
Accuracy of the test sets of the NIE models based on GAT, GCN, and MLP trained 200 times in each of the three IEEE test systems.

Figure 5.3 presents the test set accuracy of three predictive models, namely, GAT-based NIE, GCN-based NIE, and MLP-based NIE, each trained over 200 iterations across three IEEE test systems. As shown, the GAT-based NIE model consistently outperforms the other two approaches across systems of varying sizes, primarily due to its attention mechanism, which enables more effective learning of node-level dependencies. Figure 5.4 illustrates the accuracy progression of the three models during training on the IEEE 39 Bus system. The GAT-based model demonstrates faster and more stable convergence compared to GCN and MLP, highlighting its superior learning efficiency. Figure 5.5 compares the performance of the GAT-based NIE model under two observation scenarios: complete and partial node observability. Interestingly, the model achieves slightly higher accuracy under partial observation. This suggests that the information from unobserved nodes can be inferred from the observed ones, indicating that partial observability does not diminish, and may even enhance, the model's ability to capture system dynamics.

## 5.2 Prediction of Cascading Failure Propagation Paths

Complex networks are inherently susceptible to cascading failures, where a small number of initial faults can escalate into widespread, system-wide disruptions. Accurate and timely prediction of the propagation paths of such cascades is critical for enabling effective preventive and mitigative actions.

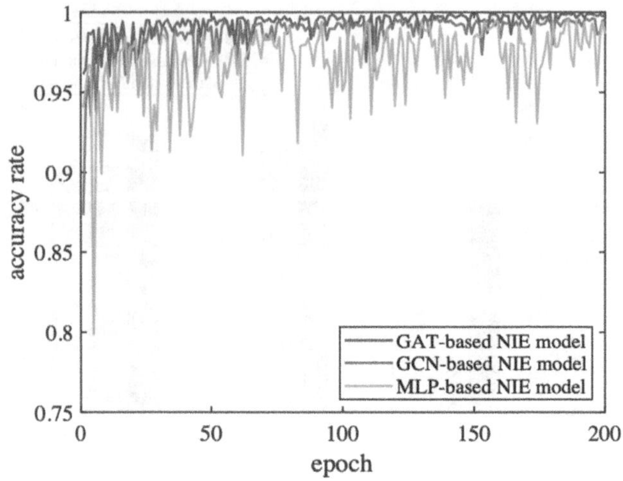


**FIGURE 5.4**
Accuracy of each model in each training cycle in the IEEE 39 Bus.

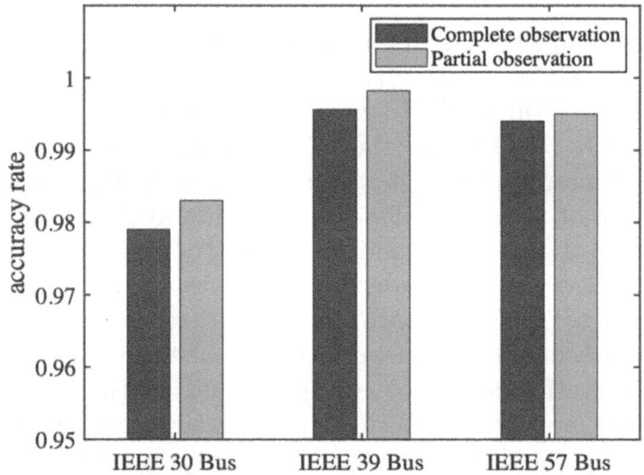


**FIGURE 5.5**
Training effect of GAT-based NIE model for complete observation and partial observation.

To address this challenge, we first develop a cascading failure simulation model that captures the cumulative impact of component overloading over time, allowing for the realistic generation of cascading failure scenarios. Using these simulated cases as ground truth, we then propose a path prediction model that integrates an attention mechanism with an LSTM neural network. The attention module enhances the LSTM's ability to focus on critical spatiotemporal patterns in the evolving network state, thereby improving prediction accuracy.

These models are evaluated on synthetic complex networks of varying scales. For benchmarking, both a standard LSTM and the attention-LSTM model are applied. Simulation results demonstrate that neural networks can predict cascading failure propagation paths both quickly and accurately, with the attention-LSTM model significantly outperforming the baseline LSTM in terms of prediction performance. This study highlights the potential of combining AI techniques with dynamic network modeling to address practical prediction challenges in complex systems.

Modern society relies heavily on the uninterrupted operation of critical infrastructure systems such as power grids, transportation networks, and communication systems. These systems can be effectively modeled as complex networks, where components are represented as nodes and their interactions as links. Over the past two decades, the study of complex networks has attracted significant attention due to their relevance in understanding system behavior and vulnerabilities. One of the most critical phenomena in complex networks is cascading failure, where a small number of initial component failures trigger a redistribution of load across the network, leading to subsequent failures and potentially large-scale outages. Such cascading events can result in severe disruptions and substantial economic losses, posing serious risks to modern society [215].

A cascading failure propagation path can be fundamentally understood as a time series comprising a sequence of component failure events [302]. Accurately reproducing and predicting these propagation paths is crucial for deepening our understanding of the mechanisms behind failure cascades. Such predictive capabilities enable the early identification of potential hazards and support the implementation of timely and effective measures to prevent large-scale network outages.

Extensive research has been conducted on predicting cascading failure propagation paths. For example, Wang et al. [256] proposed a method based on the Frequent-Pattern Growth algorithm, which predicts potential failure node sets by analyzing large volumes of simulation data and mining hidden associations among failure events. Yang et al. [288] developed a dynamic simulation-based approach that automatically selects and verifies progressive cascading failure sets. Additionally, Wang et al. [267] introduced a probabilistic analysis framework for cascading failures in power networks, relying on prior statistical probabilities and simulation to determine likely propagation paths.

While these methods have demonstrated effectiveness, they are heavily dependent on simulation-based computations, which can be time-consuming. In real-world scenarios, however, cascading failures can propagate rapidly, leaving little time for extensive simulations. This highlights the need for predictive approaches that can deliver accurate results with minimal computational delay, enabling timely intervention and mitigation.

In recent years, the rapid advancement of AI has introduced new solutions to prediction challenges, with LSTM neural networks demonstrating strong performance in sequence modeling tasks. For example, Kavinnilaa et al. [137]

applied an LSTM model to forecast stock prices, aiding investors in making more informed decisions. Cui et al. [63] developed an LSTM-based model for load forecasting in power systems, while Wang et al. [257] designed a bidirectional LSTM network to generate accurate transient stability assessments using sparse phasor measurement unit (PMU) data. For the task of cascading failure path prediction, different propagation sequences can be treated as time-series datasets. By training LSTM neural networks on these sequences, it is possible to achieve fast and accurate predictions of cascading failure paths, offering a promising approach for real-time risk mitigation in complex networks.

In this study, we present a predictive model that combines an attention mechanism with an LSTM neural network to forecast cascading failure paths in complex networks. Section 5.2.1.1 introduces a cascading failure simulation framework that accounts for the cumulative effects of component overloading over time, enabling the realistic generation of failure scenarios. These simulated cases serve as training and test data for the LSTM-based model. Section 5.2.1.2 outlines the proposed prediction strategy, where the trained attention-LSTM network is employed to forecast the complete propagation path of cascading failures. In Section 5.2.2, the model is evaluated on synthetic complex networks of varying scales. Results demonstrate that the proposed method can rapidly and accurately predict cascading failure paths, offering a practical solution for real-time risk mitigation in complex systems.

### 5.2.1 Model Description

In this section, we present a cascading failure path prediction framework for complex networks that integrates an attention mechanism with an LSTM neural network. The framework begins with a cascading failure simulation model that incorporates the cumulative effect of component overloading over time, enabling a realistic representation of failure propagation dynamics. Using this model, a large set of cascading failure scenarios is generated to form the training and testing datasets. Based on these data, we develop an artificial-intelligence-driven prediction strategy in which a trained attention–LSTM neural network is employed to forecast the entire propagation path of a cascading failure, thus providing a rapid and accurate tool for anticipating the sequence of component failures in complex networks.

#### 5.2.1.1 Cascading Failure Data Generation

A typical complex network consists mainly of nodes and connecting lines between each node. The betweenness centrality of a node is used to represent the load on the node [186], which is calculated as

$$P(\mathrm{v_i}) = \sum_{\mathrm{v_s} \neq \mathrm{v_i} \neq \mathrm{v_t}} \frac{\sigma_{\mathrm{st}}(\mathrm{v_i})}{\sigma_{\mathrm{st}}} \tag{5.14}$$

where $\sigma_{st}$ is the number of shortest paths from node $v_s$ to node $v_t$, and $\sigma_{\mathrm{st}}(\mathrm{v_i})$ is the number of paths through node $v_i$ in these paths.

Let $P_i(t), i \in \{1 \cdots n\}$ denote the power at node $i$ in the network at time $t$, and let $n$ represent the total number of nodes in the network. Each node $i$ has a predefined power limit denoted by $P_{Ci}$. If $P_i(t) > P_{Ci}$, node $i$ is considered to be overloaded. The overload rate of node $i$ at time $t$ is defined as

$$\zeta_i(t) = \frac{P_i(t) - P_{Ci}}{P_{Ci}} \tag{5.15}$$

where $\zeta_i(t)$ denotes the overload rate of node $i$ at time $t$.

In this cascading failure model, we assume that an overloaded node remains connected until its cumulative overload rate reaches a predefined threshold, at which point it is disconnected from the network. Let $\alpha_i(t)$ denote the cumulative overload rate of node $i$ at time $t$, measured from the onset of the initial failure, and can be derived as

$$\alpha_i(t) = \alpha_i(t_{i-1}) + \int_{t_0}^{t} \zeta_i(t)\mathrm{d}t \tag{5.16}$$

where $t_{i-1}$ is the time of the previous failure event in the complex network and $\alpha_i(t_{i-1})$ is node $i$'s cumulative overload rate till $t_{i-1}$. This formulation allows the model to capture the time-dependent nature of stress accumulation on network components, enabling more realistic simulation of cascading failure dynamics.

The disconnection threshold, denoted as $\alpha_{\max}$, is assumed to be uniform across all nodes in the network. At any given time, multiple components may be overloaded; however, the node that first reaches the threshold $\alpha_{\max}$ is disconnected from the network.

It is important to note that a node previously experiencing overload may return to normal operation due to power redistribution, such that $P_i(t) < P_{Ci}$. In this case, the cumulative overload rate for node $i$ is reset to zero, i.e., $\alpha_i(t) = 0$. This mechanism reflects the dynamic nature of load recovery and ensures that only sustained overloads contribute to node disconnection.

Figure 5.6 illustrates the flow chart of the cascading failure simulation model. The process begins by randomly selecting a predefined number of network components as initial failure points. Subsequently, the loads of the remaining nodes are recalculated, and overloaded nodes are identified. If overloaded nodes are present, their cumulative overload rates are updated. The node that first reaches the disconnection threshold is then removed from the network. If no overloaded nodes are detected, the cascading failure simulation terminates, indicating system stabilization.

#### 5.2.1.2 Cascading Failure Prediction Model

In this section, we introduce a cascading failure path prediction model. A complete cascading failure process may span dozens of time steps, with earlier

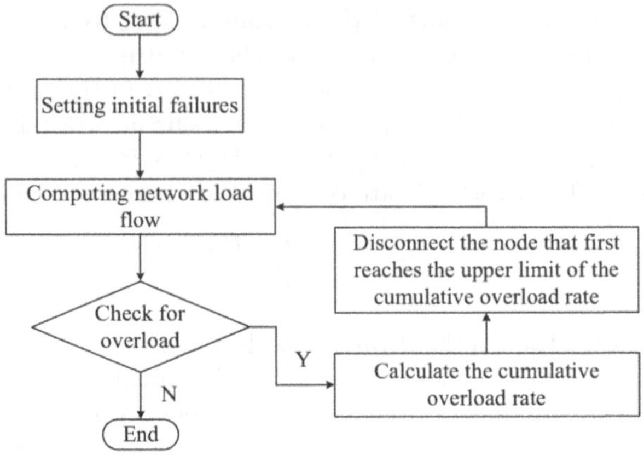


**FIGURE 5.6**
Cascading failure simulation flow chart for data generation in the prediction of the propagation paths.

failure events exerting cumulative influence on subsequent ones. Capturing these long-term dependencies is essential for accurate prediction. However, conventional recurrent neural networks (RNNs) often suffer from issues such as vanishing or exploding gradients when modeling long sequences. To overcome these limitations, we employ an LSTM neural network, which is well-suited for learning temporal dependencies and is capable of effectively predicting cascading failure propagation paths in complex networks.

#### 5.2.1.3 Prediction Method Based on LSTM

The LSTM neural network consists of memory cells and gating mechanisms that regulate the flow of information through three key gates: the input gate, the forget gate, and the output gate [233]. These gates enable the network to selectively retain or discard information, making LSTM particularly effective for modeling long-term dependencies in sequential data.

Figure 5.7 illustrates the operational structure of the LSTM cell at time $t$. The model receives three inputs: the previous cell state $C_{t-1}$, the previous hidden state $h_{t-1}$, and the current network state $\boldsymbol{x_t}$. It produces two outputs: the updated cell state $C_t$ and the current hidden state $h_t$. This architecture allows the LSTM to maintain and update contextual information over time, which is essential for accurately predicting cascading failure paths in complex networks.

Specifically, $\boldsymbol{x_t} = \{x_1, x_2, \ldots, x_n\}$ is a binary vector. If node $i$ in the network is connected at time $t$, $x_i = 0$; if it is disconnected, $x_i = 1$. The expression for the forget gate is

$$f_t = \sigma(W_f \cdot [h_{t-1}, \boldsymbol{x_t}] + b_f) \tag{5.17}$$

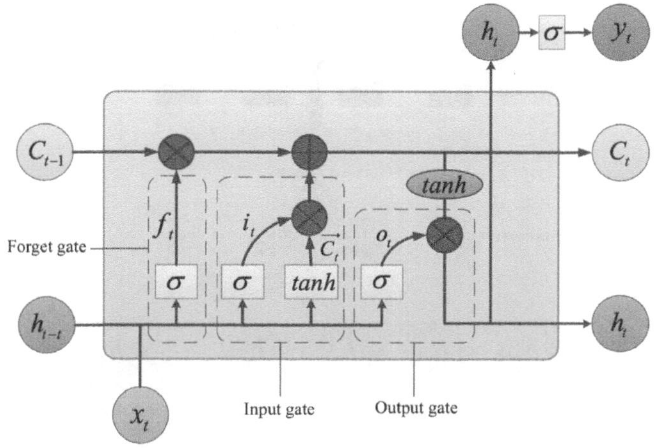


**FIGURE 5.7**
Working principle of the LSTM neural network.

where $W_f$ denotes the weight matrix and $b_f$ the bias term associated with the forget gate in the LSTM cell. The forget gate uses an activation function $\sigma$, typically a sigmoid function, to determine how much of the previous cell state should be retained. If the output of the forget gate is 1, the information is fully retained; if it is 0, the information is completely discarded. This mechanism allows the LSTM to selectively forget irrelevant information while preserving useful long-term dependencies. The input gate is

$$i_t = \sigma(W_i \cdot [h_{t-1}, \boldsymbol{x_t}] + b_i) \tag{5.18}$$

$$\overrightarrow{C_t} = \tan(W_C \cdot [h_{t-1}, \boldsymbol{x_t}] + b_C) \tag{5.19}$$

where $W_i$ and $W_C$ are the weight matrices of neurons; $b_i$ and $b_C$ are biases; $i_t$ denotes the portion of information retained by the new input; $\overrightarrow{C_t}$ denotes the information brought by the new input. The output gate is

$$o_t = \sigma(W_o \cdot [h_{t-1}, \boldsymbol{x_t}] + b_o) \tag{5.20}$$

$$C_t = f_t * C_{t-1} + i_t * \overrightarrow{C_t} \tag{5.21}$$

$$h_t = o_t * \tan(C_t) \tag{5.22}$$

where $W_o$ denotes the weight matrix of the output gate and $b_o$ denotes the bias of the output gate.

The prediction result can be obtained by processing the output of the LSTM neural network, i.e.,

$$\boldsymbol{y_t} = \sigma\left(W_y \cdot h_t\right) \tag{5.23}$$

where $W_y$ represents the weight matrix of the output, and $\boldsymbol{y_t} = \{y_1, y_2, \ldots, y_n\}$ is a binary vector that represents the state of all nodes in the network in the

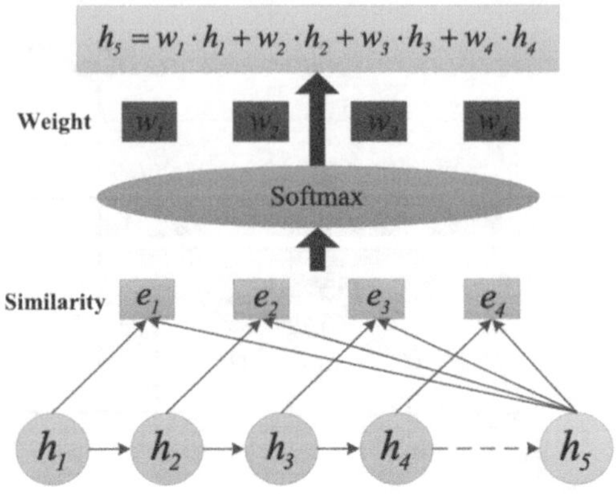


**FIGURE 5.8**
Working principle of attention-LSTM

next time step. Also, $y_n = 1$ means that node $i$ will be disconnected in the next step, and $y_n = 0$ means that node $i$ remains connected.

#### 5.2.1.4 Prediction Method Based on Attention-LSTM

In the standard LSTM prediction framework, only the hidden state output $h_{t-1}$ from the previous time step is used to compute the current cell state and output. This design can lead to earlier input information being gradually forgotten over time, making it difficult to capture long-term dependencies effectively. Moreover, the influence of previous outputs on the current prediction may vary, further complicating accurate modeling of sequential patterns. To address these limitations, an attention mechanism is introduced. This mechanism dynamically assigns weights to previous outputs based on their relevance to the current prediction, allowing the model to selectively focus on critical temporal features. By enhancing the LSTM's ability to retain and utilize important historical information, the attention mechanism significantly improves prediction accuracy. The integration of attention with LSTM is illustrated in Figure 5.8.

When the attention mechanism is combined with LSTM, the similarity between the output information and $h_{t-1}$ at different time steps is first calculated. These similarity scores are then converted into weight values using the softmax function. Next, a weighted sum of $h_{t-1}$ is computed to produce $h'_{t-1}$. Utilizing $h'_{t-1}$ allows for more accurate prediction of the cell state and output at the current time step.

The cascading failure prediction algorithm based on attention-LSTM is illustrated in Figure 5.9. Using the cascading failure simulation model

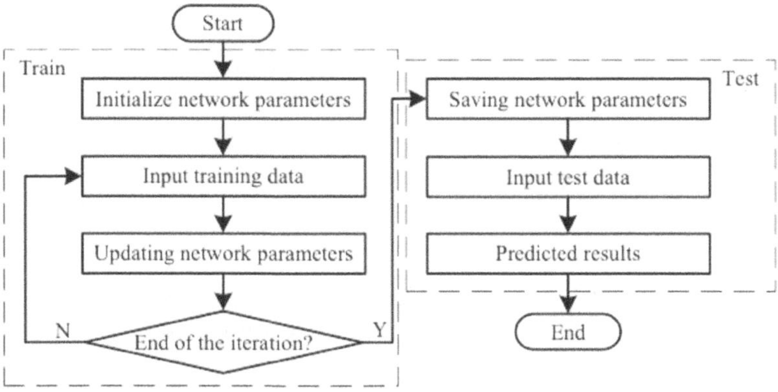


**FIGURE 5.9**
Cascading failure prediction flow chart.

described in the previous section, the training dataset is generated by simulating random initial failures. From this dataset, a single sequence representing a complete cascading failure process is randomly selected for training. The network state information at each step of the cascading failure process is sequentially fed into the neural network, allowing it to update its parameters during training. Upon completing the training process, the neural network's performance is evaluated using the test set. Once trained, the neural network can take the initial failure as input and automatically predict the network state information for each subsequent step, effectively and accurately modeling the entire cascading failure process.

### 5.2.2 Case Study

We conduct simulations to evaluate the performance of the proposed method in predicting cascading failures within complex networks. The analysis is performed on five scale-free complex network cases, each differing in size.

For case 30, 5,000 datasets are generated using the cascading failure simulation model described in Section 5.2.1.1. Each dataset represents a complete cascading failure process, with the initial failure location determined randomly for each instance. In practice, initial failures of size $N-3$ or larger are extremely rare. Therefore, the size of the initial failures in the generated data is limited to $N-3$ or smaller. Out of the total dataset, 80% is used as the training set, while the remaining 20% is randomly selected to form the test set.

Figure 5.10 shows the loss function curves on the training sets of the LSTM model and attention-LSTM model in case 30. The horizontal coordinate represents the number of training, and the vertical coordinate represents the value

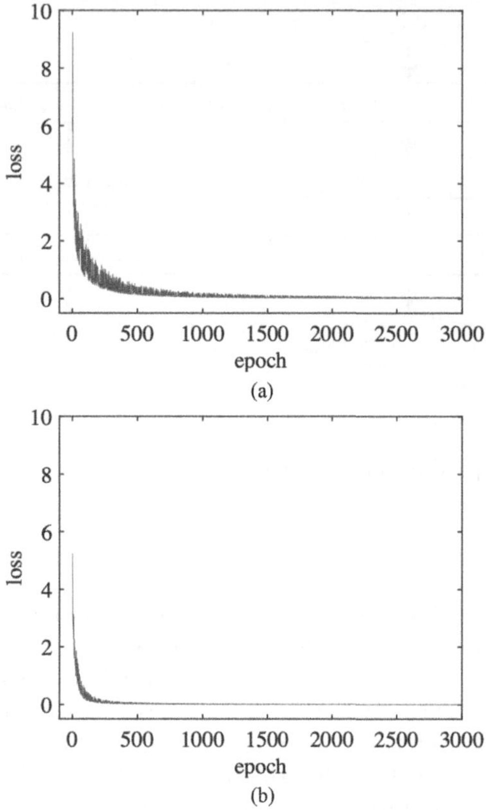


**FIGURE 5.10**
Loss curves of neural networks for (a) LSTM model; (b) attention-LSTM model.

of the loss function. The loss function is defined as

$$\text{Loss} = \sum_{t=1}^{l} \text{abs}(\boldsymbol{y}_t^{pre} - \boldsymbol{y}_t^{act}) \tag{5.24}$$

where $\boldsymbol{y}_t^{pre}$ denotes the network state information predicted at time $t$, $\boldsymbol{y}_t^{act}$ denotes the actual network state information at time $t$, and $l$ denotes the sequence length of the cascading failure.

As illustrated in Figure 5.10, the model that integrates the attention mechanism with LSTM demonstrates both faster convergence and superior final performance compared to the model that relies solely on LSTM.

Assume that the sequence length of a complete cascading failure process is $l$. Starting from an initial failure, the number of successful predictions is determined by correctly predicting the state information of all nodes at each step. When the cascading failure process concludes, the total number of successful

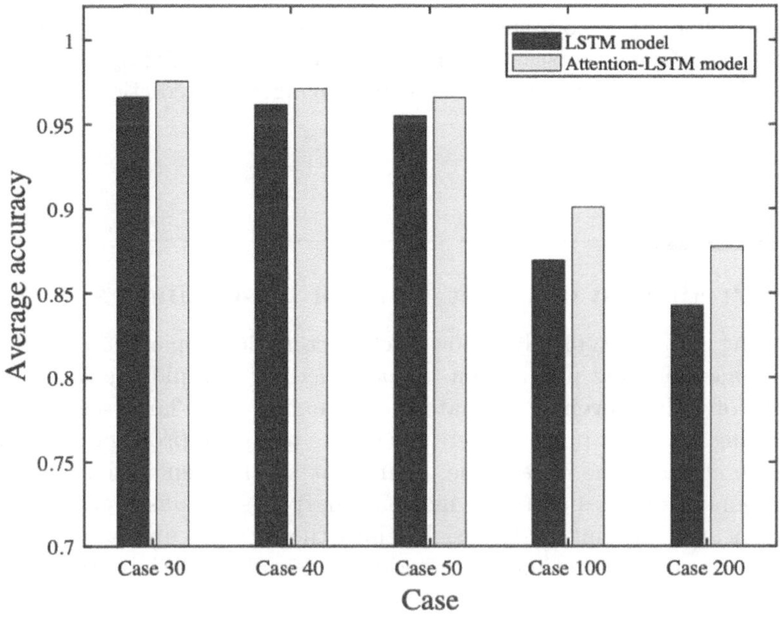


**FIGURE 5.11**
Average accuracy for all cases.

predictions is denoted as $N_{\text{success}}$. Therefore, we can define the accuracy rate of a set of predicted data as

$$\text{Accuracy rate} = \frac{N_{\text{success}}}{l} \tag{5.25}$$

Assuming that a total of $M$ sets of data are predicted, the final average accuracy rate is

$$\text{Average accuracy} = \frac{\text{Accuracy rate}}{M} \tag{5.26}$$

For the scale-free network structure, the dataset sizes are as follows: 5,000 for case 30, 12,000 for case 40, 21,000 for case 50, 30,000 for case 100, and 100,000 for case 200. From each dataset, 80% of the data is randomly selected to form the training set, while the remaining 20% is used as the test set. After training the neural network, the average accuracy on the test set is presented in Figure 5.11.

As shown in Figure 5.11, the average accuracy across all cases exceeds 85%, demonstrating the effectiveness of the proposed method in predicting cascading failure propagation paths in complex networks. However, the prediction accuracy decreases as the network size increases, indicating that larger networks present greater challenges for accurate prediction. Additionally, the neural network that combines the attention mechanism with LSTM significantly outperforms the standard LSTM network. This highlights the importance of the attention mechanism in improving the prediction of cascading failures.

## 5.3 Prediction of Onset Time of Cascading Failure

Cascading failures typically follow a characteristic propagation pattern: an initial phase of slow progression is succeeded by a rapid escalation in the number of failure events, ultimately culminating in a large-scale blackout. Identifying the onset point—where the failure begins to propagate rapidly—is therefore crucial. The onset time, defined as the moment when the number of component outages starts to increase sharply, serves as a critical robustness indicator, offering early insights into the severity and speed of failure propagation. In this section, we introduce a neural network-based multi-class classification approach to predict the onset time based on the initial state of the network and the initial failure set (i.e., the components tripped at the start of the cascade). The model architecture leverages both topological and electrical features of the system to uncover predictive patterns. Using the UIUC system as a case study, our results demonstrate that this method achieves high classification accuracy without relying on system evolution data. This provides a fast and scalable tool for assessing robustness before a cascading failure fully unfolds.

Large-scale power outages cause severe disruptions to modern society, where electricity underpins almost all economic, industrial, and social activities. Historical events highlight the devastating impact of such failures. In 2003, a blackout in North America, precipitated by ineffective protection strategies, disrupted approximately 1,800 MW of load and affected around 50 million people for more than a week [252]. Review reports on the severe outages in India (2012) [86] and the United Kingdom (2019) [83] similarly identify inappropriate or delayed protection mechanisms as primary contributors to the escalation of local faults into large-scale cascading failures. More recently, the widespread power cuts in Taiwan and Texas in 2021 again underscored the crucial role of timely and precise control strategies in preventing catastrophic fault propagation.

Over the past two decades, extensive research has been devoted to cascading failure modeling and analysis [55, 93, 162, 285] as a means to assess and, where possible, enhance the robustness of power systems against blackout scenarios. In parallel, machine learning (ML) methods have gained increasing attention for addressing problems traditionally tackled by model-based approaches, particularly in complex and uncertain operating environments.

Recent ML-driven studies have demonstrated considerable success in power system resilience assessment. Kim et al. [139] used a graph convolution network to predict optimal load-shedding ratios that prevent line overload under outage conditions. Liu et al. [160] extracted network-based features to enhance robustness using a decision-tree learning framework. Other methods have included support vector machines [304] and reinforcement learning strategies [1] to predict cascading fault chains and identify critical lines. Deep reinforcement learning has also shown promise in intelligent voltage and frequency stability control [120, 284]. These studies collectively demonstrate that ML-based methods can achieve high performance in solving complex, data-driven problems that challenge purely analytical approaches.

Building on earlier work [160, 302], this section considers onset time as a crucial indicator for early intervention in cascading failures. The onset time is defined as the moment in the propagation process when the rate of component outages increases sharply. A shorter onset time implies a more urgent need for protective action and thus greater system vulnerability. To estimate, in advance, the point at which a large number of failures will occur in rapid succession, we formulate onset time prediction as a three-class classification problem according to the degree of urgency.

The learning framework employs a neural network-based classifier that leverages:

1. Topological features of the power system extracted from the initial steady-state configuration of the UIUC 150 Bus test network [127], and
2. Electrical features obtained by comparing node and line power magnitudes before and after a small disturbance, here modeled as the tripping of $k$ selected lines in accordance with the standard $N - k$ security criterion.

Training and test data are generated from simulations that fully capture the dynamic profiles of cascading failure propagation. From these simulations, the onset time for each scenario is identified, forming ground-truth labels for the classifier. The results demonstrate that the proposed neural network-based approach can accurately predict onset time across diverse operating conditions, offering a fast and effective means to assess system vulnerability at the earliest stages of failure propagation.

### 5.3.1 Model Description

This section begins by revisiting the definition of onset time, as introduced in previous studies [160, 302], and reformulates its prediction as a multi-class classification problem. Next, we present a neural network-based classifier that integrates both topological and electrical information from a power system's initial state to categorize onset times into three urgency levels. The overall workflow of this approach is illustrated in Figure 5.12, which outlines the systematic framework for predicting the urgency level of onset time based on simulated cascading failure propagation profiles.

#### 5.3.1.1 Onset Time of Failure Propagation

A combined circuit-based AC power flow and stochastic process model is employed to generate detailed dynamic profiles of cascading failures in test power systems. An illustrative example for the UIUC 150 Bus system is presented in Figure 5.13, where the cascade is triggered by the tripping of two transmission lines.

At the start of failure propagation, the number of failed components grows slowly. At approximately $t \approx 465$ min, the system enters a rapid escalation phase in which many components fail in quick succession. The onset time, denoted $t_{\text{onset}}$, is formally defined as the first time at which the failure propagation rate increases dramatically. Operationally, it is identified when the number of newly failed components $\Delta F$ within a sufficiently short time interval $\Delta t < \Delta t_{\text{short}}$ exceeds a prescribed threshold $\Delta F_{\text{large}}$.

An early onset time indicates that protective action must be taken more promptly to arrest the cascade, implying greater vulnerability of the system to a large-scale blackout.

#### 5.3.1.2 Predictors of Neural Network-Based Classifier

Our objective is to estimate the onset time without relying on the generation of the complete failure propagation profile, which would otherwise require time-consuming, detailed cascading simulations. Instead, we aim to predict the urgency as indicated by the onset time using information available from the initial state of the network. Onset times are categorized into three classes based on their duration.

1. Urgent: $0 \leq t_{\text{onset}} < t_{c1}$
2. Relatively urgent: $t_{c1} \leq t_{\text{onset}} < t_{c2}$

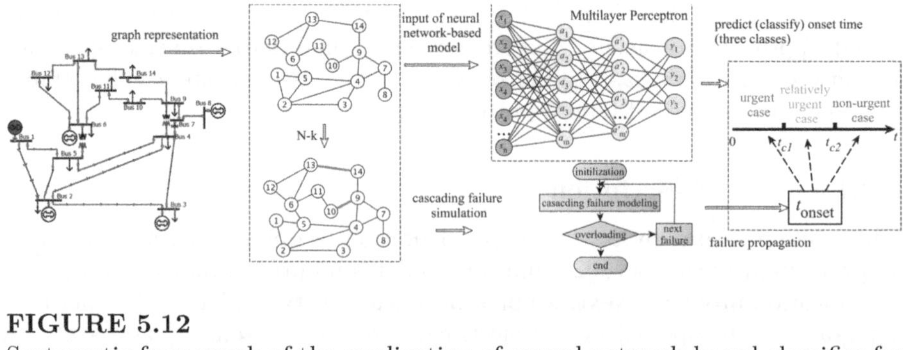


**FIGURE 5.12**
Systematic framework of the application of neural network-based classifier for predicting the degree of urgency of the onset time that is identified from failure propagation profile in a power system.

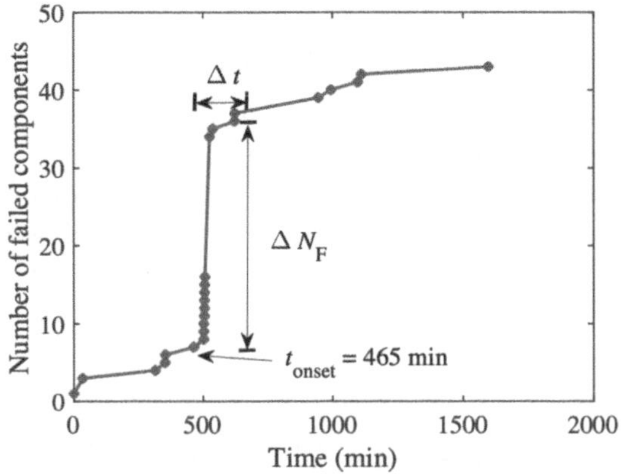


**FIGURE 5.13**
Simulated failure propagation profile of UIUC 150 Bus power network where two lines, i.e., from bus 2 to bus 14 and from bus 80 to bus 82, are initially tripped.

3. Non-urgent: $t_{\text{onset}} \geq t_{c2}$ or when no onset time is detected in the propagation profile

Here, $t_{c1}$ and $t_{c2}$ are critical time thresholds chosen from training data analysis. The classifier's task is to determine into which of these intervals the onset time for a given initial contingency will fall.

The input to the classifier is derived from the power difference matrix $\Delta M_p$, defined as

$$\Delta M_p = M_{p,N} - M_{p,N-k} \tag{5.27}$$

where $M_{p,N}$ is the $N \times N$ power matrix for the intact network (all lines present), $M_{p,N-k}$ is the corresponding matrix after removal of $k$ transmission lines, in line with the $N-k$ security criterion.

For a network represented as an $N$-node graph, $M_{p,N}(i,j)$ records the power flow from bus $i$ to bus $j$, calculated using AC power flow analysis based on the network's admittance matrix, nodal voltages, and injected currents. The diagonal element $M_{p,N}(i,i)$ represents the net power injection at bus $i$, which corresponds to the power supplied by generator buses or demanded by load buses. The difference matrix $\Delta M_p$ captures the changes in power distribution caused by the initial contingency. By definition, $\Delta M_p(i,j) = 0$ when no transmission line connects buses $i$ and $j$. While such entries carry no direct information, they provide important structural context for learning.

For each contingency scenario, node features $X_n$ are derived from the diagonal elements $\Delta M_p(i,i)$, representing changes in nodal power injection or demand. Similarly, line features $X_l$ are extracted from the non-zero off-diagonal entries $\Delta M_p(i,j)$, corresponding to existing transmission lines, and represent changes in branch power flows. The complete feature vector is then constructed as

$$X = \{X_n, X_l\} \tag{5.28}$$

which can be obtained by concatenating all selected node and line features. This vector $X$ serves as the predictor input to the neural network classifier. This formulation enables the model to exploit both structural (topological) and operational (electrical) signatures of the system's initial state, providing a robust basis for predicting whether the system's onset time indicates an urgent, relatively urgent, or non-urgent risk of rapid cascade escalation.

#### 5.3.1.3 Neural Network-Based Classifier

Using the power difference matrix $\Delta M_p$ obtained from (5.27), we develop a neural network-based classifier to predict the class of onset time. Specifically, we establish a MLP model [222,223], which has two hidden layers with $h_1$ and $h_2$ hidden units to bridge the original features and the class of the onset time.

For a specific sample, we first take the original features $X \in \mathbb{R}^{1\times d}$ as inputs, where $d$ is the dimension of features, and outputs $Y_1 \in \mathbb{R}^{1\times h_1}$ by utilizing a weight matrix $W_1 \in \mathbb{R}^{d\times h_1}$ and a bias vector $b_1 \in \mathbb{R}^{1\times h_1}$, together with a ReLU activation function, i.e.,

$$Y_1 = \text{ReLU}(XW_1 + b_1) \tag{5.29}$$

In the next layer, we utilize a similar operation and obtain $Y_2 \in \mathbb{R}^{1\times h_2}$, but take the outputs of the previous layer as inputs, i.e.,

$$Y_2 = \text{ReLU}(Y_1 W_2 + b_2) \tag{5.30}$$

where $W_2 \in \mathbb{R}^{h_1\times h_2}$ and $b_2 \in \mathbb{R}^{1\times h_2}$ are the weight matrix and bias vector, respectively. Finally, we obtain the final output as

$$Y_3 = Y_2 W_3 + b_3 \tag{5.31}$$

where $W_3 \in \mathbb{R}^{h_2\times 3}$ and $b_2 \in \mathbb{R}^{1\times 3}$. As our goal is to do a three-class classification task, $Y_3$ represents the predicted classes of the corresponding inputs in a one-hot encoded form.

### 5.3.2 Case Study

Following the $N-2$ security criterion, cascading failure scenarios in the UIUC 150 Bus test system are generated by selecting every unordered pair of distinct transmission lines as the initial failed components. These two lines are removed from the network at the start of each simulation to produce a dataset of 20503 complete failure propagation profiles for training and test the neural network–based classifier.

**TABLE 5.2**
Average Accuracy of Training Set, Test Set, and Two New Scenarios Including Changing the Original Power Parameters (Modified $N-2$) and Increasing k to 3 ($N-3$), Respectively

| Sets | Training | Test | Modified $N-2$ | $N-3$ |
|---|---|---|---|---|
| Average accuracy | 0.9417 | 0.8993 | 0.6332 | 0.8768 |

For each failure propagation case, the power difference matrix $\Delta M_p$ is computed using (5.27) with $k = 2$. Onset time detection follows the procedure outlined by Liu et al. [160], with thresholds set to $\Delta F_{\text{large}} = 10$ and $\Delta t_{\text{short}} =$ 10 min. To classify onset times into three categories, the critical time points $t_{c1}$ and $t_{c2}$ are set at $10^2$ min and $10^3$ min, respectively, in alignment with practical operational guidelines. Based on this classification, the dataset of 20,503 cases consists of 3,704 non-urgent, 14,700 urgent, and 2,099 relatively urgent samples.

The feature vector $X$ defined in Section 5.3.1.2 has a total dimension of $d = 353$, corresponding to 150 node-based features and 203 line-based features in the UIUC 150 Bus system. The neural network classifier is configured with two hidden layers containing $\{h_1, h_2\} = \{100, 50\}$ units. Training is performed for a maximum of 2,000 epochs using stochastic gradient descent and a cross-entropy loss function. To mitigate the effects of class imbalance, the loss function is weighted inversely to the number of samples in each class. The dataset is split into training and test sets in a 70:30 ratio, and all reported results are averaged over five independent train–test partitions.

The variation of accuracy with training epoch for both training and test datasets is presented in Figure 5.14. Accuracy is defined as the proportion of correctly classified samples to the total number of samples. The training accuracy exhibits a rapid increase at the start of training, followed by a gradual rise until it stabilizes around 90%. The testing accuracy follows a similar trajectory, also reaching high levels, which indicates that the model effectively avoids overfitting. Quantitative results in the second and third columns of Table 5.2 confirm that the proposed method achieves consistently high performance on both training and test datasets.

To further examine the robustness and generality of the model, two more challenging testing scenarios are considered. According to (5.27), the model constructs its input features from the difference in power parameters between the intact initial state $M_{p,N}$ and the state after removing $k$ transmission lines, $M_{p,N-k}$. We, therefore, assess the model under conditions where: (1) the original power parameters $M_{p,N}$ are modified (denoted as "Modified $N-k$"), and (2) the number of initially removed lines $k$ is increased (denoted simply as $N-k$).

For these tests, new simulation datasets are generated. In particular, two additional sets of 5,000 scenarios are created:

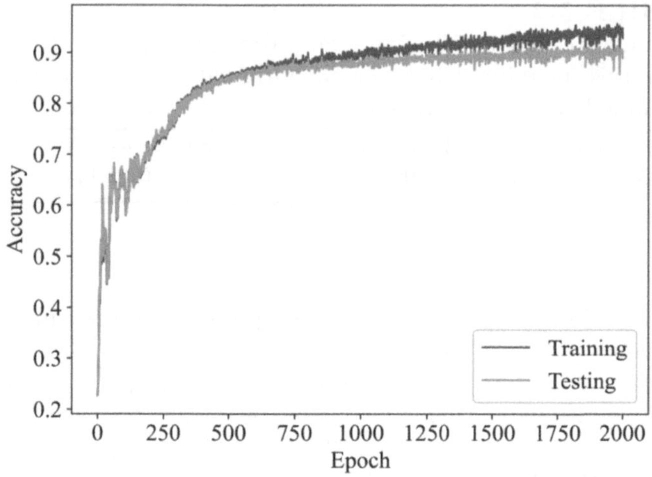


**FIGURE 5.14**
Accuracy of training set and test set with the increase of epoch.

1. Modified $N-2$: generated under the $N-2$ criterion with changes to the initial generation and load powers in $M_{p,N}$.

2. $N-3$: generated under the $N-3$ criterion without changing the initial power parameters.

The model trained on the original dataset is tested directly on these new sets without retraining. Results in the last two columns of Table 5.2 show that the classifier maintains strong performance when $k$ increases from 2 to 3, demonstrating generality with respect to attack size. However, when the original power matrix $M_{p,N}$ is modified, the average accuracy drops to 63.32%, representing a substantial decline compared to the unmodified case. This reduction is likely due to insufficient feature representation: in the training set, all samples share the same $M_{p,N}$ baseline, whereas modifying $M_{p,N}$ introduces the possibility of different original and perturbed matrices producing similar $\Delta M_p$ features. Such ambiguities can confuse the classifier, which was not trained to account for this variation. Incorporating additional features that encode variation in $M_{p,N}$ may help address this limitation in future work.

To illustrate the discriminative capacity of the learned features, the two-dimensional t-SNE projection of the output from the second hidden layer ($Y_2$) is visualised in Figure 5.15. Clear clustering patterns are visible: samples from the same onset time class are generally grouped together, while distinct classes form visibly separated clusters. This indicates that the neural network has effectively learned informative representations from the initial-state power differences across nodes and transmission lines, enabling accurate classification of onset time classes.

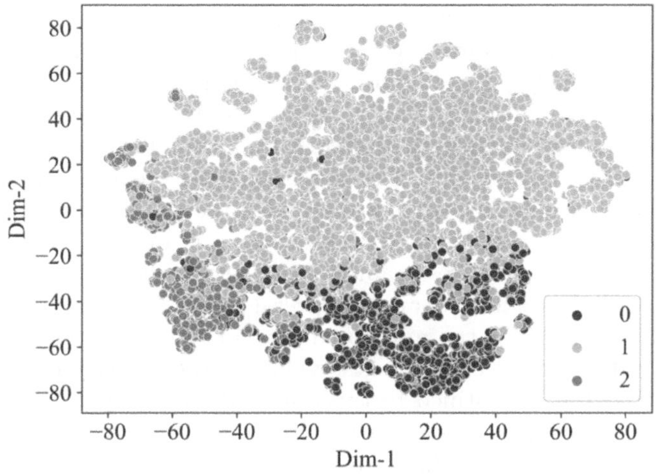


**FIGURE 5.15**
t-SNE plot of the hidden vectors of each sample in 2D space. Representations of the samples from the same class are plotted with the same color. Here, classes 0, 1, and 2 represent the non-urgent, urgent, and relatively urgent cases, respectively.

In summary, the experiments demonstrate that the proposed neural network-based classifier achieves high accuracy under standard conditions, generalizes to larger attack sizes, and learns feature representations that meaningfully separate onset time categories. Limitations arise when the baseline power matrix is altered, indicating that capturing variability in the system's initial state is a potential avenue for enhancing robustness assessment.

## 5.4 Component Criticality and Robustness Indicators via Decision Trees

In this section, our primary objective is to conduct a system-level robustness assessment of power grids, with a particular focus on identifying network-structural features that are strongly correlated with robustness. The assessment is grounded in the dynamics of cascading failure propagation, where a key observable parameter—the onset time—marks the point at which the cascade rate accelerates sharply. By jointly considering onset time and the final scale of the failed grid in each simulated cascade, we classify every component into three severity categories, each representing a distinct level of impact when that component is the initial failure. The most critical category comprises components whose failure causes both rapid and large-scale cascades, significantly degrading network robustness.

Robustness is quantified through the proportion of components in each severity category across different network scenarios. To uncover interpretable robustness indicators, we apply a decision-tree-based learning framework to a set of networks generated by small-scale edge rearrangements of the original topology (network size and degree preserved). This analysis consistently identifies three dominant features, namely, (1) average shortest path length, (2) average clustering coefficient, and (3) average effective electrical distance to the nearest generator, as the most predictive of robustness level. Subsequent robustness enhancement experiments, while not the primary focus, serve as a validation that these identified features are indeed strongly linked to system robustness. This integrated approach delivers both quantitative vulnerability metrics and qualitative insights into the structural factors influencing power grid resilience.

Power systems, as critical infrastructures supporting modern society, supply electrical energy to all sectors, including industry, government, data centers, transportation, and public services. Large-scale outages in such systems can disrupt daily life and cause severe economic losses. Historical blackouts such as the 2003 Northeast outage in the United States and Canada, which affected 50 million people for up to four days [252], as well as the blackouts in India (2012) and Taiwan (2017), have repeatedly underscored the need for effective robustness assessment of power grids.

Over the past two decades, extensive research has been devoted to cascading failure modeling for evaluating blackout risk [290]. Approaches based on historical data fitting [50, 75] offer statistical characterizations of failure propagation, whereas physics-based models treat cascades as sequences of element tripping events driven by overloads determined via DC or AC power flow [60,216,270,273,302]. Zhang et al. [302] incorporated physical power flow equations into a network-based model, reproducing universal growth patterns observed in empirical outage data.

The use of complex network theory in power system robustness studies has evolved through several stages. Early work investigated purely topological effects on robustness [8, 31, 258]. Later studies extended the analysis to weighted and directed representations [36, 62, 269], incorporating impedance and power flow direction, though often without modeling full physical redistribution [58]. The current stage emphasizes integrating component characteristics and global redistribution effects into network models [116,300], recognizing that successive failures can occur far apart geographically due to system-wide rebalancing.

Complex network metrics themselves have been explored as indicators of vulnerability. For example, Bompard et al. [38] extended betweenness centrality to better identify critical components, while Dwivedi et al. [80] applied maximum-flow concepts to detect vulnerable lines. Large-scale analysis by

Yang et al. [289] found that small vulnerable sets can dominate cascade dynamics, highlighting the value of combining structural information with electrical characteristics for robust assessment [55].

Most existing robustness assessments focus on end-state metrics—e.g., final blackout size or number of failed elements, once a cascade has run to completion. In our previous work [302], we identified a distinct onset time: the moment at which failure propagation rate accelerates sharply from a slow initial phase. Longer onset times provide operators with a larger response window for intervention [268], while short onset times greatly increase the difficulty of containment. This makes onset time, together with final blackout scale, a powerful basis for characterizing component criticality.

The main objective here is to present a system-level robustness assessment framework that identifies network-structural features highly correlated with robustness. First, based on onset time and blackout size from simulated cascades, we categorize transmission lines into three types:

1. Type I: Initial failure does not trigger a cascade;
2. Type II: Initial failure triggers a cascade with a relatively long onset time;
3. Type III: Initial failure triggers a rapid, large-scale cascade (short onset time and large blackout size).

Type III lines are of greatest concern, as their failure renders the system particularly vulnerable. The proportion of components in each category provides a quantitative vulnerability profile for the network.

Second, to extract interpretable robustness indicators, we train a decision tree on network variants generated via small-scale edge rewiring. These variants preserve size and degree distribution but alter topology; the rewiring here is not an enhancement strategy per se, but a tool to introduce structural diversity so that correlation between features and robustness can be reliably identified. The trained decision tree consistently highlights three features, namely, (1) average shortest path length, (2) average clustering coefficient, and (3) average effective electrical distance to the nearest generator, as most predictive of robustness level. In follow-up experiments, minor topological adjustments informed by these features are shown to yield networks with higher robustness, thus validating the strong correlation between the identified features and system robustness.

In this way, the focus of our study remains firmly on robustness assessment: classifying components by their potential to trigger severe cascades, and identifying key network features that serve as robustness indicators. The subsequent enhancement exercise functions primarily as a validation of the assessment results.

### 5.4.1 Model Description

This study adopts the cascading failure propagation model introduced in Section 4.1, which integrates a circuit-based DC power flow calculation with a stochastic failure progression process. The DC power flow model determines the sequence of element outages based on overload conditions, while the stochastic model captures the probabilistic timing of failures. A complete dynamic cascade profile is obtained by iteratively performing the steps described in Section 4.1, i.e., computing power redistribution after each failure, checking for overload conditions, and applying the stochastic failure timing process, until the cascade terminates.

A key observable parameter in the simulated propagation process is the onset time, defined as the point at which the rate of failure increases sharply. This parameter, together with the final blackout scale, forms the basis for a component-level vulnerability classification framework.

Recognizing the existence of a critical parameter called *onset time* in the profile of failure propagation, both from simulations [302] and historical data, we extend the methodology to quantitatively assess vulnerability, thereby assessing robustness from a reciprocal perspective. In other words, a highly vulnerable power system is not robust. The extended method consists of three main steps:

1. Detecting *onset time* in the failure propagation process;
2. Mapping *onset time* to a vulnerability index;
3. Assessing the systematic vulnerability of a power grid.

Vulnerabilities of five selected power systems will be presented and analyzed.

#### 5.4.1.1 Detection of Onset Time

The *onset time* is defined as the time when the number of failed nodes in a power grid begins to increase sharply during the propagation process. To identify the *onset time*, we apply the method developed by Goswami et al. [104], designed to detect abrupt transitions in time series.

An illustration of onset time detection is shown in Figure 5.16. Figure 5.16a plots the number of power elements (transmission lines, in this case) removed as time elapses. The number of removed power links is denoted $\mathrm{NFL}(t)$. One complete propagation process of cascading failure is characterized as a *cascading failure scenario*, an analogue of a blackout event (Figure 5.16a). From Figure 5.16a, $t_{\text{onset}} \approx 2,044$ min, meaning a fast cascading failure begins after $t = 2,044$ min in this scenario.

The detection method transforms $\mathrm{NFL}(t)$ into a recurrence matrix $\mathbf{R}$ [176], where $\mathbf{R}(i,j)$ represents the difference between $\mathrm{NFL}(t_i)$ and $\mathrm{NFL}(t_j)$. Based on this matrix (Figure 5.16b), a *differential value* $d_{\text{value}}$ is computed within a sliding window from $t_1$ to $t_2$:

$$d_{\text{value}} = \frac{\sum_{i,j=t_1}^{t_m} \mathbf{R_{ij}} + \sum_{i,j=t_m}^{t_2} \mathbf{R_{ij}}}{\sum_{i,j=t_1}^{t_2} \mathbf{R_{ij}}} \tag{5.32}$$

where $t_m$ is the mid-point between $t_1$ and $t_2$ of the current sliding window, i.e., $t_m = (t_1 + t_2)/2$. Specifically, the numerator includes two sums of the elements in two recurrence matrices, namely, one is within the time duration $[t_1, t_m]$ and the other one is within the time duration $[t_m, t_2]$. In the denominator, the summation includes all elements in the recurrence matrix within $[t_1, t_2]$. One extreme case is that if elements in $\mathbf{R}$ within the sliding time window are all the same, then $d_{\text{value}}$ is equal to 0.5.

If there are different densities of colors displayed in the heat map within a time window, as shown in Figure 5.16b, the value of $d_{\text{value}}$ is smaller than 0.5. For instance, from Figure 5.16b, when $t_m$ is around 2,000 min, the densities of colors show that both recurrence matrices corresponding to two time slots $[t_1, t_m]$ and $[t_m, t_2]$ are having smaller recurrence probabilities compared with the other recurrence probabilities outside the two recurrence matrices

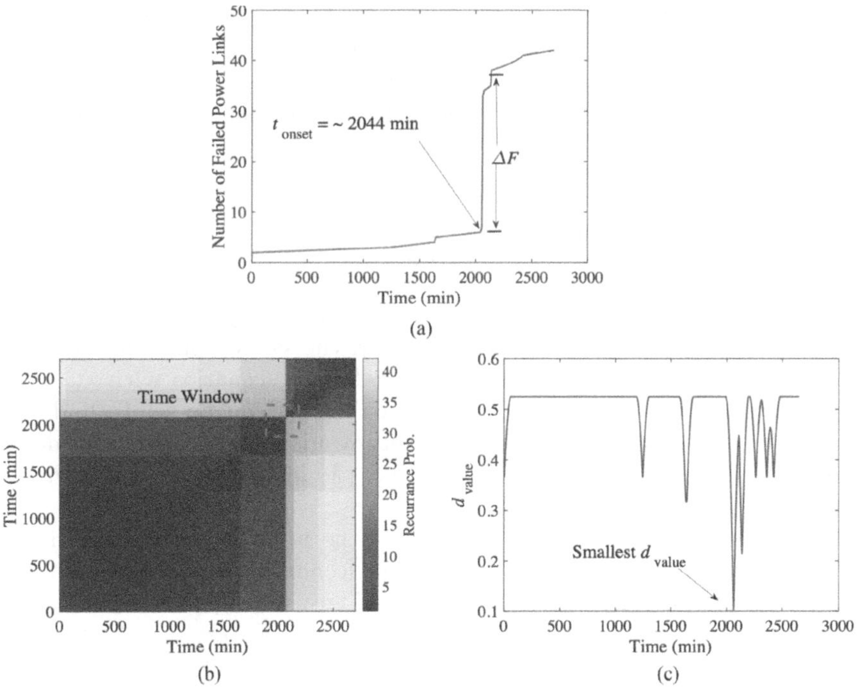


**FIGURE 5.16**
Detection of onset time using Goswami et al.'s method [104]. (a) Propagation profile of cascading failure; (b) recurrence matrix as heat map; (c) differential value indicating locations of large increments.

in the time window $[t_1, t_2]$. Thus, by plotting $d_{\text{value}}$ versus time $t_m$, and as the sliding window moves along the diagonal of the entire $\mathbf{R}$ as shown in Figure 5.16b, we identify the smallest value of $d_{\text{value}}$, which corresponds to an abrupt transition. In this example, $t_{\text{m}} \approx 2050$ min. In summary, $t_{\text{m}}$ is identified as the indicative time where $d_{\text{value}}$ is smallest, and hence provides a reference for finding the *onset time* in a cascading failure scenario. There are, however, two special scenarios worth noting. The first one is the existence of multiple abrupt transitions. In one *cascading failure scenario*, there may be more than one occurrence of rapid increment. Thus, we only focus on the first recognized largest abrupt transition corresponding to the first smallest $d_{\text{value}}$. The second scenario is related to the evaluation of the extent of the abrupt transition given by the number of failed power components within a short time period. An alternative threshold $\Delta F_{\text{th}}$ is assigned to identify whether the abrupt transition can be considered as being sufficiently large. For instance, the $\Delta F$ marked in Figure 5.16a is the value of increment in the abrupt transition, which is much larger than $\Delta F_{\text{th}}$ in the given example.

#### 5.4.1.2 Mapping Onset Time to Vulnerability

We define a *rescue time* $T_{\text{rescue}}$ as the minimum acceptable window to respond before a severe cascade. If $t_{\text{onset}} \geq T_{\text{rescue}}$, the system is considered *safe*; if $t_{\text{onset}} < T_{\text{rescue}}$, the scenario is *vulnerable* [12].

Given $N_{\text{sim}}$ simulated cascades for each initial component failure, the vulnerability index for component $i$ is:

$$V_i = \frac{N_{\text{sim}} \mid_{t_{\text{onset}} < T_{\text{rescue}}}}{N_{\text{sim}}} \tag{5.33}$$

where component $i$ is initially tripped in one *cascading failure scenario* and a total of $N_{\text{sim}}$ simulations are run. Basically, the vulnerability index $V_i$ of component $i$ is the percentage of cascading failure scenarios with $t_{\text{onset}} < T_{\text{rescue}}$. Furthermore, a threshold denoted as $V_{\text{critical}}$ can be set to find out whether the selected component would likely result in a vulnerable scenario when it is initially tripped. In particular, if $V_i \geq V_{\text{critical}}$, a vulnerable scenario is expected when component $i$ fails.

The final size of the blackout marked by the number of the failed links $\text{NFL}(t_{\text{final}})$ at the end of one *cascading failure scenario* is another key parameter reflecting the severity of a power outage. Thus, setting a threshold called $\text{NFL}_{\text{critical}}$, we expect a large-scale power outage if $\text{NFL}(t_{\text{final}}) \geq \text{NFL}_{\text{critical}}$. Specifically, a *fast* and *large-scale* cascading failure scenario can be identified by checking, respectively, $t_{\text{onset}} \geq T_{\text{rescue}}$ and $\text{NFL}(t_{\text{final}}) \geq \text{NFL}(t_{\text{critical}})$.

#### 5.4.1.3 Classification of Component Types

In order to gain a practical understanding of how vulnerable a power network is, we classify the power components in the power network into three types.

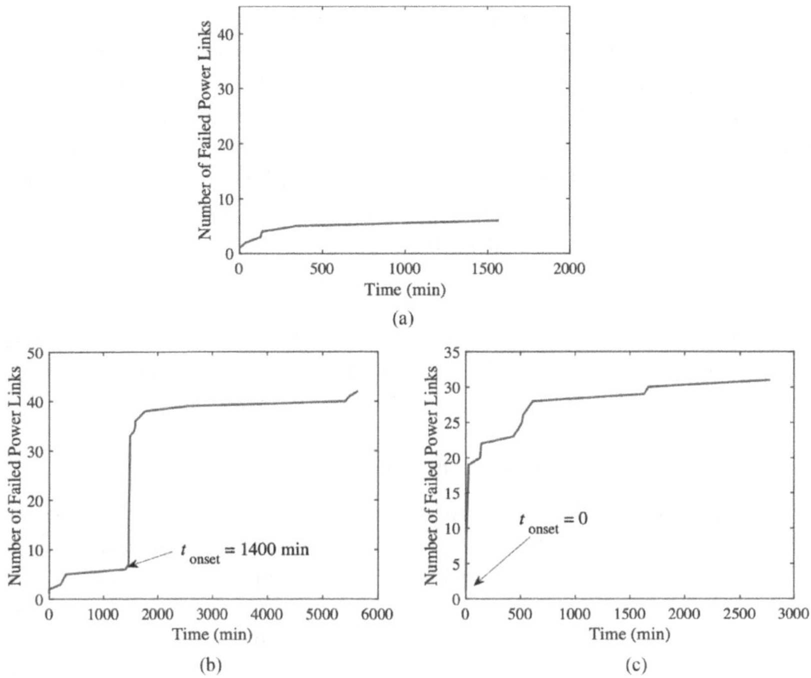


**FIGURE 5.17**
Three types of power lines observed in failure propagation. (a) Type I: no significant number of failed links; (b) Type II: significant number of failed links but relatively long onset time; (c) Type III: significant number of failed links and relatively short *onset time*.

1. Type I components: A Type I power component is a power component which, upon failure, does not lead to cascading failure of a significant scale. Here, failure of power components leading to $\mathrm{NFL}(t_{\mathrm{final}}) < \mathrm{NFL}_{\mathrm{critical}}$ is regarded as insignificant, as shown in Figure 5.17a.

2. Type II components: A Type II power component, upon failure, will lead to cascading failure of the system, meeting the criteria $\mathrm{NFL}(t_{\mathrm{final}}) \geq \mathrm{NFL}_{\mathrm{critical}}$. Moreover, the *onset time* is longer than $T_{\mathrm{rescue}}$, thus allowing recovery in time. An example is shown in Figure 5.17b.

3. Type III components: A Type III power component, upon failure, will lead to cascading failure of the system with a significant number of subsequent failed components, i.e., $\mathrm{NFL}(t_{\mathrm{final}}) \geq \mathrm{NFL}_{\mathrm{critical}}$. Moreover, the *onset time* is shorter than $T_{\mathrm{rescue}}$, making recovery difficult, as shown in Figure 5.17c.

Having established a systematic methodology for quantifying robustness through onset time–based vulnerability assessment and component classification, the next step is to investigate why certain networks exhibit higher or lower robustness. In particular, beyond identifying individual critical components, it is essential to uncover structural characteristics of the network that are closely correlated with robustness. Understanding these relationships enables us to interpret the vulnerability profile of a network in terms of its topology and may inform design or operational strategies. To this end, we employ a decision-tree–based feature extraction approach, which uses the robustness assessment results as labeled data and a set of candidate network-structural metrics as predictors, thereby allowing the identification of significant features most strongly associated with the robustness of the power grid.

#### 5.4.1.4 Decision Tree–Based Feature Extraction

A decision tree [40] is adopted to relate network-structural features to system robustness, using as the robustness indicator the proportion $P_{\mathrm{III}}$ of Type III lines in the network. A typical decision tree consists of a root node, a set of decision nodes, and a set of leaf nodes. The construction process starts at the root and continues until no decision node can be further split into leaf nodes. At each decision node, one of the candidate predictors (network features) is selected [170]. The predictor is chosen according to a split criterion, here

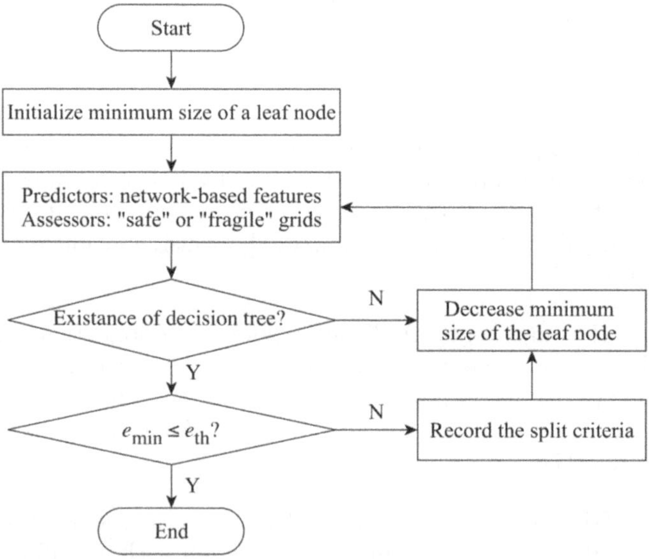


**FIGURE 5.18**
Flow chart for generating an appropriate decision tree.

defined by the node error: the fraction of misclassified samples (networks) within that node. For example, if a leaf node is labeled as representing "safe" networks, but 10% of the networks assigned to that node are actually "fragile", then the node error is 0.1. At the root node, the predictor yielding the smallest node error after splitting, e.g., Predictor A, would be selected as the basis for the first decision.

The decision tree is trained on a dataset consisting of predictors (network-structural features) and the corresponding assessors (labels). However, in order for the extracted features to be practically relevant, it is essential that the chosen predictors exhibit a clear positive or negative correlation with the assessor labels. This ensures that the split conditions on these features can be directly interpreted as actionable guidelines for modifying network structure to improve robustness. The training dataset is generated from a reference network and a set of updated networks obtained through small-scale edge rewiring [271]. The rewiring preserves each node's degree, maintains the total number of edges, fixes the line admittances, and is constrained to a rewiring fraction $p < 0.1$. For each updated network, if $P_{\text{III}}$ is smaller than in the reference network, the network is labeled as "safe" (1); otherwise, it is labeled as "fragile" (0).

The algorithm for generating a decision tree, summarized in the flow chart of Figure 5.18, proceeds as follows:
Step 1: Set the minimum number of samples per leaf. This parameter specifies the smallest allowable number of networks in a leaf node. A larger value yields a simpler tree, while a smaller value allows more detailed splits.

Step 2: Train an initial decision tree on the dataset of predictors and labels using the current minimum leaf size. Then, iteratively reduce this size and retrain until a "standard" tree is produced, i.e., one with at least one decision node and two leaf nodes.

Step 3: For the generated standard tree, compute the node error $e$ for each leaf node classified as "safe". For example, if a "safe" leaf node contains 25% "fragile" networks, then $e = 0.25$. Let $e_{\text{min}}$ denote the smallest node error across all leaf nodes. If $e_{\text{min}}$ is less than the specified threshold $e_{\text{th}}$, record the corresponding split criterion, that is, the specific increase or decrease in the predictor values, since it indicates a potentially effective modification for improving robustness. Otherwise, return to Step 2 and continue the iteration.

#### 5.4.1.5 Network-Based Features

Significant network-based features of a power grid are used in this chapter for enhancing the robustness of power grids. We have tested the significance of more than ten network-based features and eventually selected the following three.

1. Average shortest path length ($L$): the mean value of the shortest paths between each pair of nodes in the network [192], which is given in Section 2.1.2.
2. Average clustering coefficient ($\Gamma$): a local metric to measure the density of triangles in a network. A triangle is formed if two nodes that are connected to a third node are also connected to each other. Thus, for each node $i$ in a network, there is one corresponding clustering coefficient $\Gamma_i$ [192]. For the entire network, the average clustering coefficient is given in Section 2.1.2.
3. Average effective resistance (distance) to a nearest generator of all consumer nodes ($D$): a measure of the accessibility to generators of all power-consuming loads in a power network, i.e.,

$$D = \frac{1}{(N-g)} \sum_{i \in \mathcal{N} \setminus \mathcal{G}} d(i) \tag{5.34}$$

where $\mathcal{N} \setminus \mathcal{G}$ is the set of nodes excluding the generator nodes, $N$ is the total number of nodes, and $g$ is the number of generators, $d(i)$ represents resistance distance of node $i$ to its nearest generator introduced in the previous study [300].

Both *Average Shortest Path Length* and *Average Clustering Coefficient* reveal the major characteristic of the network structure of a power grid. A network exhibiting a short $L$ and a large $\Gamma$ is classified as a small-world network from the network science's viewpoint [272]. Recent studies have found that power grids displaying the small-world property have higher robustness [61]. Moreover, *Average Effective Resistance (distance) to a Nearest Generator of All Consumer Nodes* reveals the location information of generators in a power network. A smaller $D$ implies that generator distribution is more decentralized in a power network. It has been found [300] that decentralization of power source nodes exhibiting a smaller $D$ reduces the vulnerability of the power network to cascading failure.

### 5.4.2 Case Study

To evaluate the proposed robustness assessment framework, simulations are conducted on five selected power systems: (1) UIUC 150 Bus (UIUC150), (2) South Carolina 500 Bus (SC500), (3) South China 612 Bus (GD612), (4) European 1354 Bus (EURO1354), and (5) Texas 2000 Bus (TEXAS2000)

For each network, a possible cascading failure scenario is launched by initiating an initial fault in a power line. For every initial line failure, $N_{\text{sim}} = 1,000$ simulation runs are performed. The assessment parameters are set to $V_{\text{critical}} = 0.75$ and $\text{NFL}_{\text{critical}} = 10$, providing the classification condition for differentiating the three categories of power lines defined in Section 5.4.1. Under these settings: (1) A line is classified as Type III if over 75% of its

**TABLE 5.3**
Assessment Results of Selected Power Systems

| | UIUC150 | SC500 | GD612 | EURO1354 | TEXAS2000 |
|---|---|---|---|---|---|
| $N_{node}$ | 150 | 500 | 612 | 1,354 | 2,000 |
| $N_{link}$ | 203 | 584 | 852 | 1,710 | 2,667 |
| $N_{T1}/P_{T1}$ | 90 / 44% | 269 / 46% | 344 / 40% | 1123 / 66% | 827 / 31% |
| $N_{T2}/P_{T2}$ | 50 / 25% | 170 / 29% | 112 / 13% | 192 / 11% | 327 / 12% |
| $N_{T3}/P_{T3}$ | 63 / 31% | 145 / 25% | 396 / 47% | 395 / 23% | 1513 / 57% |

simulated cascades are deemed vulnerable; and (2) A cascade is considered significant when ten or more lines fail eventually.

#### 5.4.2.1 Assessment Results

The assessment results for all five networks are summarised in Table 5.3. The first two rows report each system's topological attributes: number of nodes $N_{node}$ and number of links $N_{link}$. Rows 3 to 5 give the number and percentage of each type of power lines as $N_{T1}/P_{T1}$, $N_{T2}/P_{T2}$, and $N_{T3}/P_{T3}$. As discussed in [290], the percentage $P_{T3}$ is a key indicator of vulnerability.

From Table 5.3, GD612 and TEXAS2000 each have roughly 50% of their lines classified as Type III, while in the other three systems the proportion is below one-third. This indicates that GD612 and TEXAS2000 are more prone to fast, large-scale cascading failures. Reducing the proportion of Type III lines is therefore critical to enhancing system robustness. In the next stage, we use network rewiring and the decision tree–based analysis to identify network features correlated with reductions in $P_{T3}$.

#### 5.4.2.2 Network Rewiring for Updated Topologies

Once a large power network is constructed or has evolved over time, a complete structural redesign is typically infeasible. However, operational strategies such as transmission switching [114] can modify topology, and selected rewiring of transmission lines has been shown to reduce vulnerability [28]. Here, we implement a rewiring scheme [271], with additional practical constraints:

1. A small fraction $p$ of existing edges is rewired ($p < 0.1$) to ensure economic feasibility and limit large-scale disruption;
2. The total number of edges and the degree of each node remain unchanged;
3. The electrical admittance of each line is preserved.

This produces a set of updated (candidate) networks whose robustness is then reassessed. Figures 5.19 and 5.20 illustrate the rewiring process. Three edges are removed in Figure 5.19, and three new edges are added in Figure 5.20. Post-rewiring assessment shows a reduction in Type III lines, particularly around the rewired edges, demonstrating increased robustness. Many

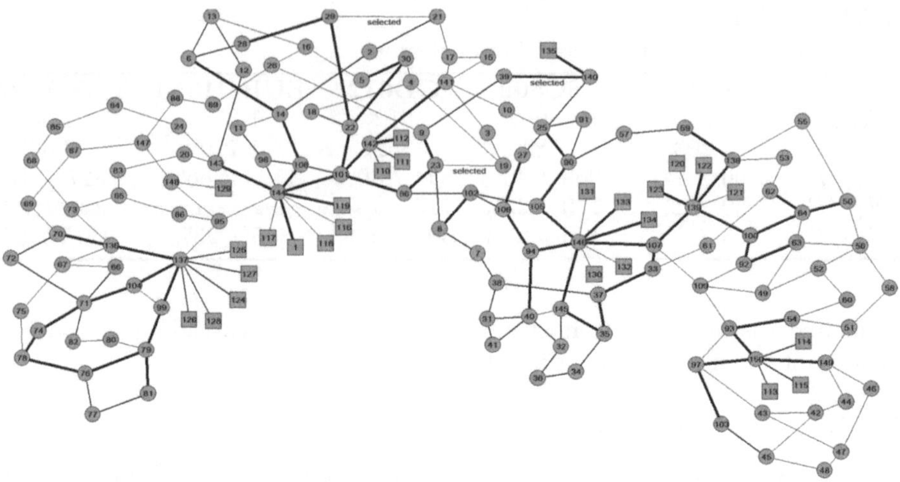


**FIGURE 5.19**
UIUC 150 Bus power system before edge modification. Numbers of Types I, II and III (thick edges) power lines are 90, 63, and 50. Edge modification involves removing edges (19, 23), (21, 29) and (39, 140).

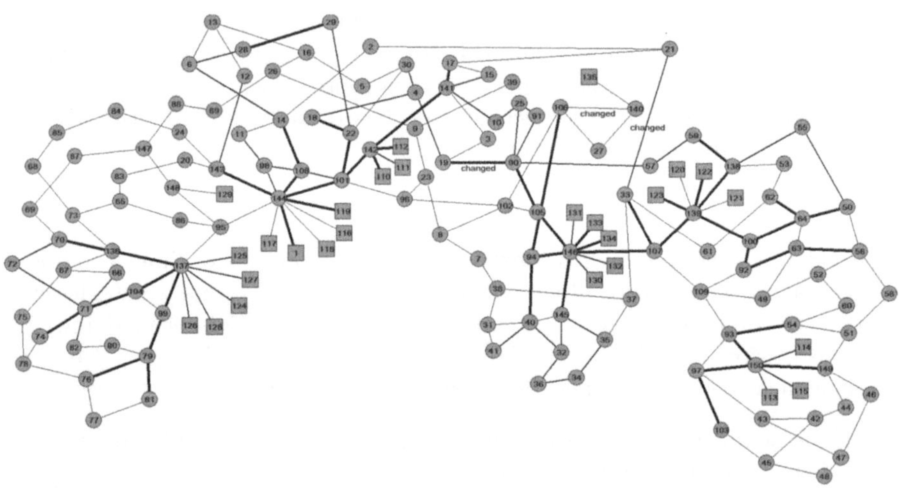


**FIGURE 5.20**
UIUC 150 Bus power system after edge modification. Numbers of Types I, II and III power lines (thick edges) are 107, 51, and 45. Edge modification involves adding edges (19, 90), (21, 33) and (106, 140).

former Type III lines become Type I or II, with only a few new Type III lines introduced.

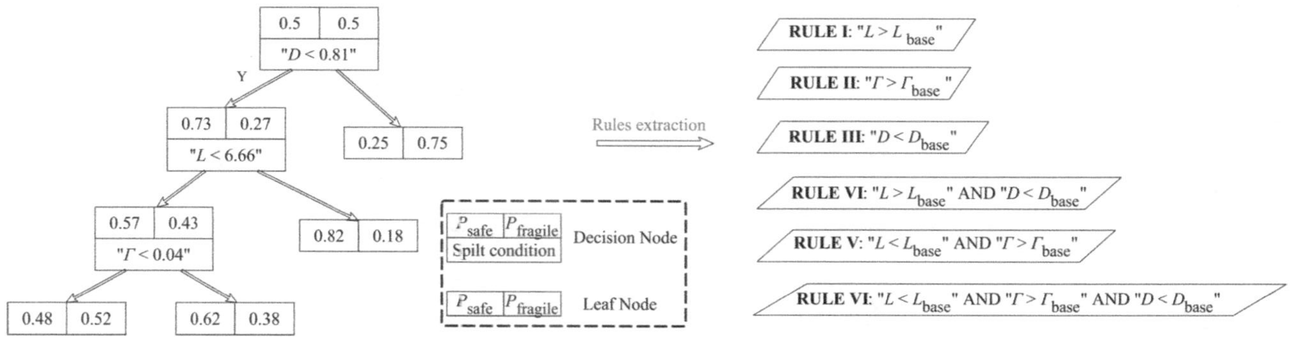


**FIGURE 5.21**
Rules derived by comparing the original and updated networks. Updated network is generated by decision tree with $e_{\mathrm{th}} = 0.25$. Split condition at decision node uses either network-based feature $L$, $\Gamma$ or $D$.

#### 5.4.2.3 Decision Tree Training and Rule Extraction

Using the algorithm described in Section 5.4.1, a decision tree is trained on the dataset of network features and "safe" or "fragile" labels derived from the rewired variants. Figure 5.21 shows an example decision tree containing three decision nodes and four leaf nodes. Each node displays the proportion of "safe" ($P_{\text{safe}}$) and "fragile" ($P_{\text{frag}}$) networks, along with the split condition.

For example, at the root node, the split condition might be $D \leq 0.81$, where satisfying this condition yields a greater fraction of "safe" networks. A set of six enhancement rules is extracted:

Rule I: $L_{\text{u}} > L_{\text{base}}$
Rule II: $\Gamma_{\text{u}} > \Gamma_{\text{base}}$
Rule III: $D_{\text{u}} < D_{\text{base}}$
Rule IV: $L_{\text{u}} > L_{\text{base}}$ and $D_{\text{u}} < D_{\text{base}}$
Rule V: $L_{\text{u}} < L_{\text{base}}$ and $\Gamma_{\text{u}} > \Gamma_{\text{base}}$
Rule VI: $L_{\text{u}} < L_{\text{base}}$ and $\Gamma_{\text{u}} < \Gamma_{\text{base}}$ and $D_{\text{u}} < D_{\text{base}}$

where $L$, $\Gamma$, and $D$ are the average shortest path length, average clustering coefficient, and average effective generator distance, respectively; subscripts "u" and "base" denote updated vs. original network.

#### 5.4.2.4 Effectiveness Verification of Enhancement Rules

To test each rule's utility, we follow the automated process shown in Figure 5.22:

1. Generate an updated network via the rewiring scheme;
2. Check compliance with the chosen rule (e.g., Rule I: is $L_{\text{u}} > L_{\text{base}}$). If not, discard and generate a new network.
3. Assess vulnerability of compliant networks by recalculating $P_{\text{T1}}, P_{\text{T2}}, P_{\text{T3}}$. Continue until a target number $N_{\text{th}}$ of "safe" networks is obtained.

The number of simulations required ($N_{\text{ast}}$) and the resulting counts of "safe" ($N_{\text{safe}}$) and "fragile" ($N_{\text{frag}}$) networks are shown in Figure 5.23. Rule VI delivers the highest efficiency (largest $N_{\text{safe}}/N_{\text{ast}}$ ratio), highlighting the effectiveness of simultaneously increasing $L$ and decreasing $D$. Rules I and II, which each modify a single feature, still eliminate over half of "fragile" networks, but the combination in Rule VI proves more impactful. From the simulation and decision tree analysis, two major observations can be made:

1. Impact of $L$: Increasing the average shortest path length, effectively weakening the small-world property, can enhance robustness against cascading failures [61].
2. Impact of $D$: Reducing the average effective distance to generators (i.e., more decentralized power., more decentralized power source distribution) consistently lowers vulnerability, aligning with the conclusions of [300].

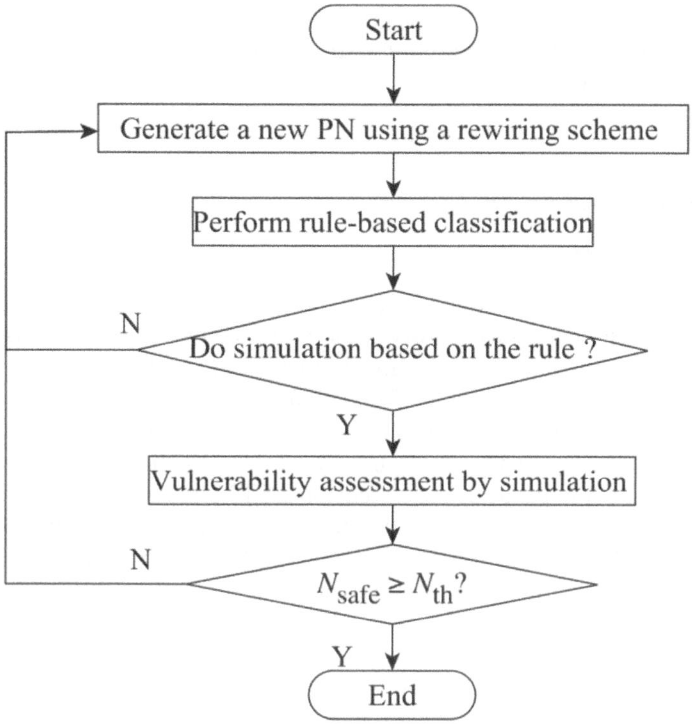


**FIGURE 5.22**
Flow chart for verifying the effectiveness of the rules for enhancement of robustness of the rewired power grid.

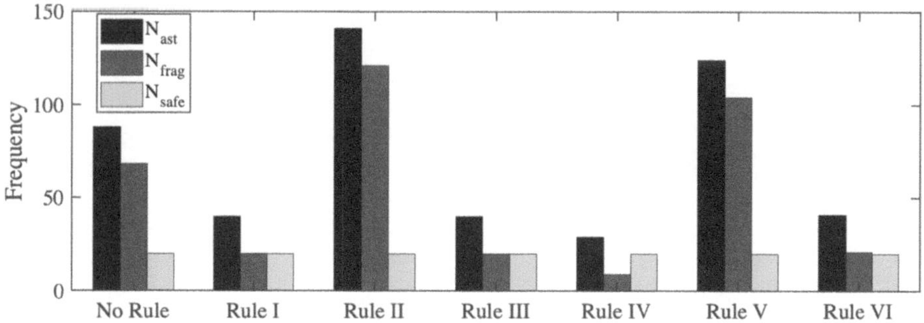


**FIGURE 5.23**
Effectiveness of rules on robustness enhancement of power networks

These observations highlight that specific topology metrics (and their combination) are strongly correlated with robustness. While the rewiring process primarily serves to validate the methodology, the extracted metrics offer valuable insights that can inform planning and operational strategies for designing more robust power grids.

## 5.5 Summary

In this chapter, we introduce a cascading failure simulation model that accounts for the cumulative effects of component overloading over time. Additionally, we present a cascading failure path prediction framework utilizing LSTM neural networks. To evaluate the effectiveness of the proposed models, we apply them to synthesized complex networks of varying scales and compare the performance of two prediction approaches: the standard LSTM model and the attention-LSTM model. Simulation results reveal that neural networks can predict the propagation paths of cascading failures both quickly and accurately, with the attention-LSTM model demonstrating superior performance compared to the standalone LSTM model. This work serves as a compelling example of the effective application of AI techniques to address practical challenges in complex network systems. We further introduce a key parameter that characterizes the propagation profile of cascading failure. This parameter is the onset time, which can be utilized for assessing the vulnerability of power grids. We present some practical classifying methods, including neural-network-based and decision-tree-based methods. It is found that the robustness of a power grid is highly correlated with its topological features, and effective metrics can be derived based on topological characteristics of the grid.

# 6

# *Robustness Enhancement of Evolving Power Grids via System Planning*

Modern power system planning prioritizes robustness by proactively designing infrastructure to withstand uncertainties, disturbances, and extreme events. Typical planning strategies include designing appropriate network topologies, hardening critical system components, deploying distributed energy resources, allocating auxiliary regulation resources, etc. These strategies are long-term and applied before the occurrence of cascading failure events. System planning strategies collectively create a "defensive depth", minimizing cascading failure risks while balancing cost and sustainability. This chapter presents a study of the effects of typical network properties on grid robustness, including network topology and accessibility to generators, and a gradient-targeted hardening strategy to enhance the power grid in the face of natural disasters.

## 6.1 Background

Power blackouts inevitably caused inconvenience to millions of users and incurred significant economic losses. Enhancing the robustness of power systems has always been a priority for engineers. Based on the cascading failure models and robustness assessment methods, research efforts have been devoted to developing power grid robustness enhancement solutions [162], which serve as long-term system planning and short-term system control and protection. System planning is applied before the occurrence of cascading failure events, e.g., optimizing network connections, adding network components, and hardening critical components [101, 262].

The topology of a power grid can influence its robustness significantly, rendering a feasible robustness enhancement strategy through the appropriate design of network connections. Since the scale-free and small-world properties were revealed and defined in networked systems [23, 272], the research of complex systems in terms of their network properties has made rapid and fruitful progress. By abstracting power stations in the power grid as nodes

DOI: 10.1201/9781003622130-6 

and transmission lines as edges, the power grid is amenable to complex network analysis [199]. Many researchers have tried to apply complex network theory to power systems, aiming to gain new insights into the power grid network topology design that would help enhance the robustness of power systems.

In early studies [7,221], real data from power grids in different regions were analyzed, with the objective of extracting structural characteristics of this man-made infrastructure. Average degree, degree distribution, and betweenness distribution are three important parameters to reveal the power grids' structural properties. Results have shown that the average degree of most of power grids is between 2 and 3, along with typical degree distribution and betweenness distribution. Cotilla-Sanchez et al. [58] compared the structural and electrical properties using the concept of "resistance distance", which is an important parameter for measuring accessibility of nodes. The functional properties of power grids, e.g., robustness, synchronization, and efficiency [187], have been explored in later studies, among which robustness has always drawn much attention.

The exact influence of network topology on its robustness should be investigated. For example, Yakup et al. [141] proposed the concept of effective graph resistance to evaluate the impact of grid topology on resisting cascading failure under malicious attacks. By increasing the effective graph resistance, the grids' robustness can be improved. Zhang et al. [300] revealed that the connection with a larger average shortest path length (ASPL) and higher clustering coefficient has lower cascading failure risks. Kaiser et al. [134] introduced a method to create such subgraphs that can stop failure propagation by adding certain edges to existing power networks. Considering the impacts of cyber coupling on cascading failure in power grids [298], Dong et al. [77] proposed a convenient and economic strategy to enhance the grids' robustness by adding information links other than transmission lines.

Grid hardening, which increases the tolerance capability of grid components, is the most direct strategy to improve the system's ability to withstand disruptions. For example, underground infrastructure can resist extreme weather events. Like many other complex networks [8], power grids exhibit the characteristic of heterogeneity [25, 111]. Heterogeneity refers to the uneven distribution of properties or characteristics among the components of a network. In the context of power grids, heterogeneity implies that a small number of critical components have a significant impact on the grid's overall performance, while the majority of components have little influence. Achieving uniform hardening across all components is costly and inefficient. Considering the heterogeneous nature of power grids, a targeted hardening approach that focuses on strengthening the critical components can be economically feasible and effective in enhancing power grid resilience [147, 282].

Accurately quantifying the criticality of components poses the foremost challenge in achieving successful targeted hardening of power grids. Extensive research has been conducted on identifying critical components in complex networks based on degree centrality [39], betweenness centrality [98], closeness centrality [224], and PageRank [41]. Additionally, numerous studies have combined some of these metrics or methods for identification purposes [21, 189]. In their work, the decision variables are the binary connection states of edges, and the objectives are to maximize global efficiency and minimize the number of disconnected edges. This type of research solely focuses on identifying the importance of critical components that cause the most damage to network connectivity, without taking into account the functional properties of power grids.

It is important to recognize that the electrical properties of a power network significantly influence its functional performance [72]. Such electrical properties cannot be adequately captured by topological quantities alone. Thus, the focus should shift towards evaluating functional performance rather than solely relying on topological measures. Some scholars use iterative and optimization model-solving methods to identify critical components. In our previous work, the largest amount of power supply achieved by quick restoration methods after the interruption of disasters was used as an important resilience metric, based on which critical network components were identified by the greedy search method [297] and intelligent optimization methods [232].

Some studies have assessed the impact of sequentially removing nodes from the network on performance metrics such as resilience [121] and robustness [261]. Other studies introduce novel metrics to evaluate the significance of components, such as risk sensitivity [166], criticality importance [227], vulnerability index [156], TrendRank value [247], and comprehensive topological potential [296]. However, these component criticality metrics fall short of being systematically proposed or directly relevant to power grid resilience, resulting in inaccurate identification results and inefficient hardening strategies in terms of enhancing power grid's resilience.

The appropriate assignment/investment of budgets on hardening system components poses the second challenge to the successful targeted hardening of power grids. By hardening only the critical components of power grids, the performance can be improved with fewer budgets [227]. A common approach to allocating reinforcement resources is to apply uniform strengthening to all nodes [184] or the identified critical nodes [246]. A more rational approach is to establish optimization models to determine the resource allocation for hardening overhead distribution lines [155, 206]. Nonetheless, in these studies, a common approach was to assign sufficient hardening resources to all the critical components to render them completely indestructible. It would be more efficient to allocate limited but adequate amounts of hardening resources based on the criticality metrics of components.

## 6.2 Improving Network Structure to Enhance Grid's Robustness

In this section, we investigate the typical network properties affecting a grid's robustness, including network topology and accessibility to generators. By deploying the cascading failure model introduced in Chapter 2, we perform simulations and generate a large number of cascading failure cases. In order to quantitatively describe a system's robustness, two robustness parameters are presented in Section 6.2.1, i.e., the percentage of unserved nodes (PUN) caused by a component's failure and the percentage of non-critical links (PNL) that will not cause severe damage. Section 6.2.2 shows robustness assessment results of some real power systems using the method described. Many factors can influence a power system's robustness, and Section 6.2.3 specifically explores the influence of network structure and the locations of generators. Simulation results show that, for a given set of numbers of generators, consumers, and transmission lines, the connection with a short ASPL can significantly reduce a power system's robustness. To explore the effects of generators' distribution in the grid, we propose, in Section 6.2.3, a new metric based on node-generator resistance distance [140] (DG) for measuring the degree of accessibility to generators of all consumers in a power network, which is shown to affect robustness significantly.

### 6.2.1 Robustness Parameters

Robustness refers to the ability of a system to tolerate faults. It can be assessed in terms of a measure that describes the ability of a power system to provide normal service to a critical percentage of clients under the condition that some components of the system fail. It is important to define appropriate metrics that can quantitatively indicate a power system's robustness. In our study, a power system is represented as an undirected graph $W$ with $n$ nodes and $m$ links. Formally, a graph $W$ is $\{N, M\}$, where $N$ is the set of all nodes and $M$ is the set of all links. Also, $G$ represents the set of generators in $W$, and $G \subseteq N$.

In the field of power system analysis, the extent of the unserved area is usually used to measure the size of a blackout [46]. Here, we introduce the *fraction of unserved area* caused by the failure of a component to indicate the importance of that component. Specifically, we define PUN($i$) as the *percentage of unserved nodes* caused by the failure of component $i$, i.e.,

$$\text{PUN}(i) = \frac{n_{\text{unserved}}(i)}{n} \tag{6.1}$$

where $n_{\text{unserved}}(i)$ is the number of unserved nodes due to component $i$'s malfunctioning. Unserved nodes are the nodes that are deprived of power in a blackout. Unserved nodes are either nodes whose power loadings exceed their capacities or nodes that exist in a subgraph containing no generators. A component that has a high PUN can seriously damage the network, and conversely, a component with a small PUN will not have much influence when it fails. The power system is more resilient to the faults occurring at components whose PUN is smaller, and we refer to these types of components as non-critical components. If a power system is resilient to the faults occurring at most of the components, i.e., most of the components of the power system have small PUNs, we can say that the system is robust.

To measure the robustness of the whole system, we use the *percentage of non-critical links* (PNL) whose PUNs are smaller than a *threshold* to indicate the ability of a network to tolerate faults. The PUN threshold is a specific percentage of nodes in the power grid. We define PNL(threshold) as the *percentage of non-critical links* for a given threshold, i.e.,

$$\text{PNL}(\text{threshold}) = \frac{1}{m} \sum_{i \in M} \delta(i) \tag{6.2}$$

where

$$\delta(i) = \begin{cases} 1, & \text{PUN}(i) < \text{threshold} \\ 0, & \text{otherwise} \end{cases}$$

A large PNL means that the power system has a large portion of links whose failure will not lead to serious damage to the grid (i.e., the PUNs remain larger than the threshold). In other words, the system can tolerate faults that occur in a high percentage of components of the system. The power system with a large PNL is highly robust.

### 6.2.2 Preliminary Study

In this section, we present the simulation results of the robustness assessment for several real power systems. The IEEE 118 Bus serves as a power flow test case, referenced in [54], while data for the Northern European Grid (NEG) is obtained from [180]. For our study, we set the generator voltages to 1 p.u., with all nodes, except for the generators, each sinking 1 p.u. of current. The admittance of each transmission line is configured to 11 p.u. Additionally, the safety margins for nodes and links are established at ($\alpha = 0.2$) and ($\beta = 0.5$). The simulations are conducted using MATLAB®, with a toolbox library developed by Lev Muchnik [188] that provides essential functions for computing complex network parameters.

Figure 6.1 shows the PUNs of all links in the IEEE 118 Bus and the NEG. It can be observed that the roles of different links in the same power system are prominently different, as they have different PUNs. Comparing Figure 6.1a with 6.1b, the percentage of non-critical links of the NEG is larger than that of the IEEE 118 Bus. In order to clearly display the robustness of the two power systems, we plot the PNLs of these two networks for different PUN thresholds. As shown in Figure 6.2, the PNLs of the NEG are always larger

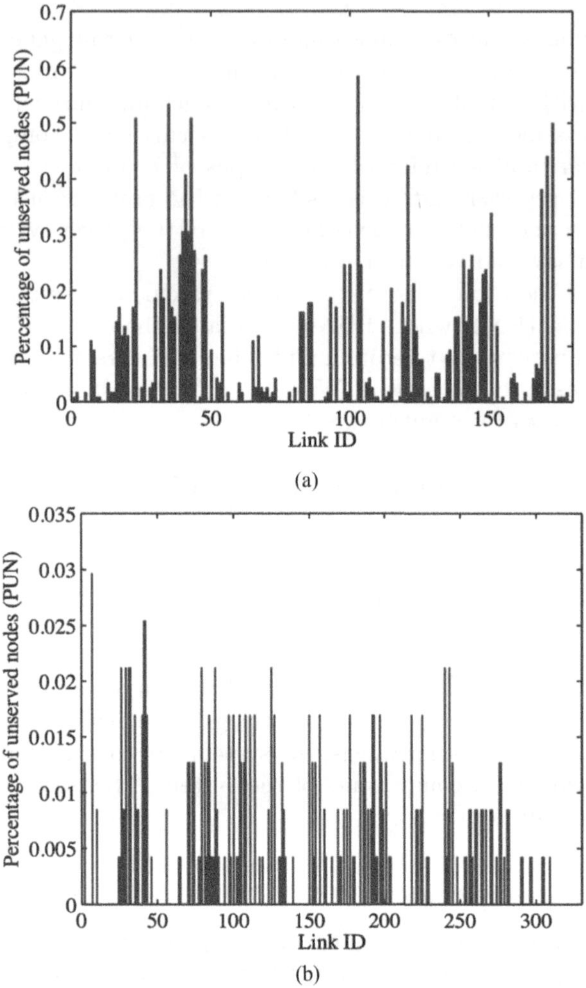


**FIGURE 6.1**
Simulation results of cascading failure and robustness assessment. (a) PUN of each link in IEEE 118 Bus; (b) PUN of each link in Northern European Grid.

than those of the IEEE 118 Bus, for the *threshold* ranging from 0 to 0.33. This means that the NEG is more robust than the IEEE 118 Bus.

The above result raises a number of important questions. Why does the NEG have better robustness than the IEEE 118 Bus? What are the factors that affect a power system's robustness, and in what way do these factors influence a power system's robustness? Is there a consolidated metric that can conveniently measure the robustness of a system? The answers to these questions will offer useful clues and design guidelines for power engineers to construct more reliable power transmission systems.

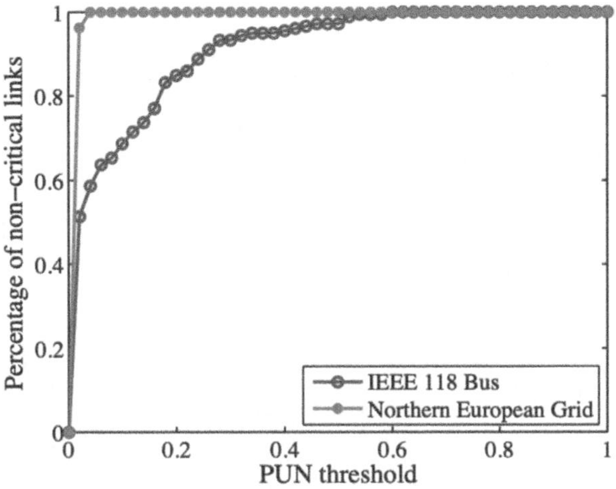


**FIGURE 6.2**
Robustness assessment of IEEE 118 Bus and Northern European Grid.

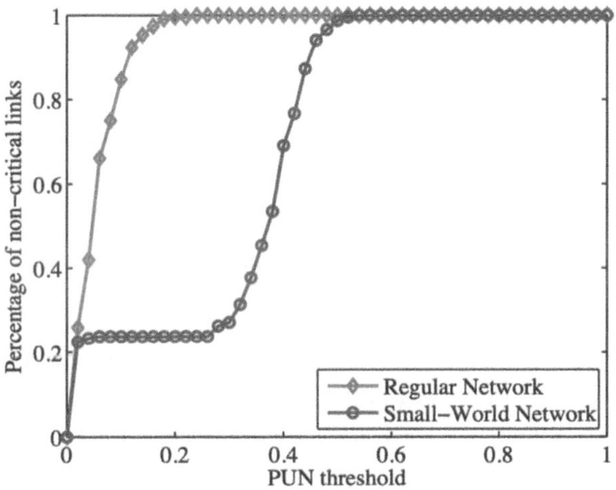


**FIGURE 6.3**
Robustness assessment of small-world and regular networks.

Table 6.1 lists the ASPL and percentage of generators (PG) of the two networks. ASPL describes the structural characteristics of a network, whereas PG gives information about generators. The NEG's ASPL is larger than the IEEE 118 Bus's, indicating that the IEEE 118 Bus displays a higher degree of small-world connectivity. The network structure can play an important role in affecting the robustness of a power system. At the same time, the NEG has a larger PG than the IEEE 118 Bus. The PG is also an important factor. Many other factors can influence the robustness of a power system as well, e.g., the

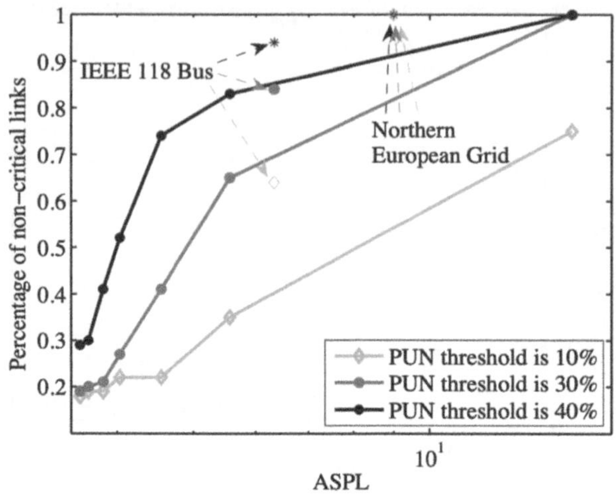


**FIGURE 6.4**
Effects of small-world connectivity on robustness of power systems.

**TABLE 6.1**
Average Shortest Path Length (ASPL) and Percentage of Generators (PG) of Networks

| | ASPL | PG(%) |
|---|---|---|
| IEEE 118 Bus | 6.33 | 8 |
| Northern European Grid | 8.99 | 50 |

locations of generators, the safety margins, and so on. It should be noted that the robustness of the two systems, as inferred from Figure 6.2, is the result of the combined influence of these factors. In the next section, we will compare the effects of various parameters systematically, aiming to develop an effective metric that can be used to assess robustness.

### 6.2.3 Effects of Network Properties

As mentioned in Section 6.2.2, the robustness of a network is affected by a number of parameters. In this subsection, we focus on network properties that determine the robustness of a network. Specifically, we consider the network structure and the availability of generators in a network. Our purpose is to derive effective guidelines that can be used by electrical engineers to determine the network structure and generator distribution in order to optimize robustness. Note that we do not consider component parameters, e.g., ratings and safety margins, which can be considered as post-design parameters and be dealt with separately after the desired network is constructed.

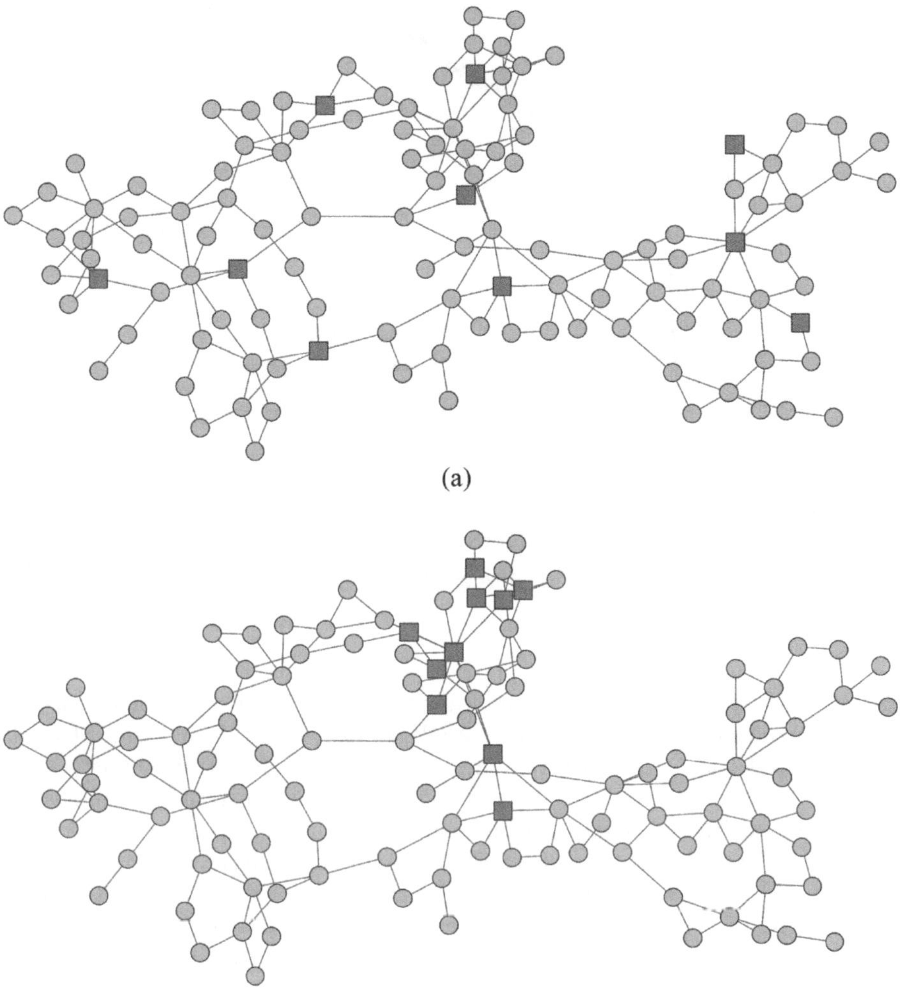

**FIGURE 6.5**
Topologies of power networks. (a) IEEE 118 Bus A; and (b) IEEE 118 Bus B. Squares represent generators.

#### 6.2.3.1 Effect of Network Structure

Phadke et al. [210] pointed out that the graph of a power system is relevant to its efficiency and robustness. Here, we investigate the influence of a grid's topology on its robustness. From Table 6.1, ASPL can be an important factor, and it also indicates that the network with short ASPL may display weak robustness. In order to focus on the effect of network structure, we generate

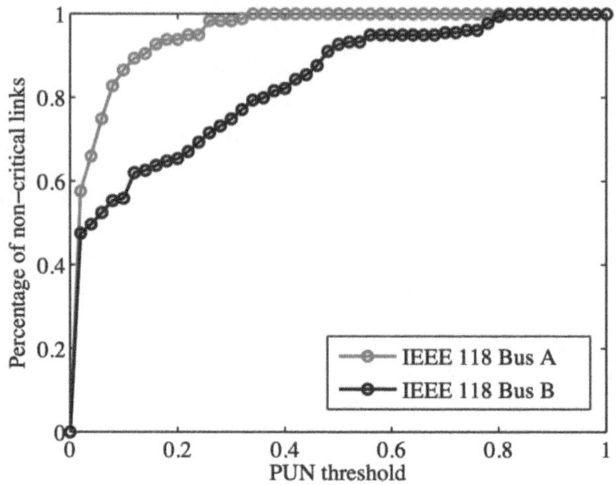


**FIGURE 6.6**
Robustness assessment of IEEE 118 Bus. Buses A and B only differ in the locations of generators, with Bus A having a more decentralized distribution of generators.

networks of specific structures for in-depth study. Small-world networks are one typical kind of networks whose ASPL is very small. Watts and Strogatz [272] showed that small-world connectivity could have significant effects on the dynamics of networked systems. To verify the effect of the connection with short ASPL, we first study the robustness of small-world networks. For instance, to probe into the effect of small-world connectivity, we construct regular and small-world networks of similar scale and identical PG nodes. Specifically, we generate a regular network of 118 nodes with an average degree of 4. The small-world network is generated by rewiring the links of the regular network with a probability $q = 0.3$. The PG is 8%. In order to scale the effects of other factors, such as the locations of the generators, we construct 100 realizations of the small-world network to get the average results. The result of the robustness assessment shown in Figure 6.3 shows that the PNLs of the regular network are much higher than those of the small-world network for PUN thresholds ranging from 0.02 to 0.60.

In order to further explore the effects of the connection with short ASPL, we generate seven groups of networks with the link-rewiring probability $q$ ranging from 0 to 0.6. Each group contains 100 realizations, similar to the group with $q = 0.3$ mentioned above. The group with $q = 0$ is essentially the regular network group. Table 6.2 lists the averaged PNLs with three thresholds, along with ASPL and PG of each group. We see that as $q$ increases, ASPL decreases. In Figure 6.4, we plot the relationship between PNL and ASPL. The lines are results derived from the seven groups of synthesized networks listed in Table 6.2, and the dots are robustness assessment results of IEEE

**TABLE 6.2**
Average Shortest Path Length (ASPL), Percentage of Generators (PG) of Networks with Different Levels of Small-World Connectivity Characterized by the Link-Rewiring Probability $q$

| $q$ | ASPL | PG | PNL(10%) | PNL(30%) | PNL(40%) |
|---|---|---|---|---|---|
| 0.0 | 15.13 | 8% | 0.75 | 1.00 | 1.00 |
| 0.1 | 5.56 | 8% | 0.35 | 0.65 | 0.83 |
| 0.2 | 4.55 | 8% | 0.22 | 0.41 | 0.74 |
| 0.3 | 4.03 | 8% | 0.22 | 0.27 | 0.52 |
| 0.4 | 3.84 | 8% | 0.19 | 0.21 | 0.41 |
| 0.5 | 3.68 | 8% | 0.19 | 0.20 | 0.30 |
| 0.6 | 3.59 | 8% | 0.18 | 0.19 | 0.29 |

Their corresponding PNLs for the threshold of PUN set at 10%, 30% and 40% are shown.

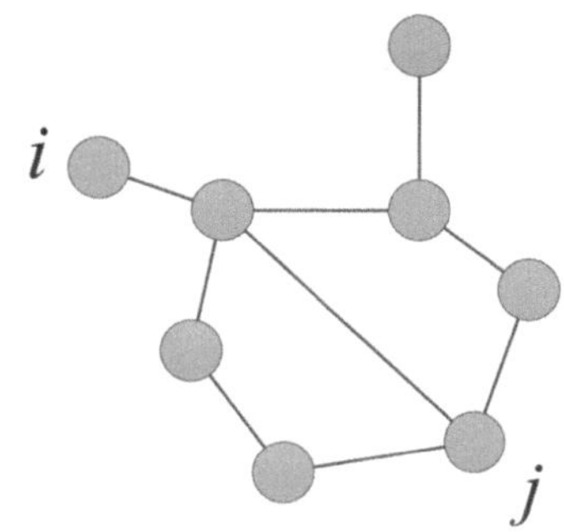


**FIGURE 6.7**
An example of electrical network.

118 Bus and NEG. It is obvious that the value of PNL will be lower if the system has a smaller value of ASPL. In other words, a smaller ASPL connectivity deteriorates the robustness of a power system. Hence, we can conclude that with equal percentages of generator nodes, transmission lines, and the same power consumption, the connection with short ASPL degrades a power system's robustness significantly when the safety margins are limited. This is consistent with the robustness assessment results for the IEEE 118 Bus and the NEG, i.e., the one with the shorter ASPL is less robust.

Several publications have studied the problem related to the influence of small-world connectivity on the robustness of a power system. For instance, small-world networks were found to be prone to cascading failures [179], and small-world networks were also found to be more readily recovered from failures and hence more robust [213]. The main reason for the different conclusions is that the assumptions of the studies are different. In the former study [213], the authors did not consider any constraints on the amount of flow that could

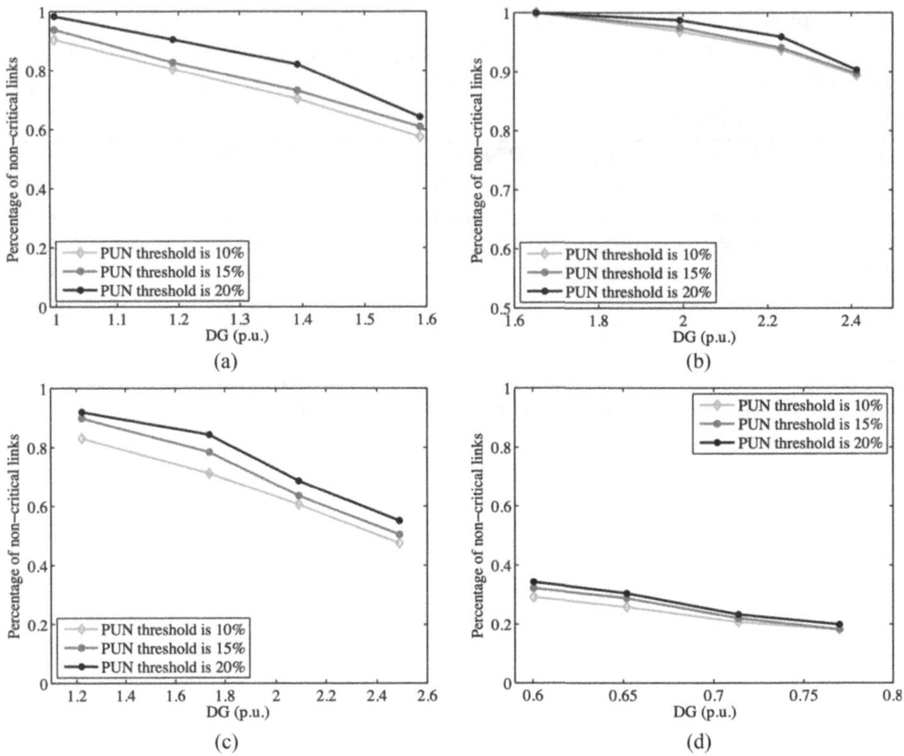


**FIGURE 6.8**
Effect of locations of generators. (a) IEEE 118 Bus; (b) Northern European Grid; (c) regular network; (d) small-world network. DG measures nodes' distance to generators. Higher DG means less decentralised distribution of generators.

be transported by any link, i.e., the capacities of the components were infinite and cascading processes were not considered. From a topological view, small-world networks have better connectivity than regular networks. Thus, the possibility of repairing them by adding new links when the network is decomposed is higher. In practice, due to economic considerations, the safety margins are often limited. Thus, it is more realistic to base our study on the practical assumption that the capacities of the components of a power system are limited.

### 6.2.3.2 Effect of Accessibility to Generators

From our simulations, power grids of the same structure can also display distinct robustness performances. We generate two power systems based on the IEEE 118 Bus, namely, IEEE 118 Bus A and IEEE 118 Bus B. Figure 6.5

shows the graph layouts of these two systems, where the rectangle nodes are generators and the circle nodes are consumers. IEEE 118 Buses A and B share the same characteristics, including network structure, PG, and safety margins, but the generators in the two networks are located differently. The robustness assessment results of these two systems are shown in Figure 6.6. From this result, we see that the IEEE 118 Bus A is more robust than the IEEE 118 Bus B. Thus, the locations of the generators significantly affect the robustness of the system.

In terms of generator distribution, the IEEE 118 Bus A is more decentralized than the IEEE 118 Bus B. Theoretically, for a given number (percentage) of available generators, a decentralized distribution of generators permits most of the consumers in the network to reach a power source within shorter distances. In other words, electrical power does not have to travel through long power lines from the generators to the consumers. To transmit the same amount of power from generators to consumers, highly decentralized locations of generators can reduce the total "traffic" volume in the transmission lines as well as the distribution nodes.

It is desirable to find a variable that quantitatively describes the location information of the generators in a network. Here, we review the concept of *resistance distance* of a power system proposed by Klein and Randić [140]. Essentially, the *resistance distance* between two nodes refers to the effective resistance between them.

Referring to Figure 6.7, when calculating the effective resistance between nodes $i$ and $j$, we set node $i$ as a voltage source with $V_i$, node $j$ as a current sink of $I_j$, and all other nodes as distribution nodes with zero sink current. $V_j$ can be readily derived. The *effective resistance* between nodes $i$ and $j$ is defined by

$$R_{ij} = \frac{V_i - V_j}{I_j} \tag{6.3}$$

When considering the electrical distance in a DC model [226,291], as the resistances of transmission lines are ignored, the equivalent metric of the resistance distance between node $i$ and $j$ can be interpreted as reactance distance $x_{ij} =$. Similarly, when all other nodes are set as distribution nodes with external injected power of 0, and the external power of nodes $i$ and $j$ are balanced, and $x_{ij} =$ can be computed with

$$x_{ij} = \frac{\theta_i - \theta_j}{P_j} \tag{6.4}$$

where $\theta_i$ and $\theta_i$ are the voltage phase angles of nodes $i$ and $j$, and $P_j$ is the injected of power of node $j$.

The minimum effective resistance of a consumer node $i$ to any nearest generator represents its shortest distance to a power source. This is a measure of the distance over which power is transmitted between the pair of nodes. Thus, the minimum effective resistance of consumer $i$ represents the *accessibility* to

**TABLE 6.3**
DG and PG of IEEE 118 Bus A and B

| | IEEE 118 Bus A | IEEE 118 Bus B |
|---|---|---|
| DG (p.u.) | 0.9977 | 1.5334 |

Percentage of generators is fixed at 8% for comparison.

power sources of this node. Specifically, we define the resistance distance of node $i$ to its nearest generator, $d(i)$, as follows:

$$d(i) = \min\{R_{is}, s \in G\} \tag{6.5}$$

For a given network structure, if the generators are evenly distributed and the PG nodes is adequately high, all consumers can reach a power source within a short resistance distance, i.e., all nodes have ready access to a power source. This will reduce the total power load imposed on the transmission lines, making the system more robust. Here, we define *average effective resistance (distance) to a nearest generator of all consumer nodes* (DG) as a measure of the accessibility to generators of all consumers, i.e.,

$$\text{DG} = \frac{1}{(n-g)} \sum_{i \in N \setminus G} d(i) \tag{6.6}$$

where $N \setminus G$ is the set of nodes excluding the generator nodes, $n$ is the total number of nodes, and $g$ is the number of generators in the network. Small DG indicates better accessibility to power sources for generators. A network has a smaller DG if its generators are more decentralized or have a sufficiently large number (percentage) of generators. Thus, in terms of basic network design, DG offers an effective measure of accessibility to power, which is the combined effect of the distribution of generators and the PG in a network. A large percentage and evenly decentralized locations of source nodes can make the parameter DG small. It is obvious that a power system could be very robust if there exists a large percentage of source nodes. We focus on the influence of the locations of the generator on a system's robustness. Table 6.3 gives the DG values of IEEE 118 Buses A and B, with the PG fixed at 8%. We see that the DG of IEEE 118 Bus A is smaller than that of Bus B, which indicates that the generators in IEEE 118 Bus A are more decentralized than in IEEE 118 Bus B.

We now study the effect of varying DG in the IEEE 118 Bus, NEG, regular, and small-world networks. For the IEEE 118 Bus system, a series of tests is performed, with generators' locations randomly chosen while keeping the same structure and fixing the PG at 8%. Then, we sort the results into five groups according to the values of DG. Ten test results are chosen in each group, and we average their PNLs and DGs. Figure 6.8a shows the PNLs with different PUN thresholds for the IEEE 118 Bus. It is obvious that the value of PNL drops significantly as DG increases. We then apply the same test procedure to assess the NEG, the regular network and the small-world network. The regular network is the same network generated in Section 6.2.3.1, and the small-world network is generated by rewiring the links of the regular network with a probability of 0.3. Figures 6.8b–d show consistent results. Thus, the metric DG defined here is an effective design parameter for guiding the power engineers to choose appropriate locations for generators in a given network structure to achieve a more robust power system.

It should be emphasized that our conclusion here has been drawn on the condition that the network structure is fixed. If the network structure is varied, small-world connectivity may also make DG very small. In that case, a small DG does not necessarily describe a decentralized distribution of the generators. In Section 6.2.3.1, we have observed that small-world connectivity can degrade a power system's robustness even though the DG value is small. The reason for this is that small resistance distances among nodes make the overall sensitivity of all components to a failure relatively high.

## 6.3 Hardening Critical Network Components

### 6.3.1 Historical Records

Due to direct exposure to wild environments, the grids are highly vulnerable to natural disasters, like earthquakes and hurricanes [6,9,107]. Between 1990 and 2020, 138 major power outages with clearly identified causes were recorded worldwide, of which 77 were attributed to natural disasters. As shown in Figure 6.9, natural disasters account for over half of these outages. Furthermore, Figure 6.10, derived from EM-DAT data, provides detailed insights into the frequency of natural disasters, the number of affected individuals, and the economic costs of damage from 2003 to 2023. These findings highlight that natural disasters not only result in significant economic losses but also pose serious risks to public safety. For instance, in October 2018, Hurricane Michael struck Florida and the southern United States [203], causing extensive damage to power transmission lines and substations. Millions of residents were left without power, with restoration efforts taking weeks in some areas. Similarly, in October 2019, Japan experienced Typhoon Hagibis and Typhoon Hibiscus, which caused widespread flooding and storm surges [185]. Severe damage to the power grid left millions temporarily without electricity. More recently, in March 2021, eastern Australia endured heavy rainfall and flooding [219], which damaged power facilities, including transmission lines and substations. As a result, hundreds of thousands of people experienced prolonged power outages lasting several weeks.

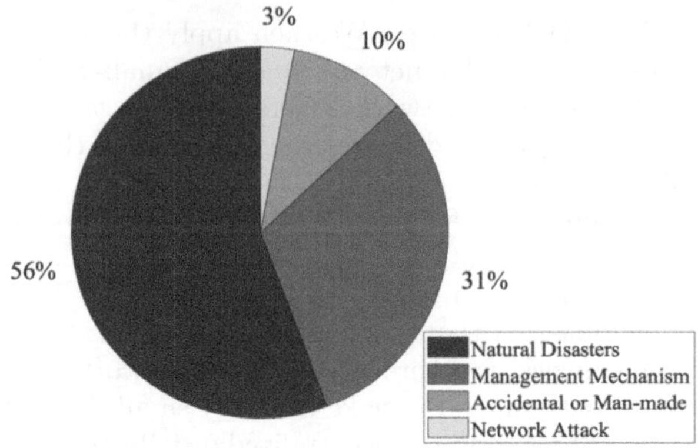


**FIGURE 6.9**
Statistics for the causes of real power outages.

Power grids are susceptible to natural disasters, which can cause physical damage to critical infrastructure and trigger cascading failures. In this section, we introduce a gradient-targeted hardening strategy designed to bolster grid resilience against such events, especially under resource constraints. This strategy integrates several essential components: a power grid resilience metric, a component resilience-criticality metric, metric value estimation, and gradient-based resource allocation for hardening. As illustrated in Figure 6.11, the framework is structured around four core elements.

1. Establishing a power grid resilience metric: We establish a general power grid resilience metric that evaluates grid robustness by integrating data from both the power network and natural disaster profiles. This metric is designed in a flexible and comprehensive form, capturing the stochastic nature of diverse natural disasters, their probabilistic impacts on system components, and the dynamic recovery behavior of power grids. By accounting for these uncertainties and recovery processes, the metric provides a more realistic and actionable assessment of grid resilience.
2. Determining the resilience-criticality of components: We define a component resilience-criticality metric for quantifying the impact of components on power grid resilience based on sensitivity analysis in statistics. The component criticality metric has the advantages of having a clear expression and being directly relevant to grid resilience.
3. Monte Carlo simulations for metric estimation: We develop a Monte Carlo–based estimation method that significantly reduces computational overhead while maintaining high accuracy in generating resilience metric values.

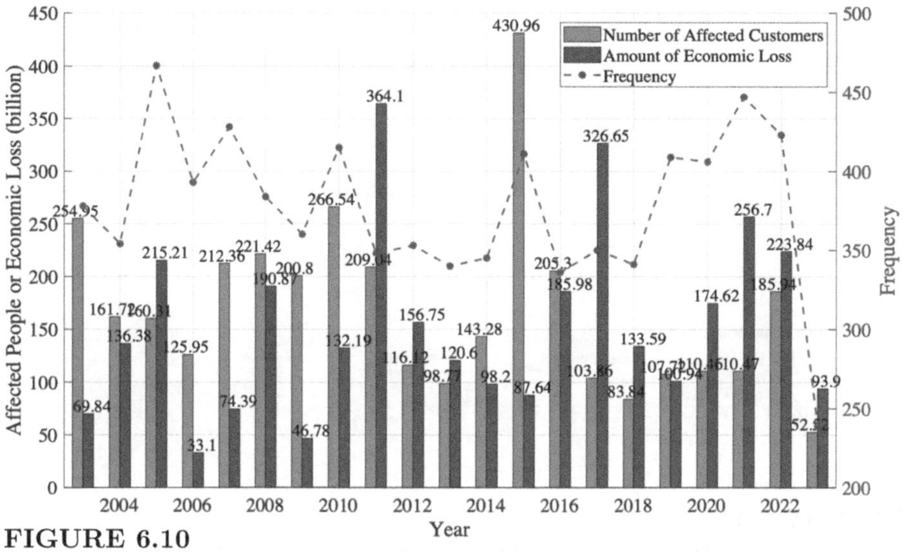


**FIGURE 6.10**
Frequency and damage extent of natural disasters.

4. Developing a gradient hardening resource allocation strategy: Based on the component resilience-criticality values, critical components are grouped into distinct clusters, each corresponding to a specific hardening level. Components within the same cluster receive an equal allocation of hardening resources. Unlike existing approaches that aim to render critical components completely indestructible, the strategy presented here enables more efficient utilization of limited resources by distributing them proportionally according to resilience-criticality. This targeted allocation enhances overall grid resilience while maintaining cost-effectiveness.

By following this strategy framework, power grid operators and planners can systematically assess the resilience of their systems, identify critical components, and develop effective strategies to enhance the power grid's ability to withstand natural disasters.

### 6.3.2 Criticality Metric of Components

Since power grid resilience is a complex property, various metrics have been proposed in the literature to quantify this property from different perspectives and with different emphases [67, 119, 133, 207]. Most existing resilience evaluation metrics focus on calculating the resulting function loss in the system recovery process. However, regarding power system planning strategies, a resilience metric that quantifies the expected function loss under the recovery context and includes the stochastic characteristics of impacts of natural disasters is more appropriate. In practice, a power grid can be affected by multiple

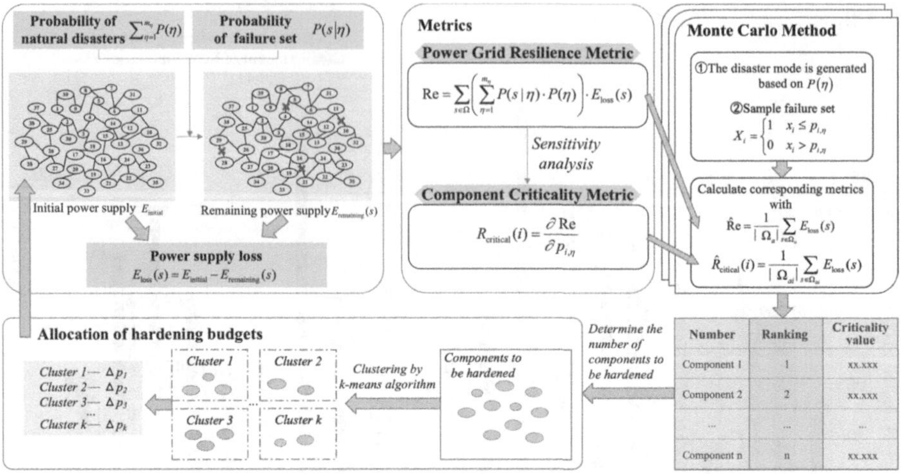


**FIGURE 6.11**
Overall framework for the gradient-targeted hardening strategy.

types of natural disasters, including floods, lightning strikes, strong winds, and icing. It is meaningful to assess and improve the power grid's resilience by incorporating the data of natural extremes.

### 6.3.2.1 Power Grid Resilience Metric

We denote a power network as $G(V, E)$, where $V$ is the set of nodes and $E$ is the set of edges. Power grid resilience characterizes its ability to tolerate extreme events, which can be quantitatively measured by the expected value of loss of power supply due to such events. Due to its exposure to various natural disasters such as earthquakes, floods, and snowstorms, the power grid is highly vulnerable. To evaluate power grid resilience in the face of strong natural disasters, we introduce the concept of risk [164] and define the power grid resilience metric as follows:

$$\mathrm{Re} = \sum_{s \in \Omega} P(s) \cdot E_{\mathrm{loss}}(s) \tag{6.7}$$

where Re is the power grid resilience metric, $s$ is a set of primary failures that are independent and physically damaged components by a disaster, $P(s)$ is the occurrence probability of failure set $s$, and $\Omega$ denotes the family of primary failure sets that may occur in the power grid. In the life cycle of grid resilience, $s$ is followed by fast functionality degradation, quick restoration, and slow restoration stages. Here, we are mainly concerned with the power supply loss associated with failure set $s$ after quick restoration, quantified by $E_{\mathrm{loss}}(s)$, when cascading failures due to relay protection actions have all been fixed. The detailed calculation of $E_{\mathrm{loss}}(s)$ is provided in Appendix 8.2.

The power grid resilience metric quantifies the expected power supply loss of a power grid due to natural disasters. Under this definition, a smaller value of Re indicates that the power grid's resilience is higher. The power grid resilience metric can be written as

$$
\begin{aligned}
\mathrm{Re} &= \sum_{s\in\Omega} P(s)\cdot E_{\mathrm{loss}}(s) \\
&= \sum_{s\in\Omega}\left(\sum_{\eta=1}^{m_\eta} P(s|\eta)\cdot P(\eta)\right)\cdot E_{\mathrm{loss}}(s) \\
&= \sum_{s\in\Omega}\left(\sum_{\eta=1}^{m_\eta}\left(\prod_{i\in s, j\in G-s} p_{i,\eta}\cdot(1-p_{j,\eta})\right)\cdot P(\eta)\right)\cdot E_{\mathrm{loss}}(s)
\end{aligned}
\tag{6.8}
$$

where $P(\eta)$ represents the probability of disaster mode $\eta$, and $P(s|\eta)$ represents the probability of failure set $s$ under disaster mode $\eta$. Also, $p_{i,\eta}$ denotes the damage probability of component $i$ under disaster mode $\eta$. $G$ represents the set of all components in the power grid, and $G - s$ is the set of all normal components that are not damaged by the natural disaster. The detailed derivation process is provided in Appendix 8.2.

#### 6.3.2.2 Component Resilience-Criticality Metric

Equation (6.8) shows that different components have varying degrees of influence on a power grid's resilience. The degree of influence of components, i.e., component resilience-criticality can be quantified by applying the sensitivity analysis method [225, 250]. Based on the afore-described grid resilience metric, we introduce a component resilience-criticality metric, which can be derived by calculating the partial derivative of $R(e)$ with respect to the damage probability, i.e.,

$$
\begin{aligned}
R_{\mathrm{critical}}(i) &= \frac{\partial \mathrm{Re}}{\partial p_{i,\eta}} \\
&= \sum_{s\in\Omega}\left(\sum_{\eta=1}^{m_\eta}\left(\prod_{\substack{k\in s, k\neq i\\ i\in s, j\in G-s}} p_{k,\eta}(1-p_{j,\eta})\right)\cdot P(\eta)\right)\cdot E_{\mathrm{loss}}(s)
\end{aligned}
\tag{6.9}
$$

where components $k$ is one of the damaged components except for components $i$.

The above equation represents the sensitivity coefficient of the proposed power grid's resilience metric to the damage probability of component $i$. A higher value of $R_{\mathrm{critical}}(i)$ indicates that the power grid's resilience is more sensitive to changes in the damage probability of component $i$. In other words, increasing the damage probabilities of components with higher resilience-criticality metric values can lead to more effective enhancements in grid resilience. Therefore, by calculating and ranking the resilience-criticality metric for all components in the power grid, we can identify the critical components that require special attention and prioritized hardening efforts.

#### 6.3.2.3 Metrics Estimation Based on Monte-Carlo Method

The calculation of a power grid's resilience metric according to its definition is analytically possible. However, due to the exponential growth of the failure set $\Omega$ as the power grid expands, this method becomes computationally infeasible. As an alternative, the Monte-Carlo method is employed [130], which is a computational technique to estimate the value of a parameter or solve complex problems through statistical sampling. It is particularly useful when analytical solutions are difficult to obtain. The key idea behind the Monte-Carlo method is that by generating a large number of random samples and performing simulations, the distribution of outputs provides an approximation of the true parameter or solution. As the number of samples increases, the accuracy of the estimate improves. The Monte-Carlo method can significantly reduce the computational loads and obtain accurate values of metrics.

Figure 6.12 shows the flow chart for estimating the power grid resilience metric with the Monte-Carlo method. After feeding the parameters of the targeted power grid, random samples of damage events are generated. The disaster mode is first generated based on $P(\eta)$. Then, for a power grid with $n$ components, its connection state can be represented by $\boldsymbol{X} = \{X_1, X_2, \ldots, X_n\}$, where $X_i$ refers to the connection state of component $i$. If $X_i = 1$, it means that component $i$ is disconnected due to damage, while $X_i = 0$ means that component $i$ is undamaged and well connected. Then, generate a set of random numbers $\boldsymbol{x} = \{x_1, x_2, \ldots x_n\}$ and compare them with $p_{i,\eta}$ to determine the damaged components with

$$X_i = \begin{cases} 1 & x_i \leq p_{i,\eta} \\ 0 & x_i > p_{i,\eta} \end{cases} \tag{6.10}$$

Given $s$, we calculate the power supply outage $E_{\text{loss}}(s)$, and the detailed calculation of $E_{\text{loss}}(s)$ is provided in Appendix 8.2. According to equation (6.8), Re can be estimated with

$$\hat{\text{Re}} = \frac{1}{|\Omega_a|} \sum_{s \in \Omega_a} E_{\text{loss}}(s) \tag{6.11}$$

where $\hat{\text{Re}}$ is the average value of $E_{\text{loss}}(s)$ of all generated samples, $\Omega_a$ is the set of random samples of damage events, and $|\Omega_a|$ denotes the number of elements of $\Omega_a$. The above process iterates until $\hat{\text{Re}}$ converges, and we obtain the estimated power grid resilience.

Similarly, the Monte-Carlo method is used to estimate the resilience-criticality metrics, as shown in Figure 6.13. After initializing the targeted power grid, damage events are sampled randomly with equation (6.10) and

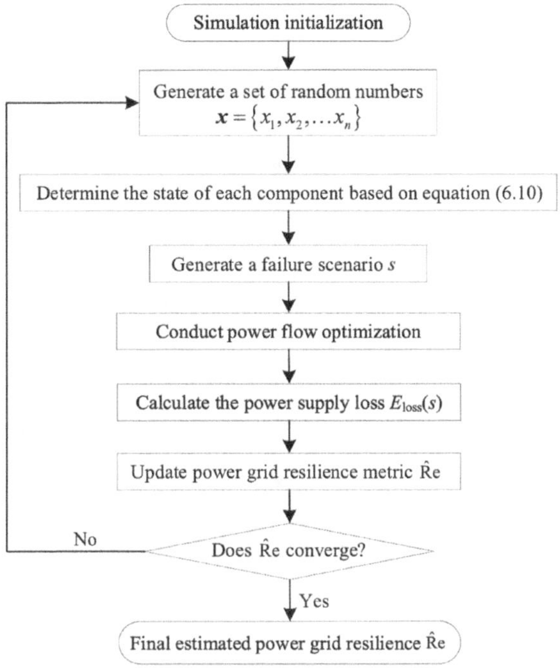


**FIGURE 6.12**
Flow chart for estimating the power grid resilience metric with Monte-Carlo method.

the power supply outage $E_{\text{loss}}(s)$ is calculated with equations (B.1)–(B.7). According to equation (6.9), $R_{\text{critical}}(i)$ can be estimated with

$$\hat{R}_{\text{critical}}(i) = \frac{1}{|\Omega_{ai}|} \sum_{s \in \Omega_{ai}} E_{\text{loss}}(s) \tag{6.12}$$

where $\hat{R}_{\text{critical}}(i)$ is the estimated value of resilience-criticality metric of component $i$, $\Omega_{ai}$ is the set of all generated samples of damage events that contain component $i$, and $|\Omega_{ai}|$ denotes the number of elements in $\Omega_{ai}$. The sample generation and the following calculation process are repeated until all components' criticality metrics converge.

### 6.3.3 Allocation of Hardening Budgets Based on Resilience-Criticality Metric

Decreasing the damage probability of critical components is an effective as well as economical approach to improving power grid's resilience. In this section, we introduce a gradient-targeted hardening strategy that divides the components

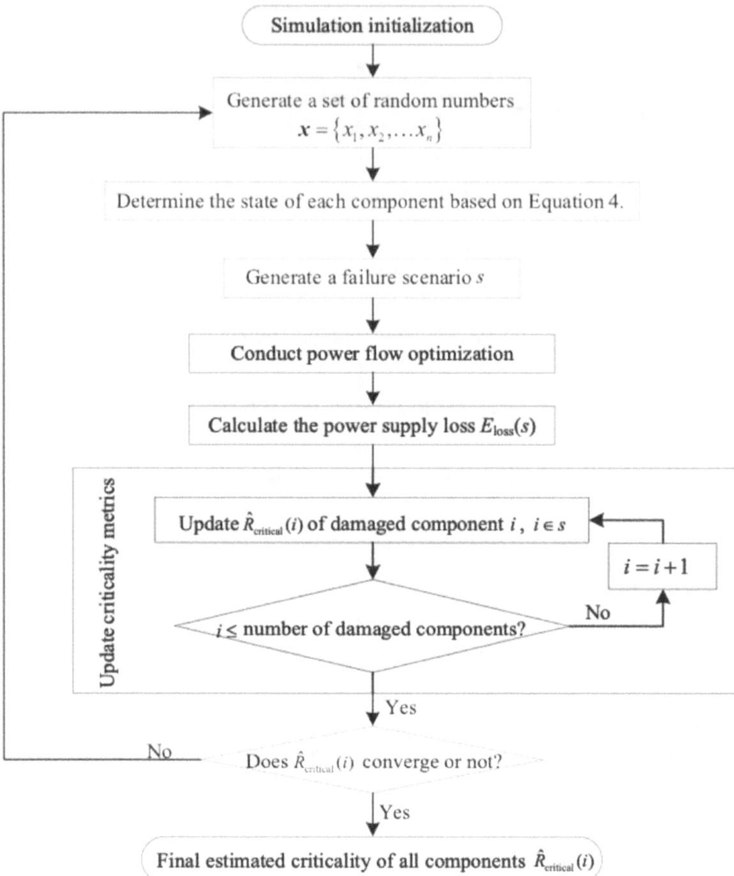


**FIGURE 6.13**
Flow chart for estimating the resilience-criticality metric by Monte-Carlo method.

into different clusters and allocates component hardening resources accordingly. This strategy aims to allocate resources more efficiently by prioritizing the hardening of critical components.

First, the k-means algorithm is used for clustering the components according to the values of their resilience-criticality. It aims to partition a given dataset into k distinct clusters, where each data point belongs to the cluster with the nearest mean (centroid). The number of clusters significantly affects the quality of the clustering result [281] as well as the effectiveness of the hardening strategy. Here, we use the elbow method to determine the optimal number of clusters. It helps to find the point of diminishing returns, where adding more clusters will not significantly improve the clustering performance. As the number of clusters increases, the degree of aggregation within each cluster increases while the sum of squared errors (SSE) decreases. SSE is a measure of clustering error and is defined as:

$$\text{SSE} = \sum_{i=1}^{k} \sum_{\text{po} \in C(i)} |po - m_i|^2 \tag{6.13}$$

where $C(i)$ represents cluster $i$ that contains the components with similar resilience-criticality metrics, po represents the resilience-criticality metric of one component in $C(i)$, $m_i$ is the centroid of $C(i)$, and SSE characterizes the effectiveness of the clustering.

---

**Algorithm 6.1: k-means for Classifying Critical Components into Clusters**

**Require: resilience-criticality metrics** $Data$; **resilience-criticality metrics number** $n$; **Clusters number** $k$; **Iteration number** $T$
**Ensure: resilience-criticality metrics of each cluster**

```
Create k initial clusters and select k individuals from Data as the initial centroids;
t=0; i=0; j=0
while t < T do
    while i < n do
        while j < k do
            calculate the distance between criticality metric of component i and centroid j
            j = j + 1;
        end while
        assign criticality metric of component i to the cluster of the nearest centroid
        i = i + 1;
    end while
    recalculating the centroid of each cluster
    t = t + 1;
end while
```

---

Given a predetermined number of clusters, denoted as $k$, and resilience-criticality metrics sorted in descending order $Data$, the specific process of clustering resilience-criticality metrics based on k-means is outlined in Algorithm 6.1. After defining the iteration number as $T$, it is crucial to select $k$ resilience-criticality metrics that are widely separated as the initial centroids. Subsequently, the distance between each data point and each centroid is calculated to determine the cluster to which the data belongs. Following the classification of all resilience-criticality metrics, the centroid of each cluster is updated. This classification process continues until the specified iteration number is reached.

Based on the classification results, we develop a gradient-targeted hardening strategy. As the resilience-criticality metrics are sorted in descending order, the components in the top cluster hold the highest importance. The strategy involves allocating additional resources to the more critical clusters, resulting in varying degrees of component hardening across different clusters. We assume that the increased survival probability of each cluster after hardening is represented by $\alpha_1, \alpha_2, \ldots, \alpha_k$. These probabilities are observed to progressively decrease. The effectiveness of this strategy in enhancing power grid's resilience will be demonstrated in Section 6.3.4.2.

### 6.3.4 Case Study

This section presents a case study to demonstrate the practical application of the Monte Carlo method for metric estimation, evaluate the effectiveness of the gradient-targeted hardening strategy in enhancing power grid resilience, and assess the accuracy of the resilience-criticality metric in identifying critical components. We begin with simulations on IEEE test cases, considering four simplified disaster modes. The frequency and occurrence of these disaster modes over a fixed time period, along with the corresponding survival probabilities of grid components, are detailed in Appendix 8.2. Subsequently, we examine the scalability and real-world applicability of the proposed method using a power grid in Texas and two historical hurricanes. In this context, comprehensive simulation analyses are conducted to validate the robustness of the approach under realistic conditions.

#### 6.3.4.1 Accuracy and Efficiency of Metric Estimation with Monte Carlo Method

To validate the effectiveness of the Monte-Carlo method in estimating these metrics, we compare the results obtained with the analytical approach and the Monte-Carlo method for the IEEE 9 Bus system. In the analytical method, the power grid's resilience and resilience-criticality metrics are calculated according to equations (6.8) and (6.9), and all potential damage events in the power grid are enumerated. On the other hand, the Monte-Carlo method utilizes equations (6.11) and (6.12) to estimate the metrics by continuously sampling damage events. Figure 6.14 shows the estimated values of the power grid's resilience and resilience-criticality metrics of a portion of components of the IEEE 9 Bus system as the number of samples increases in the Monte Carlo method. Notably, all values converge as the number of samples gets sufficiently large. The numerical results obtained from both methods are presented in Table 6.4. The error of the estimated value with the Monte-Carlo method compared to the actual value with the analytical method is within 2%. Furthermore, the criticality order of the components obtained from both methods remains consistent, further affirming the accuracy of the Monte-Carlo method.

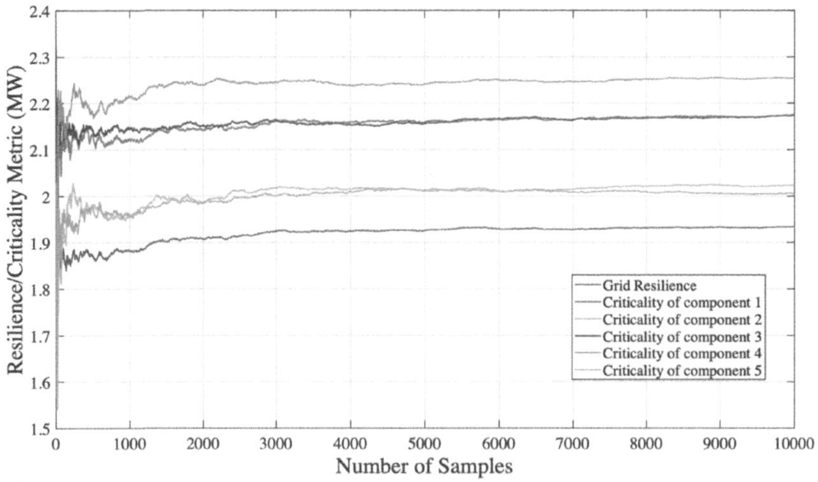


**FIGURE 6.14**
Change of estimated value of power grid's resilience metric and resilience-criticality metric with number of samples by Monte-Carlo method on IEEE 9 Bus.

It is essential to analyze the convergence rate of the Monte-Carlo method and to demonstrate the computational performance of this method. Assume that the estimated value is convergent when the estimation error is less than a given value $\epsilon$. The minimum number of samplings $N$ is derived as

$$N \geq \left(\frac{\sigma}{\epsilon}\right)^2 \tag{6.14}$$

where $\sigma$ is the standard deviation of the loss function $E(\boldsymbol{s})$ for all the Monte-Carlo samplings.

It can be seen that the expression of $N$ is strongly correlated with the component failure probability and the expression of the loss function. According to Appendix 8.2, $E(\boldsymbol{s})$ is nonlinear and extremely complex, making it difficult to directly derive the specific expression for $N$. Therefore, we obtain the minimum number of Monte-Carlo samplings $N$ for power grids of different scales through simulation, including IEEE case14, case39, case57, case118, case141 and the real-world case ACTIVSg200. Here, $\epsilon$ is set as $1 \times e^{-4}$. Figure 6.15 shows the minimum number of Monte-Carlo samplings $N$ for different grids. It can be seen that $N$ exhibits a nonlinear relationship with the power grid's scale (characterized by the number of power lines) and does not show a proportional increase as the scale of the power grid increases, indicating that the convergence rate is not primarily related to the power grid's scale but may also be related to other factors such as the power grid's topology, distribution of generators, etc. Meanwhile, $N$ of the same power grid ACTIVSg200 under different hurricane scenarios is different, which reveals

**TABLE 6.4**
Values of power grid resilience metric and resilience-criticality metric of IEEE 9 Bus obtained with analytical and Monte Carlo methods

| | | Value obtained with analytical method (MW) | Value obtained with Monte Carlo method (MW) | Error (%) |
|---|---|---|---|---|
| Power grid resilience | | 1.9374 | 1.9332 | -0.42 |
| | Component 7 | 2.5440 | 2.5537 | 0.97 |
| | Component 4 | 2.2497 | 2.2530 | 0.33 |
| | Component 8 | 2.2364 | 2.2416 | 0.52 |
| | Component 6 | 2.1990 | 2.2035 | 0.45 |
| | Component 1 | 2.1776 | 2.1749 | -0.27 |
| Component resilience-criticality | Component 3 | 2.1589 | 2.1729 | 1.40 |
| | Component 9 | 2.0661 | 2.0628 | -0.33 |
| | Component 2 | 2.0292 | 2.0237 | -0.55 |
| | Component 5 | 2.0046 | 2.0050 | 0.04 |

that the convergence rate is also related to the effects of natural disasters on the grid.

To illustrate the reduction in computational cost, we refer to Table 6.5, which compares the computational costs of the analytical method and the Monte Carlo method. The computational cost is measured by the number of calculations required for $E_{\text{loss}}$ in a failure set. For a small power grid with 20 components, the analytical method requires $2^{20}$ calculations to determine the power grid's resilience metric. In contrast, the Monte Carlo method requires only 5,201 calculations, which is significantly fewer than $2^{20}$. Although the exact expression for $N$ cannot be directly derived, simulations demonstrate that the Monte Carlo sampling method effectively reduces computational costs. Furthermore, as the size of the power grid increases, the computational cost savings achieved by the Monte Carlo method become even more pronounced. Therefore, the Monte Carlo method provides an efficient alternative to the analytical method for calculating these metrics, offering substantial reductions in computational cost.

#### 6.3.4.2 Effect of the Gradient-Targeted Hardening Strategy

Most existing hardening strategies apply a uniform level of protection to all critical components, which can lead to inefficient resource utilization. In contrast, we propose a gradient-targeted hardening strategy to enhance power grid resilience. This approach involves clustering critical components based on

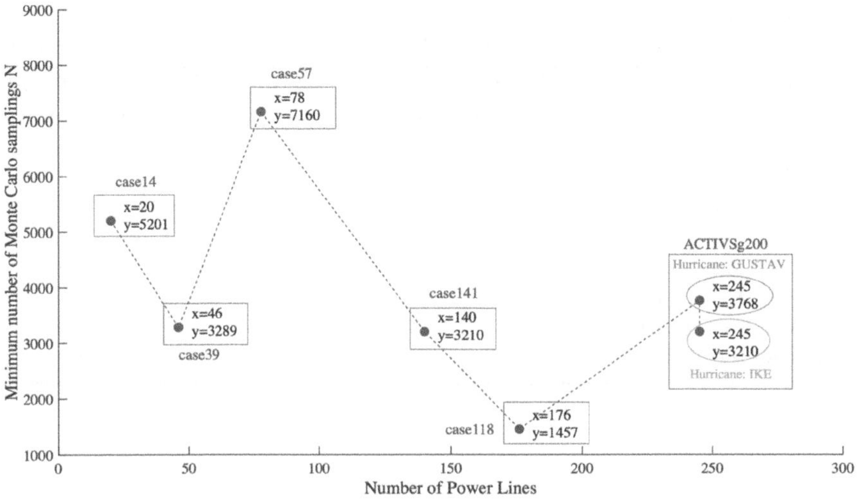


**FIGURE 6.15**
Relationship between scale of the power grid and the minimum number of Monte-Carlo samplings $N$.

their resilience-criticality values and applying differentiated hardening levels accordingly. To validate the effectiveness of this strategy, we conduct comparative analyses between non-gradient and gradient hardening approaches using IEEE 14, 39, 57, and 118 bus test cases.

In both targeted hardening scenarios, we utilize the previously defined resilience-criticality metric to prioritize power grid components and identify the top-ranked ones as critical. Hardening is modeled by increasing the survival probability of these components. In the non-gradient-targeted hardening scenario, all critical components receive an equal increase in survival probability. In contrast, the gradient-targeted hardening scenario selects the top 10 critical components and further divides them into four clusters. Components in higher-priority clusters receive a greater increase in survival probability, reflecting a larger allocation of hardening resources. To ensure a fair comparison, we assume that the resource cost for improving one unit of survival probability is uniform across all components. Additionally, the total increase in survival probability across all components is kept constant between the two strategies, thereby ensuring that both approaches utilize equivalent hardening resources. The specific survival probability increments assigned to each cluster are detailed in Appendix 8.2.

Figure 6.16 illustrates the impact of the two strategies on the enhancement of power grid's resilience. Bars with cross-hatching, right-slanted lines, and horizontal lines correspond to power grid's resilience with no component hardened, under a non-gradient-targeted hardening scenario, and under a gradient-targeted hardening scenario, respectively. From Figure 6.16, it is

**TABLE 6.5**
Number of Calculations Required by Analytical and Monte-Carlo Methods

| Cases (Hurricane) | **Number of Calculations by Monte-Carlo Method** | **Number of Calculations by Analytical Method** |
|---|---|---|
| IEEE 14 Bus (case14) | 5201 | $2^{20}$ |
| IEEE 39 Bus (case39) | 3289 | $2^{46}$ |
| IEEE 57 Bus (case57) | 7160 | $2^{78}$ |
| IEEE 141 Bus (case141) | 3210 | $2^{140}$ |
| IEEE 118 Bus (case118) | 1457 | $2^{176}$ |
| ACTIVSg200 (Gustav) | 3768 | $2^{245}$ |
| ACTIVSg200 (Ike) | 3210 | $2^{245}$ |

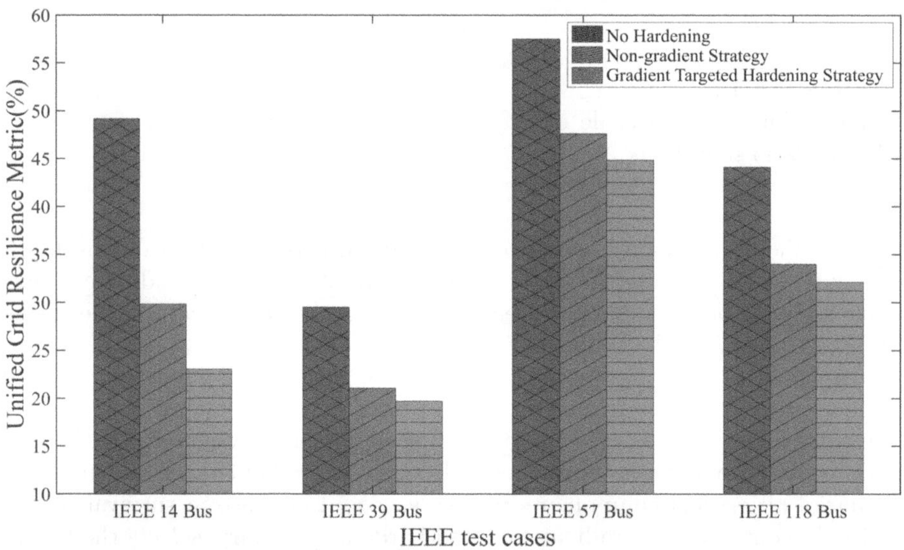


**FIGURE 6.16**
Impact of non-gradient and gradient-targeted hardening strategies on the enhancement of power grid resilience.

evident that the red bars demonstrate the smallest values of the normalized power grid's resilience compared to other colored bars across all the listed IEEE test power grid cases. This shows that the gradient-targeted hardening strategy is more effective in enhancing power grid's resilience compared to the non-gradient-targeted hardening strategy.

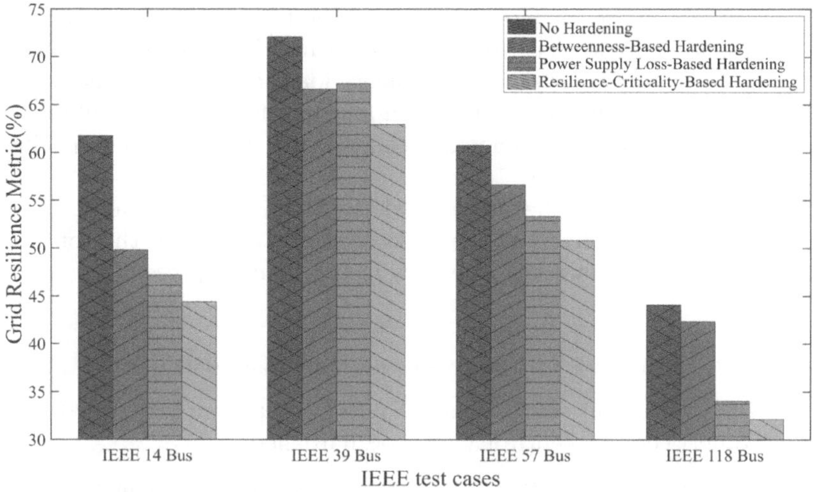


**FIGURE 6.17**
Impact of hardening different sets of components identified with different methods on the enhancement of power grid resilience.

#### 6.3.4.3 Identifying Critical Components

Numerous studies have focused on identifying critical components within power grids, aiming to effectively pinpoint those that exert a substantial influence on overall grid performance and resilience. In this section, we conduct simulations on the IEEE 14, 39, 57, and 118 Bus systems to evaluate the effectiveness of different critical component identification methods. Our objective is to determine whether the proposed resilience-criticality metric can accurately identify components that significantly influence the overall resilience of power grids. To this end, we compare three distinct approaches for identifying critical components: (1) resilience-criticality metric, (2) greedy search algorithm, and (3) betweenness centrality. Each method is applied to the test systems, and the resulting sets of critical components are analyzed to assess their impact on grid resilience.

- Resilience-criticality-based method: Sort the components in descending order according to their resilience-criticality metric ($R_{\text{critical}}$), placing those with higher values at the top of the list.

- Power supply loss-based method: Sort the components in descending order based on the power supply loss incurred by the grid due to the single removal of each component. Components whose removal results in greater power supply loss are placed at the top of the list.

- Betweenness centrality-based method: Sort the components in descending order based on their betweenness centrality, defined as the total number of shortest paths in the network that pass through each component.

To evaluate the accuracy of the three methods in identifying critical components, we assess power grid resilience following the hardening of three distinct sets of the top 10 critical components identified by each method. The gradient-targeted hardening strategy is applied uniformly across all scenarios, with survival probability increments set according to the values specified in Section 6.3.4.2.

Figure 6.17 presents the normalized resilience of the power grid under different conditions, defined as the ratio of grid resilience to the total power supply of the original system. The bars in the figure represent the following scenarios:

- Bars with cross-hatching: No components hardened
- Bars with right-slanted lines: Top ten components identified using the betweenness centrality method hardened
- Bars with horizontal lines: Top ten components identified using the power supply loss method hardened
- Bars with left-slanted lines: Top ten components identified using the resilience-criticality method hardened

As shown in Figure 6.17, the bars with left-slanted lines represent the lowest values of the normalized power grid resilience compared to the other bars across all the listed IEEE test power grid cases. This indicates that hardening critical components identified using the resilience-criticality metric is the most effective approach to enhancing power grid resilience, compared to components selected based on other metrics. Consequently, the described metric effectively quantifies the criticality of a component in terms of its impact on power grid resilience. Additionally, we present the top ten critical components identified by the three methods for the IEEE 118 Bus system. In Figure 6.18, the lines with separate arrows, sinewave, and dots correspond to the components identified by the resilience-criticality-based method, the power supply-loss-based method, and the betweenness centrality-based method, respectively. The resilience-criticality-based method identifies critical components that differ significantly from those highlighted by other methods. This highlights the need for power companies to exercise caution when selecting components to harden in order to enhance power grid resilience.

#### 6.3.4.4 Applications in Real-World Scenarios

To further demonstrate the applicability of the proposed method, we conducted simulation experiments under more realistic conditions. Given the challenges associated with accessing real-world power grid data, we employed the publicly available synthetic test case ACTIVSg200 [32]. This dataset provides a representative model of a large-scale power grid, enabling robust evaluation of our approach. Detailed information about the ACTIVSg200 system is provided in Appendix 8.2.

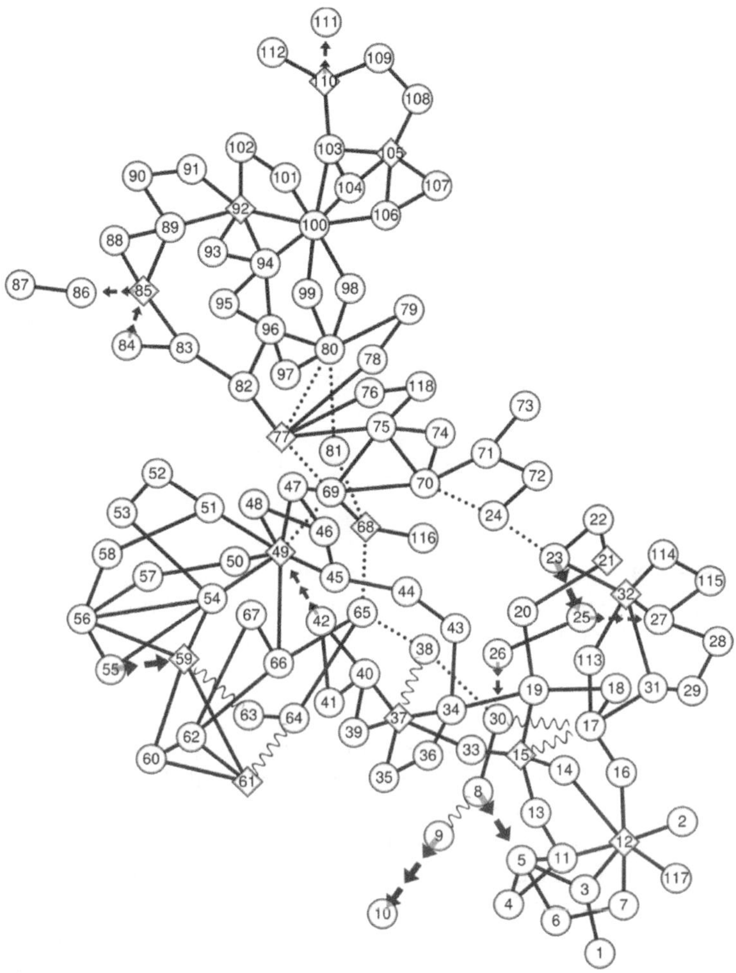

**FIGURE 6.18**
Top ten critical components on IEEE 118 Bus with different identification methods.

Different types of natural disasters exhibit distinct modeling characteristics and exert varying impacts on power grid components. To illustrate this, we focus on hurricanes as a representative example. Historical hurricane data from the past 20 years affecting the target power grid have been collected, including notable events such as Hurricanes Gustav and Ike, which made landfall in Texas in August and September 2008, respectively [26, 132]. Detailed hurricane data are provided in Appendix 8.2. Figure 6.19 displays the topology of the power grid alongside the tracks of the selected hurricanes. Using hurricane tracking data, which includes the geographic coordinates (latitude and

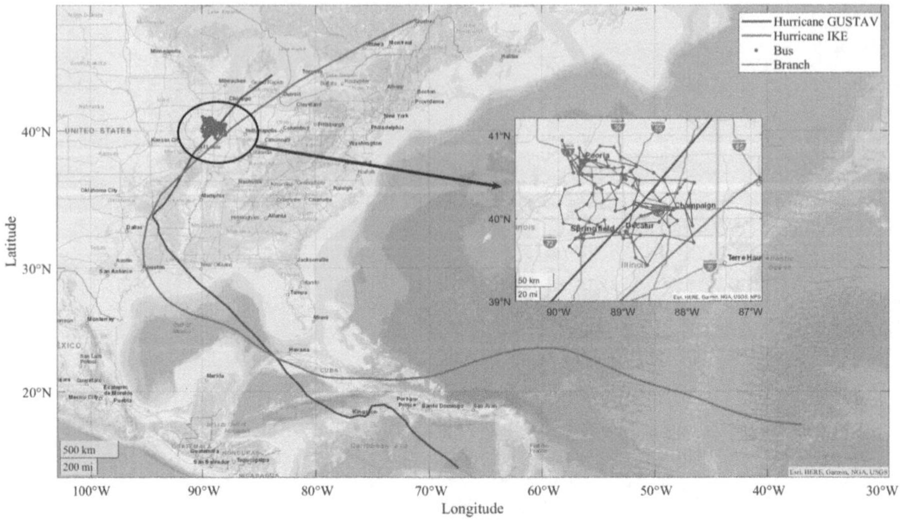


**FIGURE 6.19**
Two historical hurricanes with different pathways and intensities passed through ACTIVSg200 power grid in the last 20 years.

longitude) and wind speed at each storm center, we apply the Holland surface wind model to compute the wind field time series within a 500-km radius of each storm center, as illustrated in Figure 6.20. Based on the computed wind fields, the failure probability of a transmission line due to hurricane exposure is estimated following the methodology outlined in [246]. The detailed derivation process is provided in Appendix 8.2.

After estimating the damage probabilities of grid components, we applied the proposed method to identify and harden the top 24 critical components affected by each hurricane. The simulation results are presented in Figure 6.21. As shown, the method continues to yield strong performance under realistic conditions, demonstrating its robustness and applicability in real-world scenarios.

## 6.4 Summary

This chapter considers long-term strategies for enhancing grid's robustness in the context of system planning. The effects of network topologies and critical component enhancement are highlighted.

First, the influence of typical network properties on grid robustness is studied. A distance metric that measures the decentralization of generators

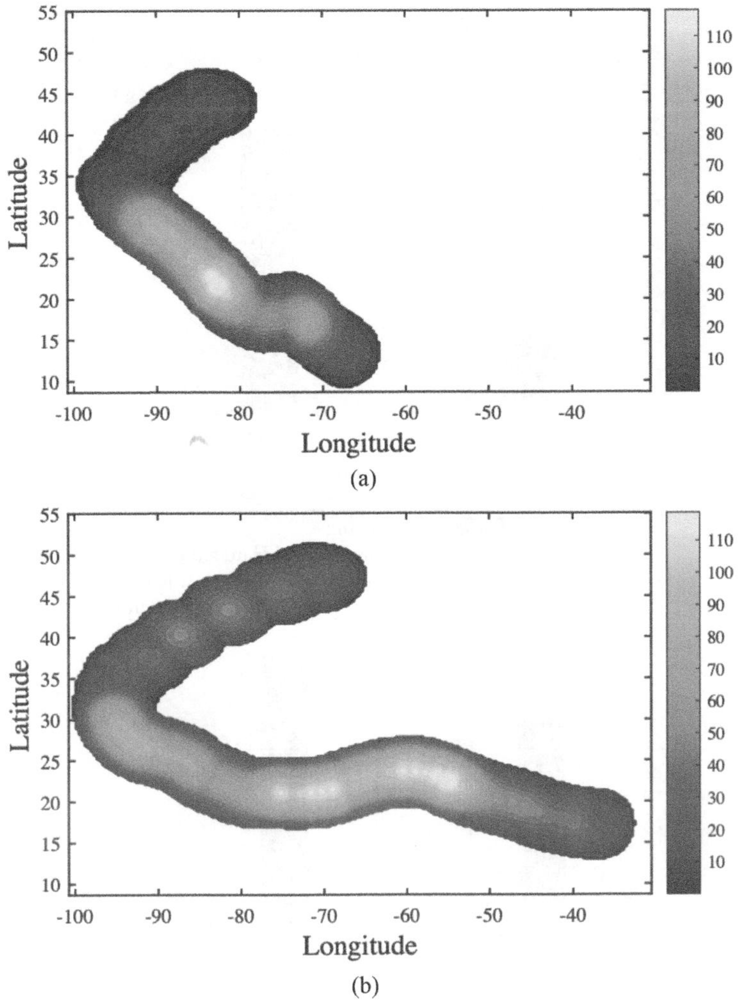


**FIGURE 6.20**
Peak wind speed estimates per grid cell simulated by the parametric wind model throughout the hurricane event.

in a network has been defined, which turns out to be an effective indicator of robustness for a given network structure. It is found that the robustness of a power system can be significantly compromised by short ASPL connectivity and consumers' accessibility to generators. The findings provide important insights for designing appropriate network topologies and for configuring generator allocation.

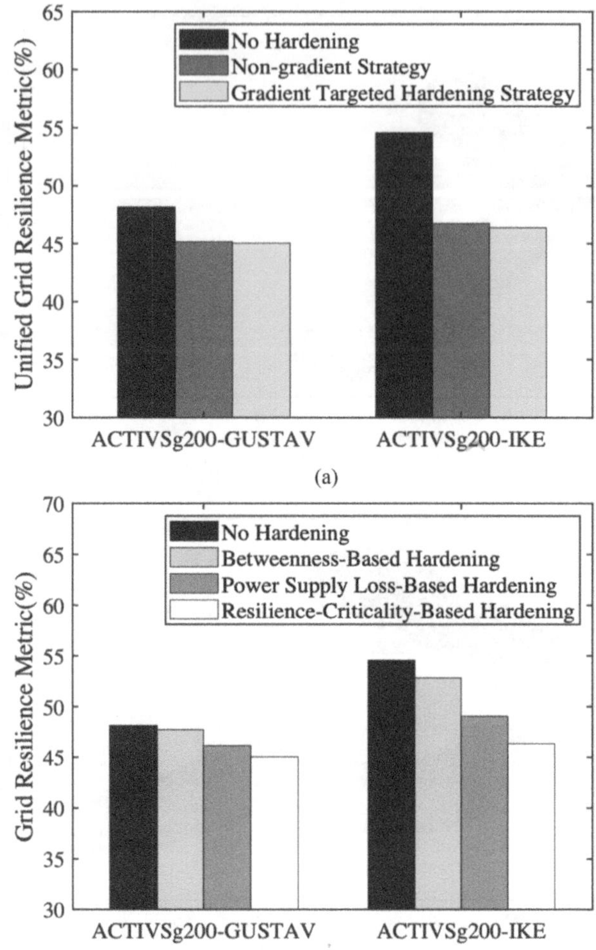


**FIGURE 6.21**
Comparison of different strategies on enhancing the resilience of ACTIVSg200. (a) Impact of non-gradient and gradient-targeted hardening strategies on the enhancement of power grid resilience. (b) Impact of hardening different sets of components identified with different methods on the enhancement of power grid resilience.

A gradient-hardening resource allocation strategy is then described to enhance power grid resilience that fuses the data of power networks and natural disasters. We have discussed a power grid resilience metric expressed in a general form to assess the grid's resilience under various types of natural disasters. Building upon this metric, we have further introduced a resilience-criticality metric that quantifies the criticality of each component. To reduce computational cost while ensuring high accuracy, we have employed the Monte-Carlo method to calculate this metric. Additionally, we have presented a gradient hardening resource allocation strategy to allocate available resources for hardening the identified critical components. We have verified the applicability of the Monte-Carlo method and have shown that this kind of strategy can effectively enhance power grid resilience more efficiently compared to existing methods.

# 7

# *Robustness Enhancement via System Control and Protection*

Cascading failure power systems can be mitigated by applying timely control and protection strategies during the failure propagation process. This chapter introduces robustness enhancement strategies in the context of system control and protection. First, an integrated cyber-physical modeling framework that simulates cascading failures is introduced. Then, the control and protection functions and their descriptions in the model are introduced–including adaptive relaying, system integrity protection schemes, and wide-area monitoring systems (WAMS)–capturing their critical role in both failure propagation and arrestment. A cascading failure study considering the wide-area protection function of remedial schemes is presented. Specifically, a novel deep reinforcement learning (DRL)-based mitigation strategy is then developed to dynamically override or augment conventional protection logic during unfolding cascades. By learning optimal actions, the DRL agent minimizes blackout size while respecting operational constraints.

## 7.1 Background

Power grids are vulnerable to cascading failure, where a small-scale local failure event triggers a chain reaction and eventually leads to widespread power outage. While system planning strategies are long-term and applied before the occurrence of cascading failure events, system control and protection strategies have a short timescale and are performed during the rapid propagation of cascading failure.

Decentralized control and protection facilities, including primary frequency responses and replay protection, act on the physical counterparts based on the local signals monitored and the commands issued by the EMS. Centralized control and protection algorithms, like the automatic generation control (AGC), the remedial action scheme (RAS), and the supervisory control and data acquisition (SCADA), analyze data and generate command signals that are transmitted through the communication network back to the respective


DOI: 10.1201/9781003622130-7

execution terminals. The various control and protection functions can effectively improve the grid's robustness and should be included in the cascading failure study.

The RAS [196] is one widely deployed method that aims to stop the abnormal condition from spreading and causing further failure by taking appropriate measures. The main components of RAS include detection devices, a decision-making algorithm, and control devices. The sensors continuously monitor the operating states of the power grid and send the necessary information to the decision-making algorithm. If abnormal operating conditions are detected, the algorithm generates remedial actions like shedding loads, tripping generators, disconnecting tie lines to create islands, etc. The command system signals are then sent to and implemented by the control devices.

One key to the success of RAS is generating appropriate remedial actions in accordance with the operating states via an effective decision-making algorithm [310]. For example, the homogeneous load shedding that reduces load at all buses in the system by an equal percentage in the cascading failure process has been investigated and implemented. Along this line, Taylor et al. [249] proposed to shed fixed amounts of loads at evenly spaced frequency intervals. In the study conducted by Das et al. [68], a comprehensive framework based on Markov decision processes was developed. This framework enables the pre-computation of the optimal percentage of loads to be shed for every possible state of the grid during a cascading failure process. While homogeneous actions are straightforward to implement, they can lead to excessive shedding of loads due to inflexible shedding of the same amounts at all buses simultaneously. To address this limitation, researchers resorted to generating heterogeneous remedial actions by introducing additional decision-making variables. For instance, each bus can have a different percentage of loads to be shed at a given time. In the context of cascading failure, the optimal power dispatch method was commonly employed to adjust the power of loads and generators, aiming to minimize changes subject to system constraints [69]. To accurately determine the location and amount of load shedding, Pahwa et al. [200] proposed a novel tree heuristic that overcomes the drawbacks of optimization fairness and scalability. This linear and simple-to-implement approach offers an effective solution. In extreme events, partitioning the power grid into smaller, self-sufficient islands by disconnecting multiple transmission lines is adopted to prevent the spread of cascading failure. Biswas et al. [33] presented two efficient heuristic methods for determining the sequence of line-switching operations needed to proactively split a power grid into smaller islands.

Timeliness is another key to the success of RAS. Fast decision-making and timely execution are essential factors for achieving successful RAS. In practice, RAS systems often rely on pre-designed actions to meet the timeliness requirement. Power engineers need to simulate target power systems and design appropriate remedial actions offline based on the expected abnormal conditions. The mappings between pre-designed actions and conditions are stored in a table, which is consulted in the real-time implementation of the

RAS. While this mechanism enables quick generation of remedial actions, it has limited effectiveness when the operating conditions vary significantly.

As power systems continue to expand, it becomes increasingly more challenging to implement effective RAS that meets the above scalability and timeliness requirements based on conventional decision-making methods. The main reason is that the increase in system scale would greatly diversify the propagation path of potential cascading failures (once triggered) [29], leading to high-dimensional problems that are intractable or non-generalizable to conventional methods. *Advanced decision-making techniques are therefore required to efficiently analyze the grid states and generate timely remedial actions.*

## 7.2 Power Grid Modeling from a Cyber-Physical Perspective

The practical power grid can be described as a cyber-physical system, as shown in Figure 7.1. Taking this perspective, we introduce an integrated modeling framework for studying cascading failure in cyber-coupled power grids.

As shown in Figure 7.1, generators, loads, and distribution stations are interconnected by transformers and transmission lines, forming a physical layer. In the physical network, a bus along with the connected generator and load is represented by a node, and the transmission lines or transformers between two buses are represented by links. Physical components interact in accordance with Kirchhoff's laws and the components' own constitutive relations. The physical network can be described by a set of algebraic and differential equations.

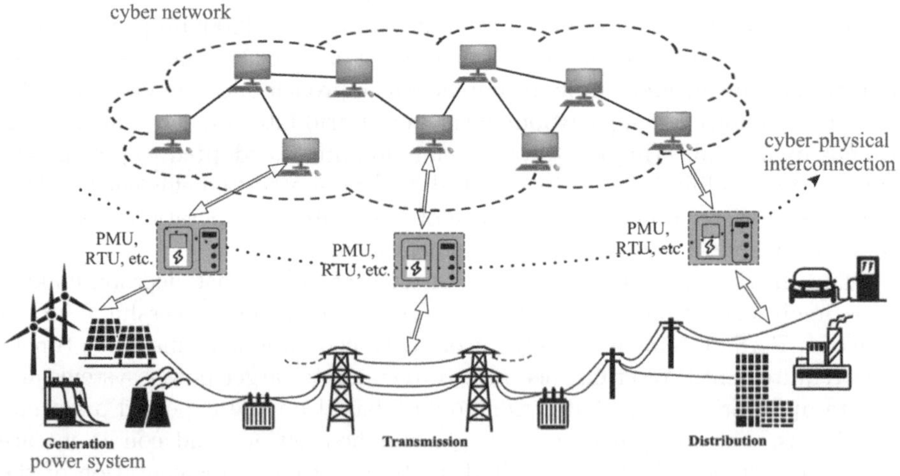


**FIGURE 7.1**
Evolving power grid tightly coupled with cyber network.

1. Each node in the physical network should satisfy the following power balance equations:

$$P_i = V_i \Sigma_{j \in N} V_j (G_{ij} \cos \theta_{ij} + B_{ij} \sin \theta_{ij}) \quad (7.1)$$

$$Q_i = V_i \Sigma_{j \in N} V_j (G_{ij} \sin \theta_{ij} - B_{ij} \cos \theta_{ij}) \quad (7.2)$$

   where $P_i$, $Q_i$, and $V_i$ are the active power, reactive power, and voltage amplitude of node $i$; $\theta_{ij}$ is the phase angle difference between nodes $i$ and $j$; $G_{ij}$ and $B_{ij}$ are the real and imaginary parts of the admittance of the link connecting nodes $i$ and $j$.

2. In addition to the above algebraic equations, the nodes have their own dynamical descriptions:

$$\frac{d\boldsymbol{x}_i}{dt} = f_i(\boldsymbol{x}_i, \boldsymbol{u}_i) \quad (7.3)$$

   where $\boldsymbol{x}_i$ represents the state vector of node $i$, $\boldsymbol{u}_i$ is the input vector, and $f_i$ is the state function of node $i$.

On top of the physical network, a cyber infrastructure is constructed to provide centralized control and protection for the physical network to enhance the performance of the whole power grid. As shown in Figure 7.1, the cyber layer is a network of measurement devices, communication lines, relays, computers, etc. The remote terminal units (RTUs) that contain measurement devices and command execution equipment are interfaces between the cyber and physical networks. Measurement devices, including various sensors, collect the real-time operating information of the physical subsystems and components. A communication network gathers and transmits the measurement data to the energy management system (EMS).

## 7.3 Control and Protection Functions

The physical layer alone is insufficient to secure the reliable and efficient operation of the power grid. In addition to obeying physical laws, these physical components are delicately controlled. Thus, a suite of control and protection functions is applied to the physical network. As shown in Figure 7.2, we divide these functions into two categories, namely, centralized and decentralized facilities.

### 7.3.1 Centralized Control and Protection Functions

As shown in Figure 7.1, a suite of system-level algorithms is executed in the EMS center to analyze data and generate command signals that are transmitted through the communication network back to the respective execution

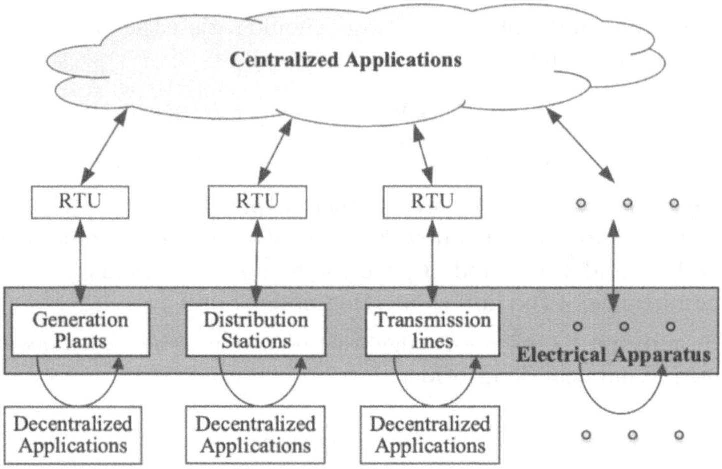


**FIGURE 7.2**
Overview of control and protection functions in cyber-coupled power grids.

terminals. For example, the AGC system adjusts the power outputs of a set of tunable generators to ensure frequency stability. The RAS can detect abnormal events in the system and take corrective actions, like cutting off a remote load/generation, to mitigate the failure cascade. It is worth noting that the newly emerging as well as promising AI-based applications also rely on the cyber network and operate in a centralized manner. Other typical applications include SCADA, state estimation (SE), and security-constrained optimal power flow (SCOPF).

In general, we use the following formula to represent centralized and protection functions:

$$[\boldsymbol{u}^c, \boldsymbol{a}^c] = \Gamma(X^\star, A^\star) \tag{7.4}$$

where $X^\star$ and $A^\star$ denote the electrical states and connection state of the physical network that are observed by the cyber network; $\Gamma$ represents the system-level algorithms executed in the EMS; $\boldsymbol{u}^c$ and $\boldsymbol{a}^c$ represent control and protection commands issued by $\Gamma$. $\boldsymbol{u}^c$ is continuous and provides references for distributed controllers. $\boldsymbol{a}^c$ is discrete and indicates the connection state of the electrical components.

### 7.3.2 Decentralized Control and Protection Functions

Decentralized control and protection facilities act on the physical counterparts based on the local signals monitored and the commands issued by the EMS. Primary frequency response is one typical example of decentralized facilities that control the power plants' output power according to the system frequency and the reference value. Normally, the power grid operates at a standard frequency and the power is balanced in the system, i.e., the sum of the power generated equals the sum of the power loadings. Component failures and disturbances can cause power mismatches in the system. The frequency deviation from the standard frequency is an important indicator of the power mismatches between the generators and the loads. This frequency-power relationship can guide the synchronous generators to collectively adjust their output power in order to maintain power balance in the system. When the frequency decreases, the generator's prime mover increases its mechanical power supplied to the rotor, and vice versa. By this simple response mechanism, the disturbances in the system are mitigated.

Protection relays are another typical example of decentralized protection facilities. Switching actions are executed when the monitored electrical signals exceed the allowable limits or when a "tripping" command from EMS is received. The aim is to prevent the corresponding components from being physically destroyed and to remove faults from the grid. Automatic under-frequency load shedding (AUFLS) is applied for under-frequency conditions with insufficient power generation. If the system frequency drops below a certain level, a predetermined set of loads is automatically disconnected to prevent frequency collapse. Usually, an AUFLS scheme contains several stages, with a series of predetermined under-frequency levels and corresponding loads. Other decentralized control facilities include over-frequency generator tripping, grid islanding, etc.

In general, we use the following formula to represent the decentralized control and protection functions:

$$[u_i, a_i] = \Phi(x_i, u_i^c, a_i^c) \tag{7.5}$$

where $x_i$ is the monitored local signal of component $i$; $u_i^c$ and $a_i^c$ are the commands delivered to component $i$ from the EMS; $\Phi$ represents local control and protection algorithms; $u_i$ and $a_i$ are control and protection actions exerted on component $i$.

## 7.4 Cascading Failure Study with Wide-Area Protection Function

In this section, we present the study of cascading failure in power systems by considering the wide-area protection function of RASs, the distributed control of primary frequency response, and the distributed protection function of protective relays.

The cascading failure model is comprehensively presented with a flow chart, where the operating characteristics of the physical network and the wide-area functions provided by the cyber network are embedded. The simulation flow chart is shown in Figure 7.3 and is briefly explained as follows.

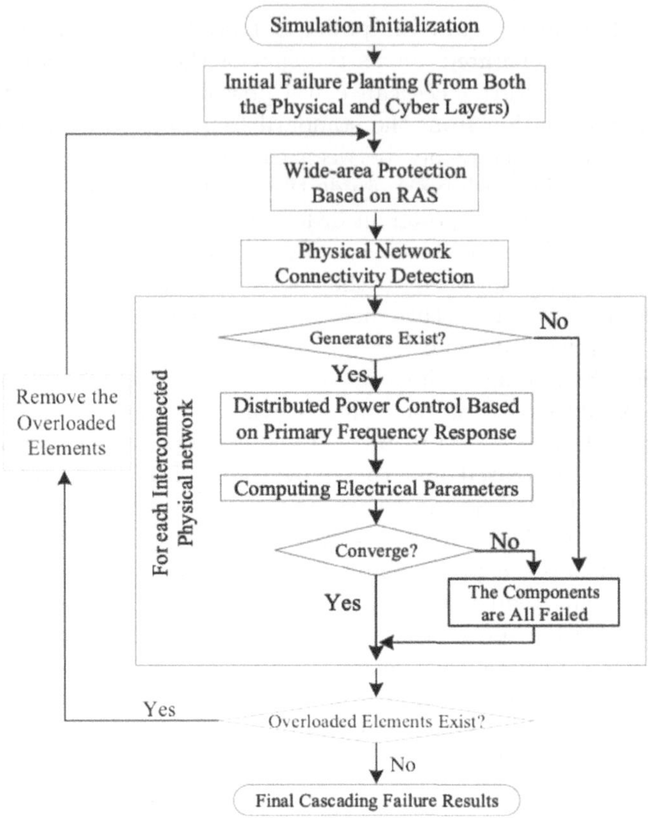


**FIGURE 7.3**
Simulation flow chart for failure propagation processes in a power system by considering the wide-area protection function of remedial action schemes, the distributed control of primary frequency response, and the distributed protection function of protective relays.

a Initialization: The power system consisting of two coupled networks and a map table that contains the coupling relations is prepared for simulation. The physical layer defines the network topology and parameters of the electrical components. The map table that contains a set of bidirectional links, $i \leftrightarrow k$, defines the correspondence between the two networks.

b Initial failure planting: Initial failures are planted in the system. Let $\zeta$ denote the set of initially failed components in the cyber-coupled power system. Unlike previous cascading studies, $\zeta$ can include initial faults from both the physical and cyber layers.

c Simulation iterations: Before performing the power flow calculation, the wide-area protection function is first implemented to take remedial actions of proactively disconnecting some connected physical components. The failure of physical components may destroy the integrity of the physical network. Thus, the connected clusters in the physical network are detected. For each cluster, if there are no generators, all components are deprived of power and fail. If there exists at least one generator, the power outputs of the generators are adjusted in a distributed manner based on the primary frequency response. The power flow distribution on the remaining network is obtained with the DC power flow model. Local protection relays remove the physical components whose power loadings exceed their corresponding operating limits. If there are no overloaded components, the simulation is terminated, and the final power outage size is obtained. Otherwise, re-iterate the process.

### 7.4.1 Centralized Protection Function of Remedial Action Schemes

In a power system, proactively disconnecting some operating components when a failure event occurs can ensure the safety of the physical network. RASs are the most typical wide-area protection function for mitigating cascading failure in power systems. RAS generates command signals of remedial actions once an abnormal condition is detected, for example, shedding loads/generators, tripping/removing electric transmission lines, and so on. In practice, a table that maps the specific abnormal conditions to the remedial action commands is made offline. These actions, to be taken after specific faults are detected, are designed by electrical engineers through extensive simulation and analysis of the power network. For simplicity, we use a simple and primitive decision-making algorithm to generate the mappings to embed the RAS function in simulating cascading failure. The decision-making algorithm used in our model can be described as follows.

For each round of iteration, the physical network's state $S(t)$ is collected as the input of the decision-making algorithm. It consists of the real-time connection status, the real-time power loadings, and the capacities of physical components. $\Gamma$ denotes the set of operating components in the physical network. The decision-making algorithm generates a command signal that proactively disconnects at most one component corresponding to $S(t)$. All possible options are enumerated, including separate disconnection of each component in $\Gamma$ and a null action. Following each option, denoted by $\gamma$, the distributed power control and the power flow calculation are performed on the remaining physical network. Then, the total number of overloaded components $\sum_{o\in\Gamma\setminus\gamma}\mu_o$ and the overall overloading rate $\sum_{o\in\Gamma\setminus\gamma}\lambda_o$ of the grid are obtained. $\mu_o$ is obtained as

$$\mu_o(t) = \begin{cases} 1, & \text{if } P_o(t+\Delta t) > P_o^{\max} \\ 0, & \text{otherwise} \end{cases} \tag{7.6}$$

where $o$ represents one component in $\Gamma \setminus \gamma$, including both links and nodes; $P_o^{\max}$ is the capacity of $o$; and $P_o(t+\Delta t)$ is the updated power loading of $o$ after option $\gamma$ is taken and the distributed power control and the power flow calculation are performed. $\lambda_o$ is obtained as

$$\lambda_o = \begin{cases} \dfrac{P_o(t+\Delta t) - P_o^{\max}}{P_o^{\max}}, & \text{if } P_o(t+\Delta t) > P_o^{\max} \\ 0, & \text{otherwise} \end{cases} \tag{7.7}$$

The option that results in the minimum number of overloaded components is selected. If multiple options yield the same minimum number, the one with the lowest overall overloading rate is chosen from among them.

### 7.4.2 Distributed Power Control Based on Primary Frequency Response

Under normal operating conditions, the power system maintains a standard frequency and a balanced power state, where total generation equals total consumption. However, the failure or removal of network components can disrupt this balance, leading to power imbalances within connected clusters. In AC power systems, deviations from the nominal frequency serve as a key indicator of such imbalances. According to the swing equation governing synchronous generators, the system frequency decreases when power demand exceeds generation, and increases when generation surpasses demand. To mitigate these imbalances, a distributed control mechanism known as primary frequency response is employed. This control strategy adjusts the mechanical power input to the generator's rotor: when frequency drops, the prime mover increases its output, and when frequency rises, the mechanical input is reduced. This dynamic response helps restore power balance and stabilize system frequency in the remaining operational network.

Here, we use the steady-state result of the primary frequency response and omit the dynamical process. The power output of a generator after the primary frequency response is obtained as

$$P_{g_i}(t+\Delta t) = (P_{N_d}(t) - P_{N_g}(t))\frac{P_{g_i}^c(t)}{\sum_{g_i \in N_g} P_{g_i}^c(t)} + P_{g_i}(t) \tag{7.8}$$

where $N_g$ denotes the set of generator nodes; $N_d$ denotes the set of load nodes in the physical network; $P_{g_i}(t)$ is the power output of generator $g_i$ before the distributed power control; $P_{N_d}(t)$ and $P_{N_g}(t)$ are the total power generated and demanded at time $t$, and $P_{g_i}^c(t)$ is the tunable power of $g_i$, i.e.,

$$P_{g_i}^c(t) = P_{g_i}^{\max} - P_{g_i}(t) \tag{7.9}$$

### 7.4.3 Local Protection by Protective Relays

Local protective relays automatically disconnect physical components whose power loading exceeds their designated safety thresholds. These switching actions are triggered when monitored electrical signals surpass allowable operational limits. The primary objective is to prevent physical damage to the

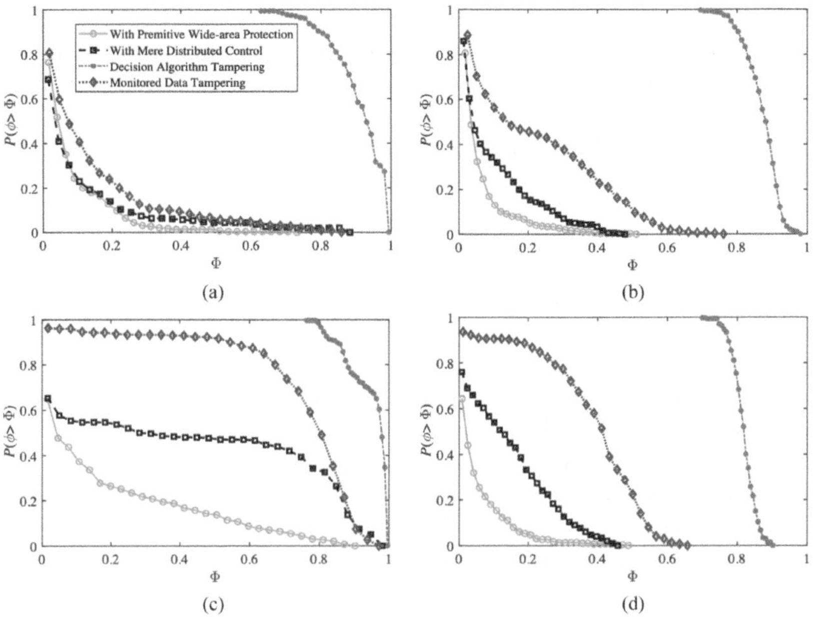


**FIGURE 7.4**
Cumulative power outage distributions of four power test cases with initial failure of three physical components, without wide-area protection, with primitive wide-area protection, with tampering the decision-making algorithm of the wide-area protection, and with tampering the monitored data. (a) IEEE 57 Bus; (b) IEEE 118 Bus; (c) IEEE 145 Bus; and (d) UIUC 300 Bus.

affected components and to isolate faults from the rest of the power grid, thereby maintaining system stability and safety.

### 7.4.4 Case Study

With the methods described above, the robustness of a power system can be quantitatively evaluated. In this section, we explore various factors that influence the robustness of cyber-coupled power systems. Through a series of case studies and simulations, we aim to analyze how different system characteristics and external conditions affect the system's ability to withstand and recover from disruptions.

#### 7.4.4.1 Effects of Cyber Coupling

We first demonstrate the robustness assessment results of power systems by considering the wide-area protection function, in contrast with the conclusions derived from the models that merely focus on the physical network. Figure 7.4 presents the cumulative power outage size distributions for the IEEE 57

Bus, IEEE 118 Bus, IEEE 145 Bus, and UIUC 300 Bus systems under various conditions. Each distribution curve is derived from 300 simulation samples, where each sample begins with the random failure of three physical components. The robustness metric used here is the cumulative power outage size distribution, denoted by $P(\phi > \Phi)$, which is the probability that the size of a power outage is higher than $\Phi$ and can be computed as

$$P(\phi > \Phi) = \frac{n(\phi > \Phi)}{|\Omega_{\mathrm{MC}}|} \tag{7.10}$$

where $n(\phi > \Phi)$ is the number of samples whose power outage size is higher than $\Phi$. Obviously, a power system with a lower distribution of $P(\phi > \Phi)$ is more robust.

The resulting power outage sizes are computed using the cascading failure model described in Figure 7.3. The curve with circle markers represent the cumulative outage distributions obtained by applying the primitive wide-area protection scheme introduced in Section 7.4.1, while the black curves correspond to simulations without any wide-area protection. Across all four test cases, the curve with circle markers consistently lie below the black curves, indicating that the integration of a cyber network and the implementation of wide-area protection based on RASs significantly enhance the robustness of the power systems. This significant difference underscores the importance of evaluating power system robustness from the perspective of interdependent networks, where the interactions between physical and cyber infrastructures play a critical role in system resilience.

#### 7.4.4.2 Effects of Enhanced Decision-Making of Wide-Area Protection Functions

While the curve with circle markers in Figure 7.4 demonstrate the benefits of implementing the primitive wide-area protection scheme; they also reveal a persistent risk of severe power outages in cyber-coupled power systems. A key limitation of the primitive decision-making algorithm described in Section 7.4.1 is its restricted action space, which allows for the proactive disconnection of at most one component upon detection of abnormal conditions. To address this limitation and further enhance system robustness, we present here an enhanced decision-making algorithm for wide-area protection. This improved approach expands the action space, enabling more flexible and effective responses to emerging threats. We evaluate the performance of this enhanced algorithm through robustness assessments of the four power test cases.

In the enhanced decision-making algorithm, the action space is enlarged to allow the proactive disconnection of at most two physical components. The command signal is generated with the following steps: Given state $S(t)$, the primitive decision-making algorithm in Section 7.4.1 is first performed to generate an initial command of the remedial action. If there exist no overloaded components with the initial command executed, the decision-making algorithm terminates. Otherwise, a second component is selected to be proactively disconnected along with the first one, constituting the final command. With the first selected component disconnected, the individual removal of all the remaining physical components is enumerated. The second component corresponds to the minimum number of overloaded components or the least overloading rate if several options have the same minimum number of overloaded components.

The power outage risk for each test case is evaluated using 300 cascading failure simulations, each triggered by the initial failure of three randomly selected physical components. The results for the four power systems are presented in Figure 7.5. The colored bars represent the average power outage sizes under three scenarios: without wide-area protection, with the primitive wide-area protection scheme, and with the enhanced wide-area protection scheme. As shown in Figure 7.5, systems without wide-area protection exhibit significantly higher outage risks, consistent with the findings discussed in Section 7.4.4.1. Moreover, the enhanced decision-making algorithm for wide-area protection demonstrates a clear advantage over the primitive scheme, effectively reducing outage risk across all four cyber-coupled power systems. It is important to note, however, that the enhanced algorithm incurs substantially higher computational costs due to its expanded action space, which allows for a broader range of control decisions. This highlights the need for developing advanced decision-making algorithms that can efficiently generate control signals while maintaining high levels of robustness in large-scale power systems.

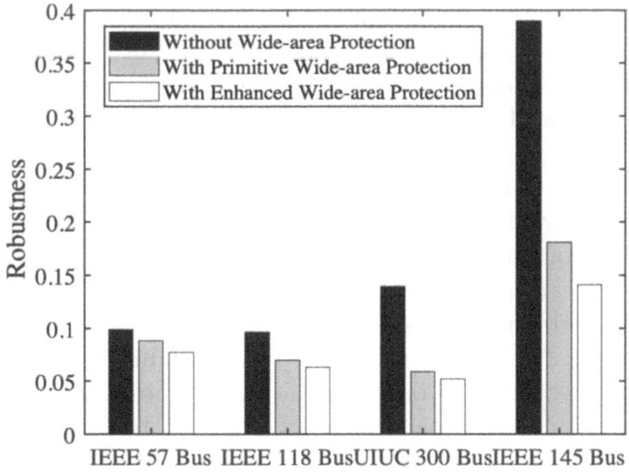


**FIGURE 7.5**
Power outage risks of four power test cases with initial failure of three physical components without wide-area protection, with primitive wide-area protection, and with enhanced wide-area protection.

## 7.5 Mitigating Cascading Failure by Deep Reinforcement Learning

In recent years, the rise of deep learning (DL) has enabled innovative approaches for improving the reliability of power systems, particularly through the analysis of large-scale and unstructured data. For example, Li et al. [154] developed a hybrid model combining random forest classification and regression techniques to assess the vulnerability of integrated power-gas systems. Wu et al. [279] introduced an influence model for predicting and screening cascading failures in large-scale DC and AC power networks. Liu et al. [169] proposed a graph convolutional network–based method capable of identifying critical components involved in cascading failures in real time. While these DL-based solutions have demonstrated strong predictive capabilities, they primarily function as "predictors" and lack integration with decision-making mechanisms such as RASs. Bridging this gap remains a key challenge for deploying intelligent, end-to-end resilience solutions in modern power systems.

One promising approach for implementing RASs is DRL, which combines DL with reinforcement learning to enable rapid, online decision-making in complex environments through trial-and-error interactions. DRL has demonstrated considerable potential across various domains. For instance, DeepMind's development of AlphaGo, based on DRL, achieved groundbreaking success by defeating human champions in the game of Go [238]. Costa et al. [57] applied DRL to solve path-planning problems in complex networks, achieving performance comparable to traditional heuristic methods. Fan et al. [92] trained DRL agents to identify critical buses in a network, whose sequential removal could significantly compromise system connectivity. These examples highlight DRL's capacity not only for prediction but also for autonomous decision-making, making it a compelling candidate for enhancing the responsiveness and resilience of power systems through intelligent RAS integration.

Researchers have also started exploring DRL applications in improving the reliability of power grids. For instance, Hao et al. [110] deployed DRL to search for the optimal operation and maintenance strategy to achieve a more reliable power supply of energy systems. Hao et al. [280] investigated a robustness enhancement method for power distribution systems that use DRL in load restoration. Kamruzzaman et al. [135] introduced a data-driven multi-intelligence framework based on DRL to enhance power grid resilience, overcoming computational and scalability issues associated with accurate system models and shunt deployment planning. However, up to now, no DRL-based RAS solutions have been developed.

Given the above, attempts have been made to bridge the gap by introducing a DRL-based RAS to mitigate the propagation of cascading failure in power grids. The main steps are summarized as follows:

1. A Markov Decision Process (MDP) framework is used to integrate the online generation of remedial actions into the cascading failure simulation model, which itself is based on a Markov chain. This formulation enables dynamic decision-making during cascading events, allowing the system to adaptively respond to evolving conditions and mitigate failure propagation.
2. An adaptation of the Proximal Policy Optimization (PPO) algorithm is used for offline training of RAS policies to solve the specified MDP.
3. Extensive numerical experiments on the IEEE 39-bus and 57-bus systems demonstrate that the proposed DRL approach can effectively generate RAS policies that slow down or even prevent the propagation of cascading failures.

In this section, the model for simulating cascading failure in power grids is briefly introduced. The detailed model description and model validation can be found in a prior work [302].

In this section, we deploy a Markov chain (MC)-based cascading failure model, taking into account the power flow distribution and the probabilistic failure mechanisms of the components. Suppose a grid has $m$ components, i.e., all the buses and transmission lines. The state space of the MC is a binary vector $\bar{\mathcal{S}} = \{0, 1\}^m$, and we use $\bar{s}_k (\in \bar{\mathcal{S}})$ to denote the state at discrete time step $k$. For $i \in \{1, 2, ..., m\}$, $\bar{s}_k^i \in \bar{\mathcal{S}}$ indicates the "connection" state of component $i$ at $k$: $\bar{s}_k^i = 1$ if $i$ is still connected to the grid, and zero otherwise. The propagation of cascading failure can then be characterized by the dynamic evolution of $\bar{s}_k$, which is composed of the state transitions of all grid components over steps $k$.

The flow chart shown in Figure 7.6a illustrates the major steps of our cascading failure simulation. To initiate the failure, a set of components is randomly disconnected, which alters the topology of the power grid, and necessitates the redistribution of power flow. If a component is consequently overloaded, the underlying disconnection/failure rates are calculated, based on which one can generate the next failure event in a stochastic manner. The subsequent failures could result in recurring changes in the grid state and hence redistribution of power flow. The simulation continues until no overloading of grid components occurs.

### 7.5.1 Problem Overview

In practice, remedial actions—such as load shedding, generator tripping, and transmission line disconnection—can be applied during cascading failures to alter their propagation paths and mitigate impacts. Figure 7.6b illustrates how these actions are incorporated into our cascading failure model. A remedial action is essentially a sequence of control commands $a_k$ generated based on the evolving grid states $s_k$. However, because cascading failures are highly nonlinear and stochastic (e.g., propagation paths can vary significantly even under

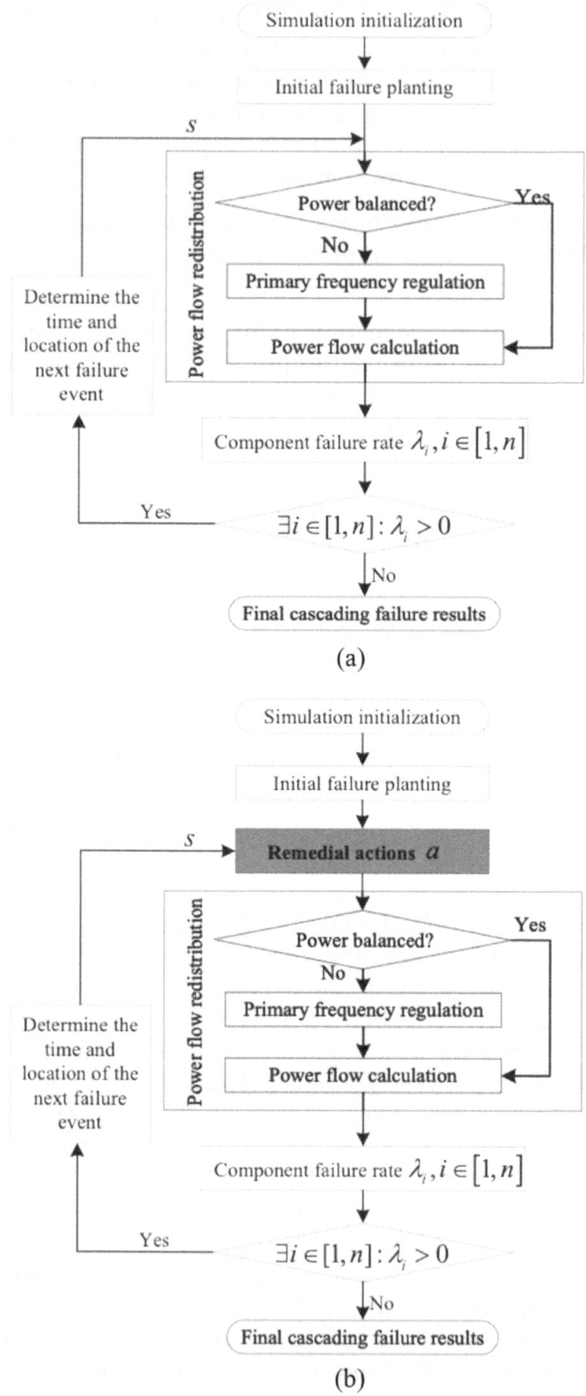


**FIGURE 7.6**
Flow charts for cascading failure simulation (a) without and (b) with incorporating remedial actions within the cascading failure process.

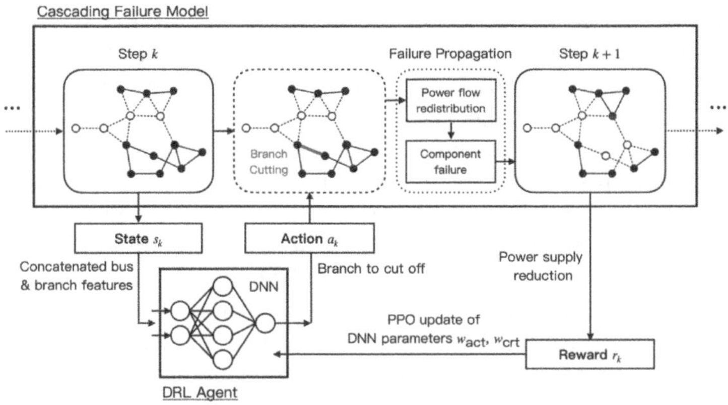


**FIGURE 7.7**
Schematic of cascading failure mitigation method with DRL-based remedial branch cutting.

the same initial conditions), determining the appropriate remedial action is extremely challenging. Moreover, the timely generation of these actions is critical, as failures can propagate rapidly.

To overcome these challenges, we present here an MDP on top of the MC-based cascading failure model to select the appropriate remedial actions. In addition to the action space and reward function, the state space is also re-designed so that sensible decisions can be made accordingly. A DRL approach is then used to solve the MDP, featuring: (1) deep neural network (DNN) agents for processing the non-linearity and uncertainty in cascading failure; and (2) efficient and stable agent training through a proximal policy optimization (PPO) implementation. The DRL agents are trained offline based on the "experiences" collected from the interactions between the DRL agent and an "environment" that implements the MDP. With proper training, the DRL agent can then be deployed in practice to determine remedial action $a_k$ based on real-time MDP states. In Figure 7.7, the schematic of this method for the mitigation of cascading failure is illustrated.

### 7.5.2 Modeling with Markov Decision Process

We define the MDP that describes cascading failure mitigation as a 4-tuple $\{\mathcal{S}, \mathcal{A}, \mathcal{T}, \mathcal{R}\}$, where $\mathcal{S}$ and $\mathcal{A}$ are the state and action spaces; $\mathcal{T}$ is a set of state transition rules, and $\mathcal{R}$ is the reward function. The elements of the 4-tuple are defined as follows:

#### 7.5.2.1 State

The state (or observation) space is defined as $\mathcal{S} = \mathcal{S}_{\text{bus}} \times \mathcal{S}_{\text{branch}}$, where $\mathcal{S}_{\text{bus}}$ and $\mathcal{S}_{\text{branch}}$ are the states corresponding to the buses and branches of the power grid. In particular, $\mathcal{S}_{\text{bus}} = \boldsymbol{\theta} \times \boldsymbol{P} \times \boldsymbol{\lambda}_{\text{bus}}$, where $\boldsymbol{\theta}$ and $\boldsymbol{P}$ are again the vectors of bus voltage phase and powers; $\boldsymbol{\lambda}_{\text{bus}}$ is the vector of bus failure rates; and similarly, $\mathcal{S}_{\text{branch}} = \boldsymbol{P}_{\text{branch}} \times \boldsymbol{\lambda}_{\text{branch}}$, where the two features are the power flows and failure rates of the branches. For each combination of component failure that is given by the binary MC state $\bar{s}_k \in \bar{\mathcal{S}}$, one can evaluate the above state features of the power grid, which practically determines how the failure may propagate. For demonstrative purposes, we hereinafter use the notation $s_k$ to denote an MDP state in $\mathcal{S}$ at decision step $k$. Assuming state $s_k$ is fully observable in real time, the goal of the MDP is hence to decide on the remedial actions based on $s_k$. All the features of $s_k$ are normalized to $[-1, 1]$ (using empirical bounds) before any further processing. A state $s_k \in \mathcal{S}$ is considered to be terminal if there is no further failure propagation.

#### 7.5.2.2 Action

The remedial action of disconnecting the transmission line is considered for the mitigation of cascading failure. Assuming that a single branch is to be cut off proactively following each event of component failure, then the action space $\mathcal{A}$ is essentially the set of remaining branches of the power grid. For any non-terminal state $s_k \in \mathcal{S}$, the action $a_k \in \mathcal{A}$ decides on which branch to cut off.

#### 7.5.2.3 Transition

By taking action $a_k$ at state $s_k$, the MDP transits to state $s_{k+1}$, at which another failure may have been triggered and altered the states of the power grid, necessitating further remedial interventions. The process of stochastic failure propagation is characterized by the MC-based cascading failure model.

#### 7.5.2.4 Reward

The reward function is defined as:

$$r(s_k, a_k, s_{k+1}) = P_k^{\text{gen}} - P_{k+1}^{\text{gen}} \tag{7.11}$$

where $P^{\text{gen}}$ denotes the total power generation (that is still active) of the power grid. The subtraction essentially measures how much power the grid lost after the transition from state $s_{k+1}$ to $s_k$, i.e., one step of failure propagation. One apparently has $r(s_k, a_k, s_{k+1}) \leq 0$.

### 7.5.3 Problem Statement

Let $\pi(a_k|s_k)$ be a stochastic policy that maps any non-terminal state $s_k$ to an action $a_k$, i.e., to determine a branch to cut off. The problem of cascading failure mitigation can then be formulated as a sequential decision-making problem:

$$\max_{\pi} J = \mathbb{E}_{\pi}[G_k] = \mathbb{E}_{\pi}\left[\sum_k \gamma^k \cdot r(s_k, a_k, s_{k+1})\right] \tag{7.12}$$

where $\gamma \in [0, 1]$ is the discount factor that controls the present value of a future reward; and $G_k$ is the discounted return (or rewards-to-go). The larger $\gamma$ is, the more "far-sighted" the DRL agent would be in taking action. The goal of equation (7.12) is hence to maximize the expectation of discounted return with respect to the agent policy $\pi$, i.e., to determine a sequence of branches to be removed so that the power grid preserves as much power supply as possible after the failure propagation stops.

### 7.5.4 Proximal Policy Optimization

One of the state-of-the-art DRL algorithms, PPO [229], is adopted to solve the MDP for cascading failure mitigation. As a policy gradient method, PPO has an actor-critic architecture, where the actor parameterizes the policy $\pi$ and interacts with the environment by inferring actions based on given states, while the critic evaluates the performance of policy $\pi$ by learning the state value functions $V$ as the training progresses. Since the state space $\mathcal{S}$ is multi-dimensional and continuous, one can resort to DNNs to represent the actor and critic, which can thus be referred to as "policy network" and "value network", respectively. Let $\omega_{\text{act}}$ and $\omega_{\text{crt}}$ be the vectors of parameters of policy and value networks, respectively. The estimators of $\pi$ and $V$ are then denoted by $\pi(\cdot; \omega_{\text{act}})$ and $V(\cdot; \omega_{\text{crt}})$.

Policy gradient methods optimize the parameters of policy network $\omega_{\text{act}}$ based on a re-formulation of the objective function derivative that is independent of the state distribution, i.e.,

$$\nabla L_{\text{PG}}(\omega_{\text{act}}) = \mathbb{E}\left[A(s_k, a_k; \omega_{\text{crt}}) \nabla \log \pi\left(a_k | s_k; \omega_{\text{act}}\right)\right] \tag{7.13}$$

where $A(s_k, a_k; \omega_{\text{crt}})$ is an estimator of the advantage function for taking action $a_k$ given the state $s_t$, and can be approximated by the value network in the form of temporal difference error:

$$A(s_k, a_k; \omega_{\text{crt}}) = r_k + V(s_{k+1}; \omega_{\text{crt}}) - V(s_k; \omega_{\text{crt}}) \tag{7.14}$$

To prevent overly large "steps" of parameter optimization and consequently poor sampling efficiency, PPO is explicitly designed with a clipped surrogate objective function given by

$$\nabla L_{\text{PPO}}(\omega_{\text{act}}) = \mathbb{E}\left[\min\left\{R_k(\omega_{\text{act}}) A(\cdot; \hat{\omega}_{\text{act}}), h(\epsilon, A(\cdot; \hat{\omega}_{\text{act}}))\right\} \nabla \log \pi\left(a_k | s_k; \omega_{\text{act}}\right)\right] \tag{7.15}$$

where the input arguments to the advantage function $A$, $s_k$ and $a_k$ are omitted for brevity; $\hat{\omega}_{\text{act}}$ is the vector of parameters of the policy network before this optimization step, $\epsilon$ is a hyper-parameter that controls how far away $\omega_{\text{act}}$ can be "updated" from $\hat{\omega}_{\text{act}}$; and $R_k(\omega_{\text{act}})$ denotes a probability ratio, i.e.,

$$R_k(\omega_{\text{act}}) = \frac{\pi(a_k|s_k;\omega_{\text{act}})}{\pi(a_k|s_k;\hat{\omega}_{\text{act}})} \tag{7.16}$$

and

$$h(\epsilon, A(\cdot;\hat{\omega}_{\text{act}})) = \begin{cases} (1+\epsilon)A(\cdot;\hat{\omega}_{\text{act}}) & \text{if } A(\cdot;\hat{\omega}_{\text{act}}) \geq 0 \\ (1-\epsilon)A(\cdot;\hat{\omega}_{\text{act}}) & \text{if } A(\cdot;\hat{\omega}_{\text{act}}) < 0 \end{cases} \tag{7.17}$$

For instance, $A(\cdot;\hat{\omega}_{\text{act}}) \geq 0$ implies that taking action $a_k$ at state $s_k$ leads to an expected return $G_k$ that is above the average, and should thus be exploited more in future iterations. However, the update of $\omega_{\text{act}}$ cannot exceed the limit of $(1+\epsilon)A(\cdot;\hat{\omega}_{\text{act}})$, thereby optimizing the networks in a stable and efficient manner.

On the other hand, the parameters of value network $\omega_{\text{crt}}$ can be updated iteratively by minimizing a mean squared error (MSE) between the returns $G_k$ and state value approximations $V(s_k;\omega_{\text{act}})$ given by the value network, i.e.,

$$\nabla L_{\text{V}}(\omega_{\text{crt}}) = \mathbb{E}_{\pi(\cdot;\omega_{\text{act}})}\left[(G_k - V(s_k;\omega_{\text{crt}}))\,\nabla V(s_k;\omega_{\text{crt}})\right] \tag{7.18}$$

Note that the subscript of the expectation operator indicates that equation (7.18) evaluates the state value functions when the DRL agent behaves consistently with policy $\pi(\cdot;\omega_{\text{act}})$.

The derivatives in equations (7.15) and (7.18) can guide the update of network parameters $\omega_{\text{act}}$ and $\omega_{\text{crt}}$ through (stochastic) gradient methods over multiple epochs and mini-batches of episodes (collected from interactions between the policy network and the environment), with an aim of improving the policy performance and estimation accuracy of state value functions.

### 7.5.5 Case Study

Extensive numerical experiments are conducted on the IEEE 39-bus and IEEE 57-bus test systems to evaluate the effectiveness of the proposed DRL-based approach for RASs. The results demonstrate that the learned RAS policies can significantly mitigate the impact of cascading failures by slowing down their propagation and, in many cases, preventing large-scale outages altogether. These case studies highlight the capability of the DRL framework to adapt to evolving system conditions and to provide timely, data-driven control actions for enhancing grid resilience.

#### 7.5.5.1 Settings

For performance evaluation, the proposed DRL approach is employed to train branch-cutting remedial policies for mitigating cascading failures in the IEEE 39 Bus and 57-bus systems. In the 39-bus system, failures are initiated by randomly disconnecting three branches, whereas in the 57-bus system, six

**TABLE 7.1**
Configuration of PPO Hyper-Parameters

| **Hyper-Parameters** | **Value** | **Hyper-Parameters** | **Value** |
|---|---|---|---|
| Total iteration ($I$) | 4e5 | Epoch number ($N^{(e)}$) | 10 |
| Mini-batch size ($N^{(b)}$) | 256 | Discount factor ($\gamma$) | 0.99 |
| Clipping range $\epsilon$ | 0.2 | Learning rates ($\alpha_{\text{act}}, \alpha_{\text{crt}}$) | 3e-4 |

branches are removed. A uniform safety margin of 0.2 is assumed for all power grid components, i.e., $C_i = 1.2 \times L_i$. The basic unit rate $a_i$ is set as 0.035 $\text{s}^{-1}$. The cascading failure simulation and the MDP model are both coded in Python.

Our PPO is implemented based on *Stable Baselines3* [214], which is a set of reliable implementations of DRL algorithms in Pytorch. The policy and the value networks are both represented by a four-layer feed-forward neural network with 256 neurons at each layer and ReLU activation. Their parameters are updated using Adam (Adaptive Moment Estimation) optimizer. Through extensive trial-and-error, the hyperparameters for agent training are determined as listed in Table 7.1.

All the numerical studies were performed on a Dell workstation with Intel Xeon(R) CPU E5-2643 v4 @ 3.40 GHz×24, 128 GB memory, and NVIDIA Quadro M4000 graphics card.

#### 7.5.5.2 Performance

In this section, the PPO policy is compared against two alternative DRL policies, which are trained under the same function approximation settings while using the basic Advantage Actor-critic (A2C) [182] and double Q-learning (DDQN) [112] algorithms, as well as two benchmark policies that are based on one-step exhaustive searches:

- Bus Load Shedding (BLS): At each decision step, BLS iterates over all the buses that are still active, and for each bus, the total failure rate of the power grid ($\sum_{i=1}^{n} \lambda_i(\cdot)$) is calculated when its load is shed. The bus that minimizes such a one-step failure rate will be selected by BLS for load shedding.

- Branch Exclusion (BE): As compared to BLS, BE iterates over all the remaining branches rather than the buses. Whereas similar to BLS, the goal of BE is to find and cut off the branch whose removal minimizes the total failure rate of the resulting power grid.

To ensure a fair comparison of policy performance, 100 initial failure scenarios are randomly generated for both the 39-bus and 57-bus systems. For each scenario, the DRL, BLS, and BE policies are independently deployed to simulate failure propagation 30 times, accounting for the inherent stochasticity of the process. The same set of random seeds is used across all policies to maintain consistency in the propagation paths. The performance of a policy for a given failure scenario is measured by the average remaining power across the 30 simulations. To evaluate the overall effectiveness of a policy, the average remaining power is further aggregated across all initial failure cases. A policy is considered more effective if it retains a higher amount of power after the failure propagation concludes.

Table 7.2 summarizes the comparison results. Among the three DRL-based policies, the PPO policy achieves the best performance in mitigating cascading failures, preserving the largest share of power supply–81.16% and 67.08% for the 39-bus and 57-bus systems, respectively. This outcome is reasonable, as PPO is a more advanced DRL algorithm explicitly designed with a clipped surrogate objective function, denoted by equation (7.15) to ensure stable and efficient agent training. In contrast, without any RAS in place, only 67.98% and 63.11% of the power supply has remained. Thus, introducing PPO-based policies yields an average improvement of 19.40% (26.32%) and 6.29% (9.48%) in preserved power for the two systems, respectively.

The BLS and BE policies deliver performance comparable to A2C and DDQN, significantly outperforming scenarios without RAS by preserving up to 79.99% and 64.21% of power for the two systems, respectively. However, they are outperformed by PPO in both failure mitigation and decision-making speed. Notably, BE performs relatively close to PPO on the 39-bus system but at the cost of a substantially longer decision time (0.36 seconds per action). In contrast, the PPO trained offline can generate remedial actions online almost instantaneously (under 0.01 seconds). Given that real-world cascading failures propagate extremely quickly (on the order of milliseconds, as evidenced by the 1996 North American blackout [81]), minimizing decision time is critical. This underscores the distinct advantage of our DRL-based approach for mitigating cascading failures using RAS.

It is also worth noting that all the policies preserve relatively fewer power supplies on the 57-bus system than on the 39-bus system – the BE policies barely improve over the no RAS scenario (i.e., 0.56% on average, and 1.43% case-by-case), even though each branch-cutting decision takes up to 0.73 seconds (on average). The results show that the one-step methods are limited by the local observations and thus incapable of addressing the highly nonlinear and stochastic nature of cascading failure, especially for large-scale power grids under simultaneous failures of multiple grid components (e.g., the failures on the 57-bus system are triggered by cutting off six branches), where the paths of failure propagation are much more variable and longer. In contrast, the DRL-based policies decide on the remedial actions step-by-step based on the evolving grid states in order to minimize the long-term impact of cascading failure. Though a careful algorithmic design is certainly entailed to better exploit such advantages for an out-performance over the benchmark policies (i.e., the PPO policies preserve 6.29% on average and 9.48% case-by-case more power supplies over no RAS scenarios, while the improvements offered by the A2C and DDQN policies are rather limited).

**TABLE 7.2**
Comparative Performance Analysis of the PPO Policies

| Power System | Policy | Rewards-to-go | | | Decision Time (second) |
|---|---|---|---|---|---|
| | | Remaining Power (%) | Improvements over No RAS | | |
| | | | On Average (%) | Case-by-Case (%) | |
| 39 Bus | No RAS | 67.98±17.48 | - | - | - |
| | PPO | 81.16±12.16 | 19.40 | 26.32 | <0.01 |
| | A2C | 68.78±16.55 | 1.18 | 2.21 | <0.01 |
| | DDQN | 78.30±13.26 | 15.18 | 20.86 | <0.01 |
| | BLS | 73.09±15.27 | 7.52 | 13.45 | 0.14 |
| | BE | 79.99±13.50 | 17.67 | 23.40 | 0.36 |
| 57 Bus | No RAS | 63.11±20.15 | - | - | - |
| | PPO | 67.08±18.56 | 6.29 | 9.48 | <0.01 |
| | A2C | 64.04±20.50 | 1.48 | 1.55 | <0.01 |
| | DDQN | 63.45±21.31 | 0.54 | 0.09 | <0.01 |
| | BLS | 64.21±20.52 | 1.75 | 5.37 | 0.33 |
| | BE | 63.47±21.01 | 0.56 | 1.43 | 0.73 |

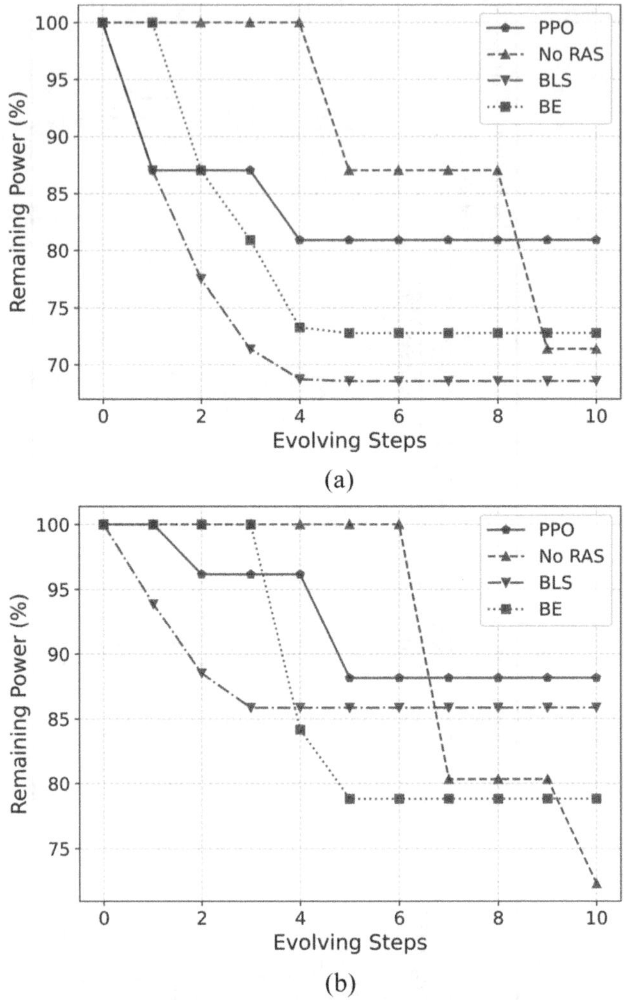


**FIGURE 7.8**
Illustration of the failure propagation on IEEE 39 Bus system. (a) Initial failure on branches 2, 23, and 33. (b) Initial failure on branches 3, 6, and 43.

#### 7.5.5.3 Stability Analysis

Figures 7.8 and 7.9 visualize the failure propagation on the IEEE 39 Bus and 57 Bus systems given different sets of initial failure branches. One can observe that when no RAS is applied, the propagation of cascading failure can easily result in abrupt reductions in power supply (e.g., in Figure 7.9a, the remaining power drops from around 92% to 36%). While with RAS policies in place, the failures propagate more slowly, and more power supplies are preserved in the end. The PPO policy is apparently the most capable of the three. Observations are also consistent with the statistical results reported in Table 7.2.

To investigate the reasons for the good performance of PPO policies, we present the grid failure rate alongside the cascading failure evolving steps in Figures 7.10 and 7.11. The grid failure rate indicates the overall overloading extent of the power grid. The figures show that, unlike the remaining power, which could stay unchanged as the failure propagates, the failure rates are constantly changing (e.g., evolving steps 1 to 4 in Figure 7.10a). In addition, high failure rates increase the likelihood of subsequent component outages. For example, when no RAS is applied, the remaining power drops sharply at step 7 in Figure 7.10b and step 15 in Figure 7.11a. Prior to these major failures, the failure rates fluctuated or remained at relatively high levels. In contrast, PPO policies proactively disconnect branches to prevent significant deterioration or sudden spikes in failure rates (as observed in the no-RAS scenario), thereby giving the power grid a greater chance to preserve more power during cascading failures.

#### 7.5.5.4 Sensitivity Analysis

To evaluate the robustness of the proposed DRL approach under varying initial failure conditions, Figure 7.12 plots the average remaining power after failure termination against the number of initially failed branches for two scenarios: (1) PPO-based RAS and (2) no RAS. The results demonstrate that PPO policies effectively and consistently mitigate the impact of cascading failures by guiding RAS actions. Moreover, as the number of initial branch failures increases, the severity of cascading failures escalates, while the relative benefit of PPO policies becomes more pronounced. For example, in Figure 7.12a, the PPO policy preserves 17.3% and 46.9% more power than the no-RAS scenario when the initial failures involve 1 and 6 branches, respectively.

## 7.6 Summary

In this chapter, we introduce the short-term control and protection strategies for enhancing the robustness of power systems. An integrated modeling framework from the cyber-physical perspective and various protection and control functions within the cascading failure are presented. A cascading failure study considering the wide-area protection function of RASs is given.

Specifically, we present a DRL-based method to generate remedial actions for the mitigation of cascading failures in power grids. This method is based on a cascading failure simulation model that incorporates power flow redistribution and the stochastic characteristics of component failures in the dynamic propagation process. The numerical studies on the IEEE 39 and 57 Bus

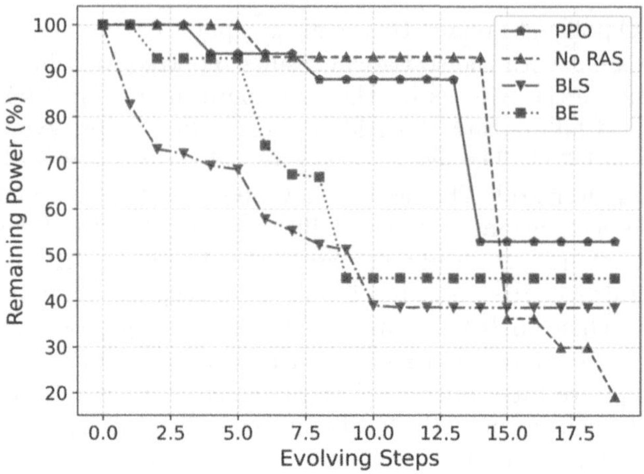


(a)

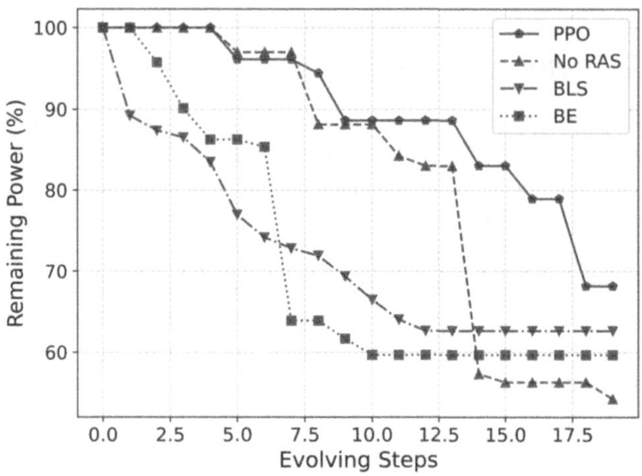


(b)

**FIGURE 7.9**
Illustration of the failure propagation on IEEE 57-bus system. (a) Initial failure on branches 2, 7, 22, 24, 28, and 47. (b) Initial failure on branches 2, 6, 17, 46, 54, and 63.

systems show that the DRL policies can generate well-timed and effective remedial actions that best suppress the failure propagation as compared to benchmark RAS policies.

The work described in this chapter explores a new approach to addressing the challenges associated with cascading failure in power grids. The integration of DRL techniques with a comprehensive cascading failure model establishes a robust framework for effective mitigation.

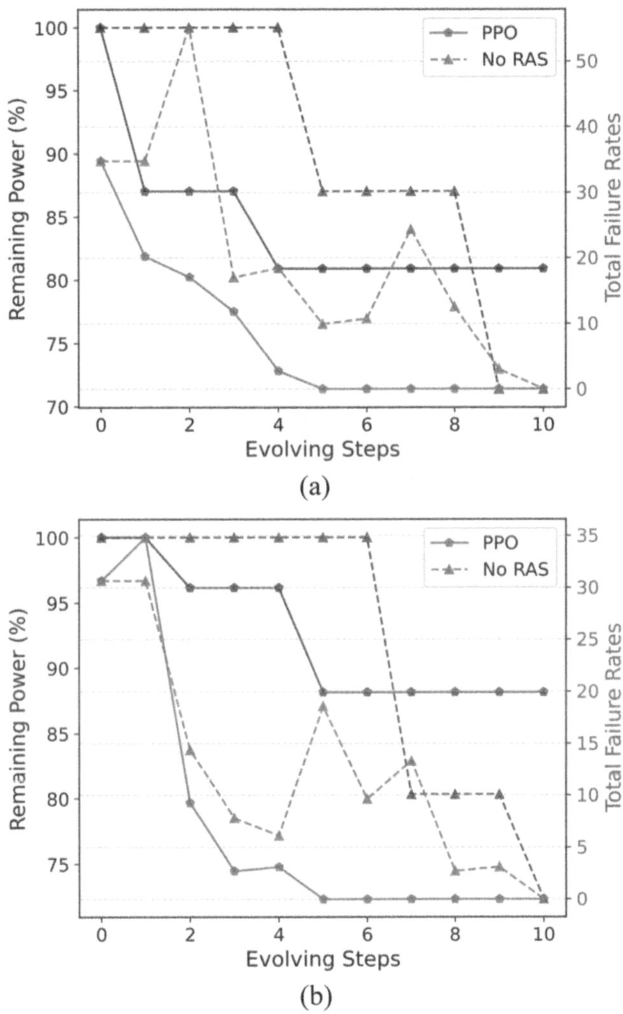


**FIGURE 7.10**
Variations in grid failure rate during the failure propagation on IEEE 39 Bus system. (a) Initial failure on branches 2, 23, and 33. (b) Initial failure on branches 3, 6, and 43.

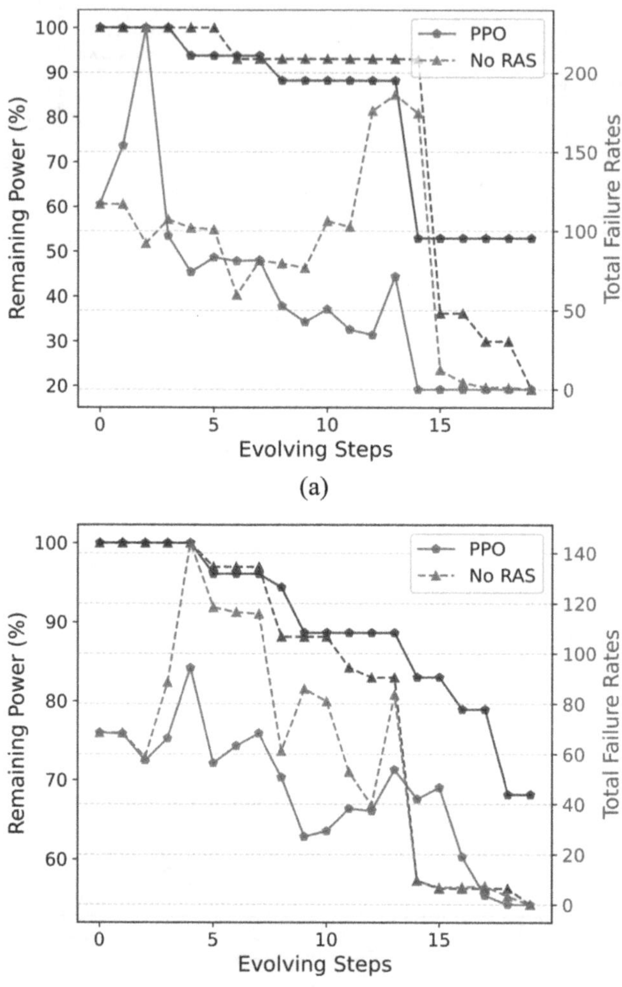


**FIGURE 7.11**
Variations in grid failure rate during the failure propagation on IEEE 57 Bus system. (a) Initial failure on branches 2, 7, 22, 24, 28, and 47. (b) Initial failure on branches 2, 6, 17, 46, 54, and 63.

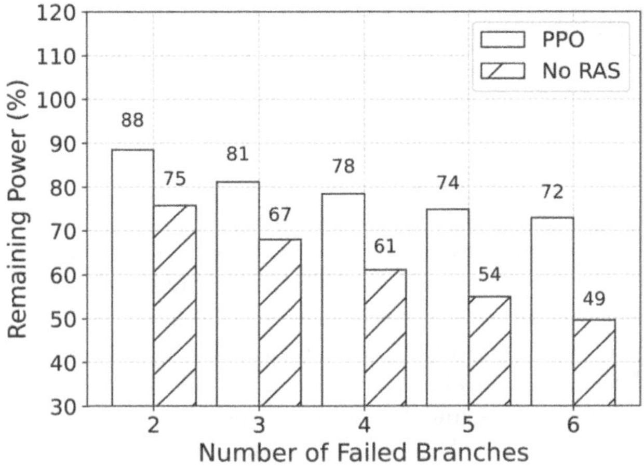


(a)

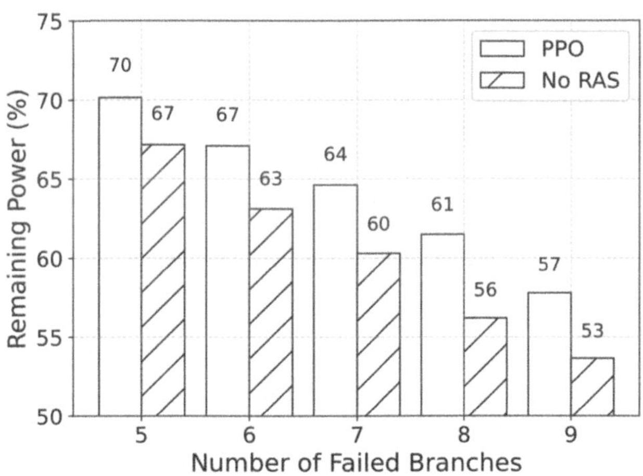


(b)

**FIGURE 7.12**
Number of failed branches that trigger the cascading failure against the average power remained after propagation terminates. (a) 39 Bus system. (b) 57 Bus system.

# 8

# *Forward Perspectives on Power System Robustness*

The robustness of future power systems will be shaped not only by technological innovation but also by strategic decisions made today by researchers, industry leaders, and policymakers. Each stakeholder plays a distinct yet interconnected role in this transformation. Researchers face the challenge of developing hybrid methodologies that combine physics-based models with advanced AI techniques. Their work includes benchmarking and validating digital twin architectures and formulating resilience metrics that accurately reflect the dynamic and evolving nature of modern power grids. Industry practitioners must embed resilience into the foundation of system design. This involves investing in cybersecurity-by-design for all operational technologies and deploying adaptive restoration and control protocols capable of responding in real time to extreme events and disruptions. Policymakers and regulators are instrumental in enabling systemic change. By establishing flexible regulatory frameworks that support AI integration, incentivize cross-sector resilience initiatives, and promote knowledge-sharing across critical infrastructure domains, they can foster an environment conducive to innovation and collaboration.

Robustness in power systems is not a fixed attribute but a continuous, adaptive process that demands both foresight and agility. The power grids of the future will do more than merely endure disruptions—they will learn from them, adapt dynamically, and evolve over time. Intelligence, automation, and sustainability will be embedded at the core of these systems. Realizing this vision requires deep collaboration across disciplines and sectors, integrating technological innovation with operational excellence and forward-looking policy. As we move forward, the path is clear: resilient, intelligent, and environmentally sustainable power grids will serve as the foundation of modern society. These systems will not only safeguard against uncertainty but also act as catalysts for progress and transformation.


DOI: 10.1201/9781003622130-8

## 8.1 Emerging Technologies and Trends

The transformation of power grids into intelligent, adaptive, and highly resilient infrastructures is being propelled by the convergence of advanced technologies, including artificial intelligence, digital simulation, autonomous control, and cyber-physical integration, as illustrated in Figure 8.1. These innovations are redefining operational capabilities and establishing new paradigms for how energy systems are designed, monitored, and restored following disruptions. This section explores several of the most impactful technological advancements and examines their practical applications across key regions, including the United States, China, and Europe.

### 8.1.1 Large Language Models for Decision Support

Large Language Models (LLMs), such as GPT-style architectures, offer advanced reasoning, natural language interaction, and rapid synthesis of complex information. In power grids, LLMs can serve as real-time decision-support assistants for operators, helping interpret grid states, analyze reports, and recommend corrective actions during disturbances. Integration with SCADA and Energy Management Systems (EMS) allows LLMs to combine live operational data with historical knowledge, providing timely, context-aware advisories. For instance, in the United States, the Pacific Northwest National Laboratory (PNNL) has been experimenting with AI-based natural language systems to

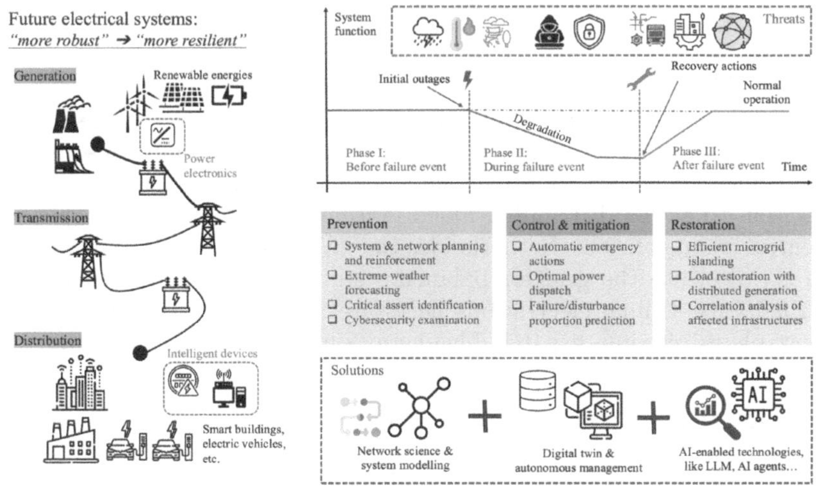


**FIGURE 8.1**
Vision of future resilient grid.

assist operators during grid stress events [198]. In Europe, the European Parliamentary Research Service (EPRS) is piloting AI-driven knowledge bases to support cross-border decisions during extreme weather [274].

### 8.1.2 Digital Twins Technologies

Digital twins are real-time, data-synchronized digital replicas of physical grid assets or network topologies. They allow operators to test scenarios, anticipate vulnerabilities, and validate strategies before applying them in the physical system. This enables predictive maintenance, early anomaly detection, and pre-validation of control actions, reducing outage risks. China's State Grid Corporation has employed digital twins platforms for major substations and urban distribution systems, incorporating IoT and AI analytics to detect faults before escalation [245]. In Europe, Germany's 50 Hz Transmission GmbH developed digital twins for offshore wind integration, simulating faults and curtailment scenarios to improve contingency responses [102].

### 8.1.3 Autonomous and Self-Healing Grids

Self-healing grids utilize sensors, automation, and AI-based decision-making to detect, isolate, and restore faults automatically, often within seconds. They greatly reduce outage duration and geographic spread, critical in high-renewable and high-demand environments. For example, Duke Energy in North Carolina has deployed self-healing distribution systems serving over one million customers, cutting outage durations by up to 75% [78]. Shanghai's urban grid pilots employ AI diagnostics and fast-acting reclosers to bring restoration times down from minutes to seconds [124].

### 8.1.4 Cross-Sector and Multi-Domain Integration

Future grids will be tightly linked with transportation, water, telecommunications, and industrial systems. EV batteries and other flexible resources can provide emergency power and balancing capacity. Cross-sector synergy allows the system to access distributed assets during crises, mitigating dependence on central plants. In the Netherlands, the "Vehicle to Grid" initiative connects EV fleets to the grid for balancing and black-start capability [82]. In Shenzhen City, China, electric bus fleets are integrated into emergency grid support plans, offering dispatchable energy during peak events [234].

### 8.1.5 AI Driven Cybersecurity Defenses

With increasing connectivity, the attack surface for cyber threats continues to grow. AI-enhanced security can detect anomalies, correlate events across different domains, and trigger rapid containment protocols. These techniques ensure that even during cyber or hybrid attacks, core operational functions of the grid remain uncompromised. In the United States, the Department of Energy's Cybersecurity for Energy Delivery Systems (CEDS) program pilots AI-based intrusion detection across operational networks [253]. In Europe, the Horizon 2020 EnergyShield project has developed a modular platform combining AI-driven anomaly detection, threat intelligence sharing, and simulation-based resilience assessment, tested across multiple national grid infrastructures [85]. The European Union Agency for Cybersecurity (ENISA) has also published detailed guidelines for securing smart grids, outlining how advanced analytics and behavioral anomaly detection can be integrated into utility cybersecurity strategies [91].

## 8.2 Challenges and Open Research Questions

Despite the promise of the aforementioned technologies, their integration into large-scale power systems faces formidable technical, operational, and governance barriers. Addressing these challenges is essential before the resilience potential outlined in Section 8.1 can be fully realized.

One critical issue is the trustworthiness of AI models used in operational decision-making. Their output can accelerate operator decision-making, but, without transparency, recommendations are difficult to validate in real time. This is particularly critical in power system operations where inappropriate control actions can have cascading consequences. Recent work in the research community, such as the comprehensive review by Machlev et al. [173], surveys explainable artificial intelligence (XAI) techniques designed specifically for energy and power system applications. This review highlights common challenges of integrating XAI into operational environments, evaluates current methodologies for improving model interpretability, and identifies research opportunities to bridge the gap between high-performance AI models and the accountability requirements of critical infrastructure.

Digital twins face scalability and interoperability bottlenecks. While highly effective at asset scale, extending them to national or multi-operator grids demands real-time synchronization of millions of data points and seamless data exchange. Europe's OneNet project illustrates the interoperability challenge, emphasizing the need for standardized data formats and model abstraction methods [88].

Autonomous, self-healing grids introduce risks if uncoordinated automation across network layers produces unintended interactions. Shanghai's pilot observed load oscillations when multiple automation schemes operated independently, leading to the implementation of human intervention thresholds [124].

Cross-sector integration, while offering resilience benefits, also increases interdependence risk. Deployments such as the Netherlands' V2G program and

Shenzhen's EV bus integration showcase the potential [82, 234]. Yet, events like Europe's 2021 gas electricity crisis underscore the need for integrated contingency planning and resilience metrics spanning sectors [96].

Finally, AI-enabled cybersecurity must counter new vulnerabilities, including adversarial manipulation of AI models themselves. The EU Horizon 2020 "EnergyShield" project has developed a modular framework combining AI-based intrusion detection, anomaly detection using physics-informed models, penetration testing, and resilience evaluation specifically for the energy sector [85]. Such an approach demonstrates that effective protection for AI-enabled control systems requires a multi-layer defense strategy integrating machine learning analytics with engineering-based anomaly detection. These findings raise urgent policy questions about governance, cyber resilience auditing, and adversarial robustness evaluation in mission-critical energy systems.

In summary, the future development of power systems will be shaped by the integration of advanced technologies, including AI, digital twins, and autonomous control systems, alongside cross-sector collaboration and enhanced cybersecurity measures. These innovations promise to transform power grids into intelligent, resilient, and adaptive systems capable of withstanding disruptions and supporting the growing demand for sustainable energy. Looking ahead, power systems will not only become more robust and intelligent but also increasingly adaptive and sustainable. The integration of AI, automation, and cross-sector collaboration will enable grids to learn from disruptions, evolve, and support the transition to a greener energy future. However, overcoming technical, operational, and governance challenges will require interdisciplinary collaboration among researchers, industry leaders, and policymakers. By addressing these barriers, the vision of resilient, intelligent, and sustainable power systems will become a reality, forming the backbone of modern society and ensuring energy security in the expectedly highly interactive, complex, yet uncertain world.

# Appendices
# A

## *Failure Propagation in the Network*

A power network is represented as an undirected graph $G$ consisting of $m$ elements. The state of $G$ is defined as $S = \{s_1, s_2, ..., s_m\}$, which is a vector containing the states of all $m$ elements. Network $G$ can have $2^m$ possible network states, and any state transition of an element will lead to a network state transition of $G$.

The dynamic propagation of cascading failures in $G$ is equivalent to the dynamic evolution of $S(t)$. Given the current state of the network, the network state transition can be described by (1) the time of the next state transition and (2) identification of the next element that will transit (be tripped).

## A.1 Basics

First, we consider the network state transitions in an infinitesimal time interval $dt$. Suppose $S(t) = N_S$, which is a specific network state among the $2^m$ possible states. Thus, $S(t + dt)$ is the network state after a duration of $dt$. Only those elements in state "0" may transit, leading to a network state transition. Let $\Omega_0$ be the set of elements in state "0", and $\Omega_1$ be the set of removed (tripped) elements. From elementary probability theory, we have the following basic results:

1) $o(dt)$ is the sum of the second and higher order terms of $dt$. Omitting $o(dt)$, the probability that no element undergoes a state transition after $dt$

can be written as

$$\begin{aligned} P[S(t+dt) = N_S \,|S(t) = N_S\,] &= \prod_{i\in\Omega_0} [1-\lambda_i(t)dt)] \\ &= 1 - \sum_{i\in\Omega_0} \lambda_i(t)dt + \sum_{x_1,x_2\in\Omega_0} \lambda_{x_1}(t)\lambda_{x_2}(t)(dt)^2 \\ &\quad - \sum_{x_1,x_2,x_3\in\Omega_0} \lambda_{x_1}(t)\lambda_{x_2}(t)\lambda_{x_3}(t)(dt)^3 + \cdots \\ &= 1 - \sum_{i\in\Omega_0} \lambda_i(t)dt + o(dt) \approx 1 - \sum_{i\in\Omega_0} \lambda_i(t)dt \end{aligned} \tag{A.1}$$

where $x_1, x_2, \ldots$ are the elements in $\Omega_0$.

2) The probability that only one element state transition (say element $k$) occurs after $dt$, i.e., only element $k$ transits, can be written as

$$\begin{aligned} P[S(t+dt) = M_S \,|S(t) = N_S\,] &= \lambda_k(t)dt \prod_{i\in\Omega_0\backslash\{k\}} [1-\lambda_i(t)dt] \\ &= \lambda_k(t)dt - \sum_{x_1\in\Omega_0\backslash\{k\}} \lambda_k(t)\lambda_{x_1}(t)(dt)^2 \\ &\quad + \sum_{x_1,x_2\in\Omega_0\backslash\{k\}} \lambda_k(t)\lambda_{x_1}(t)\lambda_{x_2}(t)(dt)^3 + \cdots \\ &= \lambda_k(t)dt + o(dt) \approx \lambda_k(t)dt \end{aligned} \tag{A.2}$$

where $x_1, x_2, \ldots$ are the elements in $\Omega_0 \setminus \{k\}$ and $M_S$ denotes the network state that only one of the "0"-state elements in $N_S$ becomes "1".

3) The probability that two or more element state transitions occur after $dt$ is given by

$$P[S(t+dt) = R_S \,|S(t) = N_S\,] = 0 \tag{A.3}$$

where $R_S$ denotes the network state that two or more of the "0"-state elements in $N_S$ become "1". From equation (A.3), there is at most one element state transition at a time.

## A.2 Extended Gillespie Method

In this section, we derive $S(t)$ using an extended Gillespie method [294], which was used for analyzing coupled chemical reactions [65, 100].

As shown in Figure A.1, the state of the power system at $t_1$ is $N_S$, i.e., $S(t_1) = N_S$. Let $Q(\tau)$ denote the probability that given $S(t_1) = N_S$, no transition occurs in $(t_1, t_1+\tau)$, i.e.,

$$Q(\tau) = P[S(t_1+\tau) = N_S \,|S(t_1) = N_S\,] \tag{A.4}$$

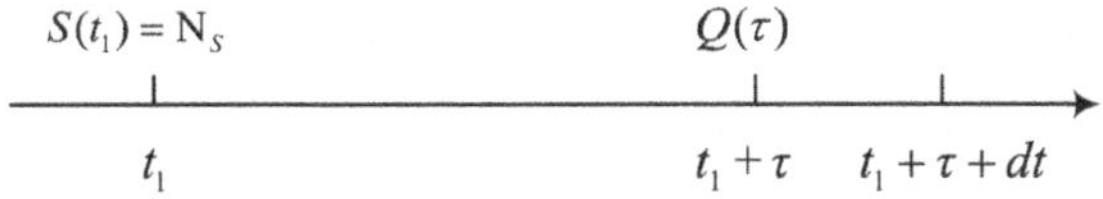


**FIGURE A.1**
Timeline of network state transitions.

Similarly, $Q(\tau + dt)$ can be written as

$$\begin{aligned} Q(\tau + dt) &= P[S(t_1 + \tau + dt) = N_S \,|S(t_1) = N_S] \\ &= P[S(t_1 + \tau + dt) = N_S \,|S(t_1 + \tau) = N_S]Q(\tau). \end{aligned} \tag{A.5}$$

Given $S(t_1) = N_S$, power flow calculation can be performed, and $\lambda_i(t_1)$ can be derived based on equation (4.5). If no state transition occurs during time interval $(t_1, t_1 + \tau)$, we have $S(t) = S(t_1)$ and $\lambda_i(t) = \lambda_i(t_1)$ for $t \in (t_1, t_1 + \tau)$. From equation (A.1), we get

$$P[S(t_1 + \tau + dt) = N_S \,|S(t_1 + \tau) = N_S] = 1 - \sum_{i \in \Omega_0} \lambda_i(t_1)dt \tag{A.6}$$

Thus, by putting equation (A.6) in equation (A.5), we get

$$Q(\tau + dt) = Q(\tau)(1 - \lambda^*(t_1)dt) \tag{A.7}$$

where $\lambda^*(t_1) = \sum_{i \in \Omega_0} \lambda_i(t_1)$. Furthermore, re-arranging equation (A.7) and taking the limit $dt \to 0$, we get

$$\begin{aligned} \frac{dQ(\tau)}{d\tau} &= \lim_{dt \to 0} \frac{Q(\tau + dt) - Q(\tau)}{dt} = -\lambda^*(t_1)Q(\tau) \\ &\Rightarrow Q'(\tau) = -\lambda^*(t_1)Q(\tau) \end{aligned} \tag{A.8}$$

The probability that nothing happens in zero time is one, i.e., $Q(0) = P\{S(t_1) = N_S \,|S(t_1) = N_S\} = 1$. Then, the analytical solution of equation (A.8) is

$$Q(\tau) = e^{-\lambda^*(t_1)\tau} \tag{A.9}$$

Let $h_i(\tau, dt)$ denote the probability of the event that given $S(t_1) = N_S$, the next transition occurs in the interval $(t_1 + \tau, t_1 + \tau + dt)$ in element $i$. There are two conditions for this event to occur. The first condition is that there is no state transition during $(t_1, t_1 + \tau)$. The second condition is that a state transition occurs in element $i$ during $(t_1 + \tau, t_1 + \tau + dt)$. Thus, $h_i(\tau, dt)$ can be written as

$$h_i(\tau, dt) = P[S(t_1 + \tau + dt) = M_S \,|S(t_1 + \tau) = N_S]Q(\tau) \tag{A.10}$$

Putting equations (A.2) and (A.9) in equation (A.10), we get

$$h_i(\tau, dt) = e^{-\lambda^*(t_1)\tau}\lambda_i(t_1)dt \tag{A.11}$$

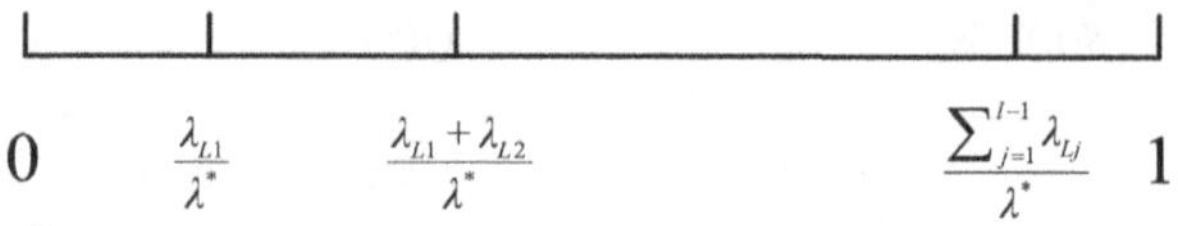


**FIGURE A.2**
Relative probability for elements in $\Omega_0$ to be first tripped given $S(t_1) = N_S$.

Let $H(\tau, dt)$ denote the probability that the next transition occurs in the time interval $(t_1 + \tau, t_1 + \tau + dt)$, given $S(t_1) = N_S$. It is readily shown that

$$H(\tau, dt) = \sum_{i \in \Omega_0} h_i(\tau, dt) = \lambda^*(t_1)e^{-\lambda^*(t_1)\tau}dt \tag{A.12}$$

Further, let $\tau$ denote the time interval between two adjacent network state transitions, and $f(\tau)$ denote the *state transition probability density function* (PDF):

$$f(\tau) = \lim_{dt \to 0} \frac{H(\tau, dt) - H(\tau, 0)}{dt} = \lambda^*(t_1)e^{-\lambda^*(t_1)\tau} \tag{A.13}$$

i.e.,

$$f(\tau) = \lambda^*(t_1)e^{-\lambda^*(t_1)\tau} \tag{A.14}$$

The cumulative probability density function that the next transition occurs before time $t_1 + \tau$, given $S(t_1) = N_S$, can be written as

$$F(\tau) = \lambda^*(t_1) \int_0^{\tau} e^{-\lambda^*(t_1)t}dt = 1 - e^{-\lambda^*(t_1)\tau} \tag{A.15}$$

Note that one can also get $F(\tau)$ from $F(\tau) = 1 - Q(\tau)$.

Equations (A.14) and (A.15) show that $\tau$ follows an exponential distribution and that the network transition rate is $\lambda^*(t_1)$. Here, $\lambda^*(t_1)$ is the sum of the element state transition rates of all the working elements in the network, and is determined by the sum of the extents of overloading of all the overloaded elements. The time interval $\tau$ is expected to be short when $\lambda^*(t_1)$ is large, i.e., the network state transition (cascading process) occurs very rapidly. Thus, the physical meaning of $\lambda^*(t_1)$ can be interpreted as the overload stress of the entire power system.

In order to include this characteristic in our model, we take the following steps to determine the time of the next network state transition, given $S(t_1) = N_S$:

1. A random number $z_1$ is generated uniformly in (0,1).
2. Let $F(\tau) = z_1$, and $\tau$ is derived as

$$\tau = \frac{ln(1 - z_1)}{-\lambda^*(t_1)} \tag{A.16}$$

### A.2.1 Order of State Transition

A number of working elements (elements in $\Omega_0$) can possibly undergo a state transition. In our analysis, we allow only one element to be removed (tripped) at a time. Pfitzner et al. [209] pointed out that the order in which overloaded lines are tripped influences the cascade propagation significantly. Here, we study the order in which element state transitions take place.

In our stochastic model, any overloaded element in $\Omega_0$ may be tripped first. From a probabilistic viewpoint, the element with a higher $h_i(\tau, dt)$ will more likely be tripped first. Thus, we define the *relative probability* for element $i$ ($i \in \Omega_0$) to be tripped first as:

$$rf_i = \frac{h_i(\tau, dt)}{H(\tau, dt)} = \frac{\lambda_i(t_1)}{\lambda^*(t_1)} \tag{A.17}$$

where $\lambda^*(t_1) = \sum_{i\in\Omega_0} \lambda_i(t_1)$. Our model can incorporate this tripping order using the following steps:

1. A random number $z_2$ is generated uniformly in (0, 1).
2. Suppose there are $l$ overloaded elements in $\Omega_0$. With no loss of generality and for ease of referral, let these overloaded elements be elements $L1, L2, \ldots, Lj, \ldots, Ll$. Figure A.2 shows the relative probability of an overloaded element in $\Omega_0$ to be first tripped, given that $S(t_1) = N_S$.
3. The $j$th element in $\Omega_0$ is selected to be tripped according to

$$\sum_{k=0}^{j-1} \frac{\lambda_{Lk}}{\lambda^*} \leqslant z_2 < \sum_{k=0}^{j} \frac{\lambda_{Lk}}{\lambda^*} \tag{A.18}$$

where $\lambda_{L0} = 0$.

# B

# *Derivation of Power Grid Resilience Metric*

The power grid resilience metric, Re, quantifies the expected power supply loss of a power grid due to natural disasters. Under this definition, a smaller value of Re indicates that the power grid resilience is higher.

## B.1 Consequences of Damage Events

The failure set $s$ can significantly affect the resilience of the power grid. Our focus here is on the resilience measures. The life cycle of power grid's resilience is depicted in Figure B.1 [301], encompassing five key stages: (1) Initial State: Power grids operate properly, and operators estimate the ability of power grids to survive disasters and develop appropriate preventive measures. (2) Violent Degradation State: Equipment of Power grids is physically damaged by extreme events. A cascading failure occurs in the power grid, causing rapid degradation of performance to the lowest level due to the action of protection relays or the proactive removal of overloaded components. (3) Quick Restoration State: Power restoration is quickly achieved by correctly opening circuit breakers, restarting offline components, and adjusting the operating mode of

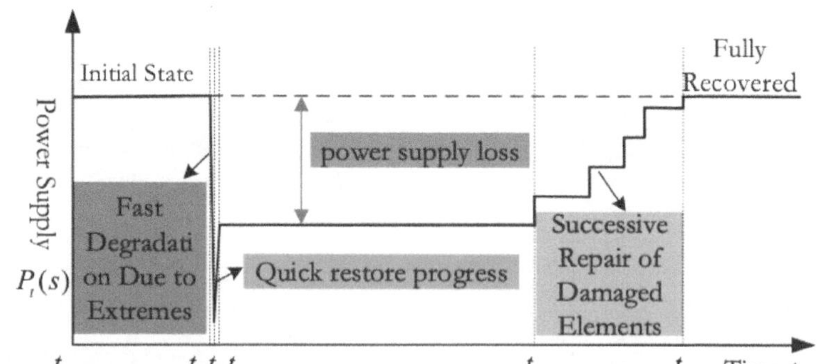


**FIGURE B.1**
Life cycle of power grid resilience.

the remaining grid. (4) Post-disturbance Degraded State: The grid's performance reaches a stable state after the disaster, where the failure does not spread further. However, except for repairing physically damaged electrical components, the power grid's performance cannot be further improved. (5) Repair State: Power grids are fully recovered by successive repair of physically damaged components.

The power supply slumps due to the occurrence of a natural disaster that acts on the power grid at time $t_1$, which is fully restored until the power grid is completely repaired at time $t_5$. Here, the power supply loss is used to quantify the destructiveness caused by natural disasters to the power grid, which is expressed as the altitude difference between the initial maximum power supply and the remaining maximum power supply after failures during the resilience life cycle of the power grid. The power supply loss can be obtained by

$$E_{\text{loss}}(s) = E_{\text{initial}} - E_{\text{remaining}}(s) \tag{B.1}$$

where $E_{\text{initial}}$ denotes the initial amount of power supply and $E_{\text{remaining}}(s)$ denotes the remaining maximum amount of power supply of the power grid after the occurrence of failure set $s$.

The remaining maximum power supply to the network can be obtained by solving the DC-OPF problem [44]. The objective function is

$$\max \quad E_t(s) = \sum_{i \in N_d} -\omega_i E_i \tag{B.2}$$

$$\text{s.t.} \quad \boldsymbol{A\theta} = \boldsymbol{E} \tag{B.3}$$

$$E_{ij} = a_{ij}(\theta_i - \theta_j) \tag{B.4}$$

$$-C_{ij} \leq E_{ij} \leq C_{ij} \tag{B.5}$$

$$-C_i \leq E_i \leq 0, i \in N_d \tag{B.6}$$

$$0 \leq E_i \leq C_i, i \in N_g \tag{B.7}$$

where $E_t(s)$ is the maximum power supply provided by the grid at time $t$; $\omega_i$ is the weight of the load node $i$, representing the importance level of the load; $\theta_i$ is the phase angle of the node $i$ and $\boldsymbol{\theta}$ is the vector of $\theta_i$; $a_{ij}$ is the derivative of the edge connecting the node $i$ and the node $j$; $\boldsymbol{A}$ is the vector of $a_{ij}$; $E_i$ is the active power injected into the power grid from node $i$; $C_i$ is the power limit of node $i$; $E_{ij}$ is the power flow across the edge passing through node $i$ and node $j$; $C_{ij}$ is the power flow limit of edge $(i, j)$; $N_d$ is the set of load nodes, and $N_g$ is the set of generator nodes.

Equations (B.3) and (B.4) denote the constraints of electrical laws. Equation (B.5) denotes the upper and lower limit constraints of the line flow. Equation (B.6) denotes the consumption power constraint for load nodes. Equation (B.7) denotes the output power constraint for generator nodes.

## B.2 Probability of Damage Events

Natural disasters are one of the key factors to consider when evaluating the resilience of power grids. Different types of natural disasters, with varying probabilities, can cause varying degrees of damage to the electricity grid [211]. It is more meaningful to improve the resilience of the power grid in all types of natural extremes.

Specifically, our metric involves summing up the product of the probability of a disaster mode, the probability of the primary failure set under that particular disaster mode, and the corresponding loss of functions. The specific disaster type, level, and location jointly determine a particular mode under which the component failures are considered independent. Assuming that there are $m$ modes, the probability of occurrence of failure set $s$ caused by a natural disaster can be written as

$$P(s) = \sum_{\delta=1}^{m} P(s|\delta) \cdot P(\delta) \tag{B.8}$$

where $P(\delta)$ represents the probability of disaster mode $\delta$, and $P(s|\delta)$ represents the set of probability of failure $s$ in disaster mode $\delta$.

The survival probability of a certain component refers to the possibility that it can function properly when the power grid is damaged by natural disasters, while the damage probability refers to the possibility that it cannot function properly. The sum of the survival probability and damage probability of a certain component is 1. Assume that the damage probabilities of all the grid components are mutually independent, given a specific natural disaster. Therefore, the probability of occurrence of failure set $s$ can be expressed based on the probabilities of component damages [171]. Then, one can write

$$P(s \mid \delta) = \prod_{i \in s, j \in G-s} p_{i,\delta}\left(1 - p_{j,\delta}\right) \tag{B.9}$$

where $p_{i,\delta}$ denotes the damage probability of component $i$ under disaster mode $\delta$, $G$ represents the set of all components in the power grid, and $G-s$ is the set of all normal components that are not damaged by a natural disaster. Thus, the power grid's resilience metric can be rewritten as

$$\begin{aligned}
\mathrm{Re} &= \sum_{s \in \Omega} P(s) \cdot E_{\mathrm{loss}}(s) \\
&= \sum_{s \in \Omega} \left( \sum_{\delta=1}^{m} P(s|\delta) \cdot P(\delta) \right) \cdot E_{\mathrm{loss}}(s) \\
&= \sum_{s \in \Omega} \left( \sum_{\delta=1}^{m} \left( \prod_{i \in s, j \in G-s} p_{i,\delta} \cdot (1 - p_{j,\delta}) \right) \cdot P(\delta) \right) \cdot E_{\mathrm{loss}}(s)
\end{aligned} \tag{B.10}$$

# C

# *Parameter Settings in Monte Carlo Case Study*

Referring to the case study of Monte Carlo method presented in Chapter 6, the number and frequency of disaster modes observed over a fixed time duration and the corresponding survival probability of components are provided in Table C.1.

**TABLE C.1**
Number and Frequency of Disaster Modes within a Fixed Time Duration

| Disaster Mode | Number of Occurrences | Occurrence Frequency | Survival Probability of Components |
|---|---|---|---|
| 1 | 19 | 0.14 | 0.7 |
| 2 | 92 | 0.66 | 0.6 |
| 3 | 24 | 0.17 | 0.5 |
| 4 | 4 | 0.03 | 0.4 |

The specific values of the increase of survival probability corresponding to each cluster are shown in Table C.2.

**TABLE C.2**
Increase of Survival Probability of each Cluster

| Cluster ID | Increase of Survival Probability |
|---|---|
| Cluster 1 | 0.25 |
| Cluster 2 | 0.2 |
| Cluster 3 | 0.1 |
| Cluster 4 | 0.05 |

# D

# *Real-World Data Sources*

## D.1 Power Grid Data

The power grid data used in Chapter 6 is sourced from IEEE test cases, including IEEE 14 Bus (case14), 39 Bus (case39), 57 Bus (case57), 118 Bus (case118), 141 Bus (case141) and the real-world case ACTIVSg200. The first five cases are obtained from the MATPOWER toolbox [309], which provides a comprehensive collection of synthetic datasets for power system studies. The MATPOWER toolbox is publicly available at https://matpower.org. Specifically, case ACTIVSg200 is synthetic but captures key characteristics of real power grids covering several regions in Texas, including Peoria, Champaign, Decatur, and Springfield. We not only used electrical data but also used geographic location information to match with hurricane data. These data include the spatial distribution of power generation and demand. It consists of 200 buses with geographic locations, 245 branches (transmission lines and transformers), and spans three voltage levels, which are publicly available at https://electricgrids.engr.tamu.edu/electric-grid-test-cases/activsg200.

## D.2 Hurricane Data

Historical hurricane cases that have impacted case ACTIVSg200 over the past 20 years were collected. The selected hurricanes include Gustav and Ike, which caused landfall in Texas in August [132] and September 2008 [26], respectively. The hurricane tracking data include the time of occurrence, location (latitude and longitude), and wind speed.

Hurricane Gustav data is available at https://www.nhc.noaa.gov/data/tcr/AL072008_Gustav.pdf.

Hurricane Ike data is available at https://www.nhc.noaa.gov/data/tcr/AL092008_Ike.pdf.

## D.3 Impacts of Hurricane on Power Grid

Here, we followed the approach outlined in reference [246]. To model wind-induced failures of transmission lines, we treat the failure process as a Poisson process, with a time-dependent rate $r(v_t)$ that varies according to the wind speed encountered by the line. As a result, the probability of transmission line $i$ failing due to wind is expressed as

$$p_i(v_\tau) = r(v_t)\tau \tag{D.1}$$

where $\tau$ represents the time step.

To ensure our calibration produces reasonable values, we have incorporated the physical constants from Winkler et al. [275] to derive the failure model:

$$r\left(v_{\mathrm{t}}\right) = r^{\mathrm{brk}} \frac{F_k^{\mathrm{wind}}\left(v_{\mathrm{t}}\right)}{F^{\mathrm{brk}}} \tag{D.2}$$

where $F^{\mathrm{brk}}$ represents the breaking force. The calibration is then carried out using $r^{\mathrm{brk}}$, the failure rate when the wind field matches the breaking force. Here, we set $r_{\mathrm{brk}} = 0.002\,\mathrm{h}^{-1}$, which we consider reasonable, as most transmission lines do not fail even during several hours of intense storms.

The complete wind force equation, based on the ASCE guidelines [276], is given by

$$F^{\mathrm{wind}}(l, z, v) = Qk_z I_{\mathrm{FW}} G_{\mathrm{WRF}}(l, z) C_{\mathrm{f}} A_{\mathrm{C}} v^2 \tag{D.3}$$

where $v$ denotes the 3-second gust wind speed at height $z$ and $G_{\mathrm{WRF}}$ is the wind gust response factor defined as

$$G_{\mathrm{WRF}}(l, z) = \frac{1}{k_v^2}\left[1 + 13.23\sqrt{\kappa}\left(\frac{z_0}{z}\right)^{1/\alpha}\sqrt{1 + 0.8l/L_{\mathrm{s}}}\right] \tag{D.4}$$

All the symbols mentioned above are explained in Table D.1, along with their typical values.

**TABLE D.1**
Typical wind and Transmission Line Parameter Values

| Parameter | Symbol | Value | Comment/ Meaning |
|---|---|---|---|
| Air density parameter | Q | 0.613 kg/m$^3$ | Value recommended by [276] |
| Terrain exposure coefficient | $k_z$ | 1.17 | Value for open terrain with scattered obstructions |
| Importance factor | $I_{FW}$ | 1.00 | Return period of 50 years |
| Wind gust response factor | $G_{\mathrm{WRF}}$ | From (D.4) | Accounts for peak dynamic wind loads |
| Surface drag coefficient | $\kappa$ | 0.005 | Value for open terrain with scattered obstructions |
| Height | $z_0$ | 10.06 m | Average height between the ground level and 10.06 m |
| Effective height | $z$ | 21.3 m | Height of wire attachment points |
| Power law exponent of sustained wind | $\alpha$ | 7.0 | Value for open terrain with scattered obstructions |
| Wind speed conversion factor | $K_v$ | 1.43 | Ratio of 3-s gust wind speed to 10-min sustained wind speed |
| Line span length | $l$ | 161 m | Distance between towers |
| Turbulence scale | $L_s$ | 67.056 m | Value for open terrain with scattered obstructions |
| Force coefficient | $C_f$ | 0.9 | Accounts for circular wire shape |
| Surface area normal to wind direction | $A_C$ | $l \cdot 0.03$ m | Wire cross-section (0.03 m is the diameter of the "Cardinal" conductor) |

# References

[1] M. Abedi, M. R. Aghamohammadi, and T. A. Mohammad. SVM based intelligent predictor for identifying critical lines with potential for cascading failures using pre-outage operating data. *Int. J. Electr. Power Energy Syst.*, 136:107608, 2022.

[2] M. Abedini, M. Sanaye-Pasand, M. Davarpanah, and R. Iravani. A loss-of-field detection relay based on rotor signals estimation. *IEEE Trans. Power Deliv.*, 33(2):779–788, 2018.

[3] J. Adams, C. Carter, and S.-H. Huang. ERCOT experience with sub-synchronous control interaction and proposed remediation. In *IEEE PES Transmission & Distribution Conference & Expo*, Orlando, FL, USA, pp. 1–5, 2012.

[4] M. Adnan and M. Tariq. Cascading overload failure analysis in renewable integrated power grids. *Reliab. Eng. Syst. Saf.*, 198:106887, 2020.

[5] S. Ahmed, N. Sarker, A. Khairuddin, M. Ghani, and H. Ahmad. A scheme for controlled islanding to prevent subsequent blackout. *IEEE Trans. Power Syst.*, 18(1):136–143, 2003.

[6] M. M. Alam, Z. Zhu, B. Eren Tokgoz, J. Zhang, and S. Hwang. Automatic assessment and prediction of the resilience of utility poles using unmanned aerial vehicles and computer vision techniques. *Int. J. Disaster Risk Sci.*, 11:119–132, 2020.

[7] R. Albert, I. Albert, and G. L. Nakarado. Structural vulnerability of the North American power grid. *Phys. Rev. E*, 69(2):25–103, 2004.

[8] R. Albert, H. Jeong, and A.-L. Barabási. Error and attack tolerance of complex networks. *Nature*, 406(6794):378–382, 2000.

[9] G. G. Ali, I. H. El-adaway, C. Sims, J. S. Holladay, and C.-F. Chen. Reducing the vulnerability of electric power infrastructure against natural disasters by promoting distributed generation. *Nat. Hazards Rev.*, 24(2):04022052, 2023.

[10] O. Alizadeh Mousavi, R. Cherkaoui, and M. Bozorg. Blackouts risk evaluation by Monte Carlo simulation regarding cascading outages and system frequency deviation. *Electr. Power Syst. Res.*, 89:157–164, 2012.

[11] P. M. Anderson, C. F. Henville, R. Rifaat, B. Johnson, and S. Meliopoulos. *Power System Protection.* New York, USA: McGraw-Hill, 2021.

[12] G. Andersson, P. Donalek, R. Farmer, N. Hatziargyriou, I. Kamwa, P. Kundur, N. Martins, J. Paserba, P. Pourbeik, J. Sanchez-Gasca, R. Schulz, A. Stankovic, C. Taylor, and V. Vittal. Causes of the 2003 major grid blackouts in North America and Europe, and recommended means to improve system dynamic performance. *IEEE Trans. Power Syst.*, 20(4):1922–1928, 2005.

[13] S. D. Antón, D. Fraunholz, C. Lipps, S. Pohl, M. Zimmermann, and H. D. Schotten. Two decades of SCADA exploitation: A brief history. In *Proceedings of an IEEE Conference on Applications of Information, Networks, and Security*, Miri, Malaysia, pp. 98–104, 2017.

[14] S. Arianos, E. Bompard, A. Carbone, and F. Xue. Power grid vulnerability: A complex network approach. *Chaos*, 19(1):013119, 2009.

[15] M. Assante. Confirmation of a coordinated attack on the Ukrainian power grid, SANS Industrial Control Systems (ICS) Security Blog, 2016. [Online]. Available: https://ics.sans.org/blog/2016/01/09/confirmation-of-a-coordinated-attack-on-the-ukrainian-power-grid.

[16] M. H. Athari and Z. Wang. Impacts of wind power uncertainty on grid vulnerability to cascading overload failures. *IEEE Trans. Sustain. Energy*, 9(1):128–137, 2018.

[17] M. H. Athari and Z. Wang. Stochastic cascading failure model with uncertain generation using unscented transform. *IEEE Trans. Sustain. Energy*, 11(2):1067–1077, 2020.

[18] Australian Energy Market Commission. System restart services, standards and testing, 2020. Available: [Online] https://www.aemc.gov.au/rule-changes/system-restart-services-standards-and-testing.

[19] Australian Energy Market Operator. Black system South Australia 28 September 2016, 2017. Available: [Online] https://apo.org.au/sites/default/files/resource-files/2017-03/apo-nid74886.pdf.

[20] Australian Energy Market Operator. 2024 Electricity statement of opportunities, 2024. Available: [Online] https://www.aemo.com.au/-/media/files/electricity/nem/planning_and_forecasting/nem_esoo/2024/2024-electricity-statement-of-opportunities.pdf.

[21] T. Aziz, Z. Lin, M. Waseem, and S. Liu. Review on optimization methodologies in transmission network reconfiguration of power systems for grid resilience. *Int. Trans. Electr. Energy Syst.*, 31(3):e12704, 2021.

[22] R. Baldick, B. Chowdhury, I. Dobson, Z. Dong, B. Gou, D. Hawkins, H. Huang, J. H., D. Kirschen, F. Li, J. Li, Z. Li, C.-C. Liu, L. Mili, S. Miller, R. Podmore, K. Schneider, K. Sun, D. Wang, Z. Wu, P. Zhang, W. Zhang, and X. Zhang. Initial review of methods for cascading failure analysis in electric power transmission systems IEEE PES CAMS task force on understanding, prediction, mitigation and restoration of cascading failures. In *Proceedings of IEEE Power & Energy Society General Meeting*, Pittsburgh, PA, USA, pp. 1–8, 2008.

[23] A. L. Barabási and R. Albert. Emergence of scaling in random networks. *Science*, 286(5439):509–512, 1999.

[24] W. Barcelo and P. Rastgoufard. Dynamic economic dispatch using the extended security constrained economic dispatch algorithm. *IEEE Trans. Power Syst.*, 12(2):961–967, 1997.

[25] D. Barth, T. Mautor, D. Watel, and M.-A. Weisser. Configuring an heterogeneous smartgrid network: Complexity and approximations for tree topologies. *J. Glob. Optim.*, 89(1):223–257, 2024.

[26] R. Berg. Tropical cyclone report: Hurricane Ike. Technical report, National Hurricane Center, 2009. Available: [Online] https://www.nhc.noaa.gov/data/tcr/AL092008_Ike.pdf.

[27] H. Bevrani. *Robust Power System Frequency Control*. New York, USA: Springer, 2009.

[28] A. Beygelzimer, G. Grinstein, R. Linsker, and I. Rish. Improving network robustness by edge modification. *Physica A*, 357(3-4):593–612, 2005.

[29] J. Beyza and J. M. Yusta. The effects of the high penetration of renewable energies on the reliability and vulnerability of interconnected electric power systems. *Reliab. Eng. Syst. Saf.*, 215:107881, 2021.

[30] Z. Bie, Y. Lin, G. Li, and F. Li. Battling the extreme: A study on the power system resilience. *Proc. IEEE*, 105(7):1253–1266, 2017.

[31] E. I. Bilis, W. Kröger, and C. Nan. Performance of electric power systems under physical malicious attacks. *IEEE J. Syst.*, 7(4):854–865, 2013.

[32] A. B. Birchfield, T. Xu, K. M. Gegner, K. S. Shetye, and T. J. Overbye. Grid structural characteristics as validation criteria for synthetic networks. *IEEE Trans. Power Syst.*, 32(4):3258–3265, 2017.

[33] S. Biswas and V. A. Centeno. Determining an operation sequence for proactive islanding of the power grid. In *Proceedings of IEEE Power Energy Society Innovative Smart Grid Technologies Conference*, pp. 1–5, New Orleans, LA, USA, 2022.

[34] S. Biswas, J. D. Follum, E. S. Andersen, and J. S. Banning. Big data analysis of synchrophasor data: Outcomes of research activities supported by DOE FOA 1861. Technical report, Pacific Northwest National Laboratory (PNNL), Richland, WA, United States, 2022.

[35] F. Blaabjerg, Y. Yang, K. A. Kim, and J. Rodriguez. Power electronics technology for large-scale renewable energy generation. *Proc. IEEE*, 111(4):335–355, 2023.

[36] E. Bompard, L. Luo, and E. Pons. A perspective overview of topological approaches for vulnerability analysis of power transmission grids. *Int. J. Crit. Infrastruct.*, 11(1):15–26, 2015.

[37] E. Bompard, R. Napoli, and F. Xue. Analysis of structural vulnerabilities in power transmission grids. *Int. J. Crit. Infrastruct. Prot.*, 2(1):5–12, 2009.

[38] E. Bompard, E. Pons, and D. Wu. Extended topological metrics for the analysis of power grid vulnerability. *IEEE Syst. J.*, 6(3):481–487, 2012.

[39] P. Bonacich. Factoring and weighting approaches to status scores and clique identification. *J. Math. Sociol.*, 2(1):113–120, 1972.

[40] L. Breiman. *Classification and Regression Trees.* New York, USA: Routledge, 2017.

[41] S. Brin and L. Page. The anatomy of a large-scale hypertextual web search engine. *Comput. Netw. and ISDN Systems*, 30(1-7):107–117, 1998.

[42] S. V. Buldyrev, R. Parshani, G. Paul, H. E. Stanley, and S. Havlin. Catastrophic cascade of failures in interdependent networks. *Nature*, 464(7291):1025–1028, 2010.

[43] Y. Cai, Y. Cao, Y. Li, T. Huang, and B. Zhou. Cascading failure analysis considering interaction between power grids and communication networks. *IEEE Trans. Smart Grid*, 7(1):530–538, 2016.

[44] M. B. Cain, et al. History of optimal power flow and formulations. *Fed. Energy Regul. Commission Rep.*, 1:1–36, 2012.

[45] B. A. Carreras, P. Colet, J. M. Reynolds-Barredo, and D. Gomila. Assessing blackout risk with high penetration of variable renewable energies. *IEEE Access*, 9:132663–132674, 2021.

[46] B. A. Carreras, D. E. Newman, I. Dobson, and A. B. Poole. Evidence for self-organized criticality in a time series of electric power system blackouts. *IEEE Trans. Circ. Syst. I Reg. Papers*, 51(9):1733–1740, 2004.

[47] H. Cetinay, S. Soltan, F. A. Kuipers, G. Zussman, and P. Van Mieghem. Comparing the effects of failures in power grids under the AC and DC power flow models. *IEEE Trans. Netw. Sci. Eng.*, 5(4):301–312, 2018.

[48] H. Chen, H. Jin, J. Sun, D. Deng, and X. Liao. Analysis of large-scale topological properties for peer-to-peer networks. In *Proceedings First IEEE/ACM International Symposium on Cluster Computing and the Grid*, Chicago, IL, USA, pp. 27–34, 2004.

[49] L. Chen, F. Hu, S. Wang, and J. Chen. Cyber-physical system fusion modeling and robustness evaluation. *Electr. Power Syst. Res.*, 213:108654, 2022.

[50] Q. Chen, C. Jiang, W. Qiu, and J. D. McCalley. Probability models for estimating the probabilities of cascading outages in high-voltage transmission network. *IEEE Trans. Power Syst.*, 21(3):1423–1431, 2006.

[51] X. Chen, H. Mao, N. Cheng, L. Ma, Z. Tian, Y. Luo, C. Zhou, H. Li, Q. Wang, W. Kong, and J. Fan. Climate change impacts on global photovoltaic variability. *Appl. Energy*, 374:124087, 2024.

[52] Z. Chen, J. Wu, Y. Xia, and X. Zhang. Robustness of interdependent power grids and communication networks: A complex network perspective. *IEEE Trans. Circuits Syst. II Express Briefs*, 65(1):115–119, 2018.

[53] S. Chib and E. Greenberg. Understanding the metropolis-hastings algorithm. *Amer. Stat.*, 49(4):327–335, 1995.

[54] R. Christie. 118 bus power flow test case. Available: [Online] https://labs.ece.uw.edu/pstca/pf118/pg_tca118bus.htm.

[55] C. C. Chu and H. H. C. Iu. Complex networks theory for modern smart grid applications: A survey. *IEEE J. Emerg. Sel. Top. Circuits Syst.*, 7(2):177–191, 2017.

[56] S. Corsi, M. Pozzi, C. Sabelli, and A. Serrani. The coordinated automatic voltage control of the Italian transmission grid-part I: reasons of the choice and overview of the consolidated hierarchical system. *IEEE Trans. Power Syst.*, 19(4):1723–1732, 2004.

[57] P. Costa, J. Rhuggenaath, Y. Zhang, and A. Akcay. Learning 2-opt heuristics for the traveling salesman problem via deep reinforcement learning. *Proc. Asian Conf. Mach. Learn.*, 129:465–480, 2020.

[58] E. Cotilla-Sanchez, P. D. Hines, C. Barrows, and S. Blumsack. Comparing the topological and electrical structure of the North American electric power infrastructure. *IEEE Syst. J.*, 6(4):616–626, 2012.

[59] S. E. Crawford, R. A. Shugayev, H. P. Paudel, P. Lu, M. Syamlal, P. R. Ohodnicki, B. Chorpening, R. Gentry, and Y. Duan. Quantum sensing for energy applications: Review and perspective. *Adv. Quantum Technol.*, 4(8):2100049, 2021.

[60] P. Crucitti, V. Latora, and M. Marchiori. Model for cascading failures in complex networks. *Phys. Rev. E*, 69:045104, 2004.

[61] L. Cuadra, S. Salcedo-Sanz, J. Del Ser, S. Jiménez-Fernández, and Z. Geem. A critical review of robustness in power grids using complex networks concepts. *Energies*, 8(9):9211, 2015.

[62] P. Cuffe and A. Keane. Visualizing the electrical structure of power systems. *IEEE Syst. J.*, 11(3):1810–1821, 2017.

[63] C. Cui, M. He, F. Di, Y. Lu, Y. Dai, and F. Lv. Research on power load forecasting method based on LSTM model. In *Proceedings of 2020 IEEE 5th Information Technology and Mechatronics Engineering Conference (ITOEC)*, Chongqing, China, pp. 1657–1660, 2020.

[64] L. Cui, S. Kumara, and R. Albert. Complex networks: An engineering view. *IEEE Circuits Syst. Mag.*, 10(3):10–25, 2010.

[65] D. T. Gillespie. Exact stochastic simulation of coupled chemical reactions. *J. Comput. Phys.*, 81(25):2340–2361, 1977.

[66] Y. Dai, R. Preece, and M. Panteli. Risk assessment of cascading failures in power systems with increasing wind penetration. *Electr. Power Syst. Res.*, 211:108392, 2022.

[67] L. Das, S. Munikoti, B. Natarajan, and B. Srinivasan. Measuring smart grid resilience: Methods, challenges and opportunities. *Renew. Sustain. Energy Rev.*, 130:109918, 2020.

[68] P. Das, R. A. Shuvro, K. Povinelli, F. Sorrentino, and M. M. Hayat. Mitigating cascading failures in power grids via Markov decision-based load-shedding with DC power flow model. *IEEE Syst. J.*, 16(3):4048–4059, 2022.

[69] A. E. David, B. Gjorgiev, and G. Sansavini. Quantitative comparison of cascading failure models for risk-based decision making in power systems. *Reliab. Eng. Syst. Saf.*, 198:106877, 2020.

[70] D. del Giudice, A. Brambilla, D. Linaro, and F. Bizzarri. Modular multilevel converter impedance computation based on periodic small-signal analysis and vector fitting. *IEEE Trans. Circuits Syst. I Reg. Papers*, 69(4):1832–1842, 2022.

[71] Department of Energy's Oak Ridge National Laboratory. Protecting electric grid health with drone-based power line inspection, 2024. Available: [Online] https://www.ornl.gov/news/protecting-electric-grid-health-drone-based-power-line-inspection.

[72] J. Dias, A. N. Montanari, and E. E. Macau. Power-grid vulnerability and its relation with network structure. *Chaos*, 33(3):033122, 2023.

[73] L. Ding, L. Xie, and X.-K. Xu. Rich-club impact on cascading failures in interdependent power and communication networks. *IEEE J. Emerg. Sel. Topics Circuits Syst.*, 12(1):115–123, 2022.

[74] V. Do, H. McBrien, N. M. Flores, A. J. Northrop, J. Schlegelmilch, M. V. Kiang, and J. A. Casey. Spatiotemporal distribution of power outages with climate events and social vulnerability in the USA. *Nat. Commun.*, 14(1):2470, 2023.

[75] I. Dobson. Estimating the propagation and extent of cascading line outages from utility data with a branching process. *IEEE Trans. Power Syst.*, 27(4):2146–2155, 2012.

[76] S. Dong, Y. Zhao, Y. Jiang, and C. Yang. Research on fault early warning technology of key operation nodes in power communication network. In *Proceedings of IEEE International Conference on Energy Internet (ICEI)*, Southampton, UK, pp.223–228, 2021.

[77] Z. Dong, M. Tian, X. Li, J. Lai, and R. Tang. Mitigating cascading failures of spatially embedded cyber–physical power systems by adding additional information links. *Reliab. Eng. Syst. Saf.*, 225:108559, 2022.

[78] Duke Energy. Self-healing technology improves reliability for customers, 2021. Press Release. Available: [Online] https://news.duke-energy.com/releases/duke-energy-deploys-self-healing-technology-across-north-carolina.

[79] E. Dusabimana and S.-G. Yoon. A survey on the micro-phasor measurement unit in distribution networks. *Electronics*, 9(2):305, 2020.

[80] A. Dwivedi and X. Yu. A maximum-flow-based complex network approach for power system vulnerability analysis. *IEEE Trans. Ind. Inform.*, 9(1):81–88, 2013.

[81] E. C. Eakeley, et al. 1996 system disturbances. Technical report, North American Electric Reliability Corporation, 2002.

[82] ElaadNL. Vehicle-to-grid in the Netherlands: Smart charging for a sustainable future, 2021. Available: [Online] https://elaad.nl/en/group/research/.

[83] Electricity System Operator National Grid. Technical report on the events of 9 August 2019, 2019. Available: [Online] https://www.ofgem.gov.uk/sites/default/files/docs/2020/01/9_august_2019_power_outage_report.pdf.

[84] Electricity System Operator National Grid. 540 MW of wind turbine generation loss due to unexpected and insufficient ride-through performance, 2023. Available: [Online] https://www.nerc.com/globalassets/programs/event-analysis/lessons-learned/ll20231102_wind_generation_failure_to_ride-through.pdf.

[85] EnergyShield Project Consortium. Final results and modular platform for cybersecurity in the energy sector. Technical report, EU Horizon 2020, 2022.

[86] Enquiry Committee. Report on the grid disturbance on 30th July 2012 and grid disturbance on 31st July 2012. Technical report, Ministry of Power, India, 2012.

[87] ENTSO-E. Project inertia-phase II: Recovering power system resilience in case of system splits for a future-ready decarbonised system, 2025. Available: [Online] https://www.entsoe.eu/2025/01/23/entso-e-releases-the-latest-work-from-project-inertia-which-studies-the-evolution-of-the-inertia-levels-in-the-long-term-horizons-in-the-continental-europe-synchronous-area-and-the-challenges-emerging-from-their-decrease/.

[88] ENTSO-E and OneNet Project Consortium. Definition of integrated and fully coordinated markets for the procurement of grid services by DSOs and TSOs, 2023. Available: [Online] https://onenet-project.eu/wp-content/uploads/2023/04/D3.2_OneNet_v1.0.pdf.

[89] P. Erdös and A. Rényi. On random graphs. *Publ. Math.*, 6(26):290–297, 1959.

[90] EU. Regulation (EU) 2019/941 of the European parliament and of the council of 5 June 2019 on risk-preparedness in the electricity sector and repealing directive 2005/89/EC, 2019. Available: [Online] https://eur-lex.europa.eu/eli/reg/2019/941/oj/eng.

[91] European Union Agency for Cybersecurity (ENISA). Smart grid threat landscape and good practice guide, 2013. Available: [Online] https://www.enisa.europa.eu/sites/default/files/publications/Smart

[92] C. Fan, L. Zeng, Y. Sun, and Y.-Y. Liu. Finding key players in complex networks through deep reinforcement learning. *Nat. Mach. Intell.*, 2:317–324, 2020.

[93] J. Fang, J. Wu, Z. Zheng, and C. K. Tse. Revealing structural and functional vulnerability of power grids to cascading failures. *IEEE J. Emerg. Sel. Top. Circ. Syst.*, 11(1):133–143, 2020.

[94] J. Feltes and C. Grande-Moran. Down, but not out: A brief overview of restoration issues. *IEEE Power Energy Mag.*, 12(1):34–43, 2014.

[95] M. Finkelstein. *Failure Rate Modelling for Reliability and Risk.* Berlin, Germany: Springer-Verlag, 2008.

[96] S. Fleming. Lessons for the energy transition from the 2021 energy crisis, 2021. World Economic Forum. Available: [Online] https://www.weforum.org/stories/2021/10/energy-transition-risks-crisis/.

[97] H. Foretic, I. Asensio, Bermejo, V. Kopustinskas, and G. Fulli. State of the art of regional coordination centres (RCCs) and their impact on security of electricity supply in Europe. Publications Office of the European Union, 2024.

[98] L. C. Freeman, et al. Centrality in social networks: Conceptual clarification. *Soc. Networks*, 1:238–263, 2002.

[99] X. Gao, M. Peng, and C. K. Tse. Cascading failure analysis of cyber–physical power systems considering routing strategy. *IEEE Trans. Circuits Syst. II Exp. Briefs*, 70(1):136–140, 2023.

[100] D. T. Gillespie. A general method for numerically simulating the stochastic time evolution of coupled chemical reactions. *J. Comput. Phys.*, 22(4):403–434, 1976.

[101] B. Gjorgiev and G. Sansavini. Identifying and assessing power system vulnerabilities to transmission asset outages via cascading failure analysis. *Reliab. Eng. Syst. Saf.*, 217:108085, 2022.

[102] H. T. GmbH. Digital twin offshore integration program, 2022. Available: [Online] https://www.50hertz.com/en/Grid/Grid-development/Digital-Twin-Offshore.

[103] H. Gooi, D. Mendes, K. Bell, and D. Kirschen. Optimal scheduling of spinning reserve. *IEEE Trans. Power Syst.*, 14(4):1485–1492, 1999.

[104] B. Goswami, N. Boers, A. Rheinwalt, N. Marwan, J. Heitzig, S. F. M. Breitenbach, and J. Kurths. Abrupt transitions in time series with uncertainties. *Nat. Commun.*, 9(1):48, 2018.

[105] J. J. Grainger and W. D. Stevenson. *Power System Analysis.* Englewood Cliff, USA: McGraw Hill, 1994.

[106] M. Z. Gunduz and R. Das. Cyber-security on smart grid: Threats and potential solutions. *Comput. Networks*, 169:107094, 2020.

[107] V. C. Gungor and F. C. Lambert. A survey on communication networks for electric system automation. *Comput. Netwroks*, 50(7):877–897, 2006.

[108] H. Guo, C. Zheng, H. H.-C. Iu, and T. Fernando. A critical review of cascading failure analysis and modeling of power system. *Renew. Sustain. Energy Rev.*, 80:9–22, 2017.

[109] G. Gurrala and I. Sen. Power system stabilizers design for interconnected power systems. *IEEE Trans. Power Syst.*, 25(2):1042–1051, 2010.

[110] Z. Hao, F. Di Maio, and E. Zio. A sequential decision problem formulation and deep reinforcement learning solution of the optimization of O&M of cyber-physical energy systems (CPESs) for reliable and safe power production and supply. *Reliab. Eng. Syst. Saf.*, 235:109231, 2023.

[111] B. Hartmann, G. Ódor, I. Papp, K. Benedek, S. Deng, and J. Kelling. Dynamical heterogeneity and universality of power. grids. *Sustain. Energy Grids Netw.*, 39:101491, 2024.

[112] H. v. Hasselt, A. Guez, and D. Silver. Deep reinforcement learning with double q-learning. In *Proceednings pf the Thirtieth AAAI Conference on Artificial Intelligence, AAAI'16*, Phoenix, AZ, USA, pp. 2094–2100. AAAI Press, 2016.

[113] H. He and J. Yan. Cyber-physical attacks and defences in the smart grid: A survey. *IET Cyber-Phys. Syst. Theory Appl.*, 1(1):13–27, 2016.

[114] K. W. Hedman, S. S. Oren, and R. P. O'Neill. A review of transmission switching and network topology optimization. In *Proceedings of the IEEE Power & Energy Society General Meeting*, Detroit, MI, USA, pp. 1–7, 2011.

[115] B. Heller, R. Sherwood, and N. McKeown. The controller placement problem. In *Proceedning of the 1st Workshop Hot Topics Software-Defined Networking (SDN)*, pp. 7–12, New York, NY, USA, 2012. ACM.

[116] P. D. H. Hines, I. Dobson, and P. Rezaei. Cascading power outages propagate locally in an influence graph that is not the actual grid topology. *IEEE Trans. Power Syst.*, 32(2):958–967, 2017.

[117] G. Ho, A. Sharma, M. Javed, V. Paxson, and D. Wagner. Detecting credential spearphishing attacks in enterprise settings. In *Proceedings of the USENIX Security Symposia*, Vancouver, BC, Canada, pp. 469–485, 2017.

[118] J. Homer. Power companies must adapt to climate change now. *Nature*, 620:9, 2023.

[119] E. Hossain, S. Roy, N. Mohammad, N. Nawar, and D. R. Dipta. Metrics and enhancement strategies for grid resilience and reliability during natural disasters. *Appl. Energy*, 290:116709, 2021.

[120] R. R. Hossain, Q. Huang, and R. Huang. Graph convolutional network-based topology embedded deep reinforcement learning for voltage stability control. *IEEE Trans. Power Syst.*, 36(5):4848–4851, 2021.

[121] G. Hou, et al. Resilience assessment and enhancement evaluation of power distribution systems subjected to ice storms. *Reliab. Eng. Syst. Saf.*, 230:108964, 2023.

[122] A. M. Howlader and T. Senjyu. A comprehensive review of low voltage ride through capability strategies for the wind energy conversion systems. *Renew. Sustain. Energy Rev.*, 56:643–658, 2016.

[123] F. Hu, L. Chen, and J. Chen. Robustness evaluation of complex power grids containing renewable energy. *Int. J. Electr. Power Energy Syst.*, 132:107187, 2021.

[124] Z. Hu and L. Wang. Analysis of power grid fault processing based on 110 kV self-healing system. In *Proceedings of the International Conference on Advanced Electrical Energy Systems (AEES)*, Shanghai, China, pp. 225–229, 2023.

[125] M. S. Ibrahim, W. Dong, and Q. Yang. Machine learning driven smart electric power systems: Current trends and new perspectives. *Appl. Energy*, 272:115237, 2020.

[126] M. D. Ilic, L. Xie, U. A. Khan, and J. M. F. Moura. Modeling of future cyber–physical energy systems for distributed sensing and control. *IEEE Trans. Syst., Man, Cybern.*, 40(4):825–838, 2010.

[127] Illinois Center for a Smarter Electric Grid. UIUC 150-bus system, 2016. Available: [Online] http://icseg.iti.illinois.edu/synthetic-power-cases/uiuc-150-bus-system.

[128] M. Inc. Intelligent security: Using machine learning to help detect advanced cyber attacks, 2016. Available: [Online] https://info.microsoft.com/rs/157-GQE-382/images/EN-GB-CNTNT-WhitePaper-Microsoft-Security-GEP-MECH.pdf.

[129] International Energy Agency. Global energy review, 2021. Available: [Online] https://www.iea.org/reports/global-energy-review-2021.

[130] F. James. Monte Carlo theory and practice. *Rep. Prog. Phys.*, 43(9):1145, 1980.

[131] P. Javanbakht and S. Mohagheghi. A risk-aware generation dispatch including wind power for a power grid subjected to hurricanes. *Int. Trans. Electr. Energy Syst.*, 25(11):2982–3003, 2015.

[132] L. B. John and B. K. Todd. Tropical cyclone report: Hurricane Gustav. Technical report, National Hurricane Center, 2009. Available: [Online] https://www.nhc.noaa.gov/ data/tcr/AL072008_Gustav.pdf.

[133] F. H. Jufri, V. Widiputra, and J. Jung. State-of-the-art review on power grid resilience to extreme weather events: Definitions, frameworks, quantitative assessment methodologies, and enhancement strategies. *Appl. Energy*, 239:1049–1065, 2019.

[134] F. Kaiser, V. Latora, and D. Witthaut. Network isolators inhibit failure spreading in complex networks. *Nat. Commun.*, 12(1):1–9, 2021.

[135] M. Kamruzzaman, J. Duan, D. Shi, and M. Benidris. A deep reinforcement learning-based multi-agent framework to enhance power system resilience using shunt resources. *IEEE Trans. Power Syst.*, 36(6):5525–5536, 2021.

[136] V. Karyotis and M. Khouzani. *Malware Diffusion Models for Modern Complex Networks: Theory and Applications*. San Francisco, USA: Morgan Kaufmann, 2016.

[137] J. Kavinnilaa, E. Hemalatha, M. S. Jacob, and R. Dhanalakshmi. Stock price prediction based on LSTM deep learning model. In *Proceedings of the International Conference on System, Computation, Automation and Networking*, Puducherry, India, pp. 1–4, 2021.

[138] M. Khodayar, G. Liu, J. Wang, and M. E. Khodayar. Deep learning in power systems research: A review. *CSEE J. Power Energy Syst.*, 7(2):209–220, 2021.

[139] C. Kim, K. Kim, P. Balaprakash, and M. Anitescu. Graph convolutional neural networks for optimal load shedding under line contingency. In *Proceedings of the IEEE Power & Energy Society General Meeting*, pp. 1–5, 2019.

[140] D. J. Klein and M. Randić. Resistance distance. *J. Math. Chem.*, 12(1):81–95, 1993.

[141] Y. Ko, M. Warnier, P. V. Mieghem, R. E. Kooij, and F. Brazier. The impact of the topology on cascading failures in a power grid model. *Physica A*, 402:169–179, 2014.

[142] L. Kong, Y. Xue, L. Qiao, and F. Wang. Review of small-signal converter-driven stability issues in power systems. *IEEE Open J. Circ. Syst.*, 9:29–41, 2022.

[143] B. Kroposki, B. Johnson, Y. Zhang, V. Gevorgian, P. Denholm, B.-M. Hodge, and B. Hannegan. Achieving a 100% renewable grid: Operating electric power systems with extremely high levels of variable renewable energy. *IEEE Power Energy Mag.*, 15(2):61–73, 2017.

[144] P. Kundur, J. Paserba, V. Ajjarapu, G. Andersson, A. Bose, C. Canizares, N. Hatziargyriou, D. Hill, A. Stankovic, C. Taylor, T. V. Cutsem, and V. Vittal. Definition and classification of power system stability IEEE/CIGRE joint task force on stability terms and definitions. *IEEE Trans. Power Syst.*, 19(3):1387–1401, 2004.

[145] R. M. Lee, M. J. Assante, and T. Conway. Analysis of the cyber attack on the Ukrainian power grid, 2016. Available: [Online] https://media.kasperskycontenthub.com/wp-content/uploads/sites/43/2016/05/20081514/E-ISAC_SANS_Ukraine_DUC_5.pdf.

[146] J. Leskovec and A. Krevl. SNAP Datasets: Stanford large network dataset collection, 2014. Available: [Online] http://snap.stanford.edu/data.

[147] G. Li, L. Zhang, Y. Wang, and Z. Kang. Critical node identification method of power grid based on the improved entropy weight method. *Electronics*, 12(11):2439, 2023.

[148] M. Li, Z. Yu, T. Xu, J. He, C. Wang, X. Xie, and C. Liu. Study of complex oscillation caused by renewable energy integration and its solution. *Power Syst. Technol.*, 41(4):1035–1042, 2017.

[149] M. J. Li and C. K. Tse. Quantification of cascading failure propagation in power systems. *IEEE Trans. Circ. Syst. I, Reg. Papers*, 71(8):3717–3725, 2024.

[150] M. J. Li and C. K. Tse. Where should inverter-based resources be located in power networks? *IEEE Trans. Circuits Syst. I Reg. Papers*, 71(3):1456–1464, 2024.

[151] M. J. Li, C. K. Tse, and M. Yi. The impact of inverter-based resources (IBRs) on cascading failures in power systems. *IEEE Trans. Power Syst.*, 39(4):5946–5955, 2024.

[152] N. Li, C. Zhao, and L. Chen. Connecting automatic generation control and economic dispatch from an optimization view. *IEEE Trans. Control. Netw. Syst.*, 3(3):254–264, 2016.

[153] Q. Li, X. Zhang, J. Guo, X. Shan, Z. Wang, Z. Li, and C. K. Tse. Integrating reinforcement learning and optimal power dispatch to enhance power grid resilience. *IEEE Trans. Circ. Syst. II, Expr. Briefs*, 69(3):1402–1406, 2022.

[154] S. Li, T. Ding, W. Jia, C. Huang, J. P. Catalão, and F. Li. A machine learning-based vulnerability analysis for cascading failures of integrated power-gas systems. *IEEE Trans. Power Syst.*, 37(3):2259–2270, 2021.

[155] Z. Li, Z. Shi, G. Ruan, Y. Lin, and J. Zhao. Resilience-oriented operation of power distribution networks with line hardening and comprehensive reconfiguration measures. In *Proceednings pf the IEEE International Conference on Communications, Control, and Computing Technologies for Smart Grids (SmartGridComm)*, Glasgow, United Kingdom, pp. 1–6. IEEE, 2023.

[156] X. Lian, T. Qian, Z. Li, X. Chen, and W. Tang. Resilience assessment for power system based on cascading failure graph under disturbances caused by extreme weather events. *Int. J. Electr. Power Energy Syst.*, 145:108616, 2023.

[157] D. Linaro, D. del Giudice, A. Brambilla, and F. Bizzarri. Application of envelope-following techniques to the simulation of hybrid power systems. *IEEE Trans. Circuits Syst. I Reg. Papers*, 69(4):1800–1810, 2022.

[158] B. Liu, Z. Li, X. Chen, Y. Huang, and X. Liu. Recognition and vulnerability analysis of key nodes in power grid based on complex network centrality. *IEEE Trans. Circuits Syst. II Express Briefs*, 65(3):346–350, 2018.

[159] D. Liu and C. K. Tse. Cascading failure of cyber-coupled power systems considering interactions between attack and defense. *IEEE Trans. Circuits Syst. I Regul. Pap.*, 66(11):4323–4336, Nov. 2019.

[160] D. Liu, C. K. Tse, and X. Zhang. Robustness assessment and enhancement of power grids from a complex network's perspective using decision trees. *IEEE Trans. Circ. Syst. II, Expr. Briefs*, 66(5):833–837, 2019.

[161] D. Liu, C. K. Tse, and X. Zhang. Tradeoff between robustness and functionality in cyber-coupled power systems. *IEEE Syst. J.*, 16(1):499–509, 2022.

[162] D. Liu, X. Zhang, and C. K. Tse. A tutorial on modeling and analysis of cascading failure in future power grids. *IEEE Trans. Circuits Syst. II Exp. Briefs*, 68(1):49–55, 2021.

[163] D. Liu, X. Zhang, and C. K. Tse. Effects of high level of penetration of renewable energy sources on cascading failure of modern power systems. *IEEE J. Emerg. Sel. Top. Circ. Syst.*, 12(1):98–106, 2022.

[164] H.-C. Liu, L. Liu, and N. Liu. Risk evaluation approaches in failure mode and effects analysis: A literature review. *Expert Syst. Appl.*, 40(2):828–838, 2013.

[165] J. Liu, T. Yin, M. Shen, X. Xie, and J. Cao. State estimation for cyber–physical systems with limited communication resources, sensor saturation and denial-of-service attacks. *ISA Trans.*, 104:101–114, 2018.

[166] X. Liu, Y. Yan, Y. Chen, Q. Tian, J. Li, W. Zheng, and T. Li. Fast risk assessment of distribution grid with iterative inference on probabilistic graph. *Energy Rep.*, 8:810–817, 2022.

[167] Y. Liu, R. Fan, and V. Terzija. Power system restoration: A literature review from 2006 to 2016. *J. Mod. Power Syst. Clean Energy*, 4(3):332–341, 2016.

[168] Y. Liu, Y. Wang, P. Yong, N. Zhang, C. Kang, and D. Lu. Fast power system cascading failure path searching with high wind power penetration. *IEEE Trans. Sustain. Energy*, 11(4):2274–2283, 2020.

[169] Y. Liu, N. Zhang, D. Wu, A. Botterud, R. Yao, and C. Kang. Searching for critical power system cascading failures with graph convolutional network. *IEEE Trans. Control Netw. Syst.*, 8(3):1304–1313, 2021.

[170] W.-Y. Loh and Y.-S. Shih. Split selection methods for classification trees. *Stat. Sin.*, 7(4):815–840, 1997.

[171] L. Ma, M. Khazaali, and P. Bocchini. Component-based fragility analysis of transmission towers subjected to hurricane wind load. *Eng. Struct.*, 242:112586, 2021.

[172] R. Ma, Y. Zhang, M. Han, J. Kurths, and M. Zhan. Synchronization stability and multi-timescale analysis of renewable-dominated power systems. *Chaos*, 33(8):082101, 2023.

[173] R. Machlev, L. Heistrene, M. Perl, K. Levy, J. Belikov, S. Mannor, and Y. Levron. Explainable Artificial Intelligence (XAI) techniques for energy and power systems: Review, challenges and opportunities. *Energy AI*, 9:100169, 2022.

[174] J. Machowski, J. W. Bialek, J. Bialek, and J. R. Bumby. *Power System Dynamics and Stability*. Chichester, U.K.: Wiley, 1997.

[175] L. A. Maglaras and J. Jiang. Intrusion detection in SCADA systems using machine learning techniques. In *Proceedings of the Scientific Information, Confirmation*, London, UK, pp. 626–631, 2014.

[176] N. Marwan, M. C. Romano, M. Thiel, and J. Kurths. Recurrence plots for the analysis of complex systems. *Phys. Rep.*, 438(5-6):237–329, 2007.

[177] B. Mather. Power electronics as an enabler of the current and future grid: Control instead of copper. Technical report, National Renewable Energy Laboratory, Golden, CO United States, 2024.

[178] W. Q. Meeker, L. A. Escobar, and F. G. Pascual. *Statistical Methods for Reliability Data.* Hoboken, USA: Wiley, 2022.

[179] S. Mei, X. Zhang, and M. Cao. *Power Grid Complexity.* Berlin Heidelberg, Germany: Springer, 2011.

[180] P. J. Menck, J. Heitzig, J. Kurths, and H. J. Schellnhuber. How dead ends undermine power grid stability. *Nat. Comm.*, 5:3969, 2014.

[181] R. Mitchell and I.-R. Chen. A survey of intrusion detection techniques for cyber-physical systems. *ACM Comput. Surv.*, 46(4):55, 2014.

[182] V. Mnih, A. P. Badia, M. Mirza, A. Graves, T. Lillicrap, T. Harley, D. Silver, and K. Kavukcuoglu. Asynchronous methods for deep reinforcement learning. In M. F. Balcan and K. Q. Weinberger, editors, *Proceedning of the 33rd The International Conference on Machine Learning (ICML*, vol. 48, pp. 1928–1937, New York, New York, USA, 20–22 Jun 2016. PMLR.

[183] N. Modi, M. V. Escudero, K. Aramaki, X. Zhou, and P. Partinen. High inverter-based resource integration: The experience of five system operators. *IEEE Power Energy Mag.*, 22(2):78–88, 2024.

[184] J. Montoya Rincon, J. Gonzalez, and M. P. Jensen. Evaluation of power transmission lines hardening scenarios using a machine learning approach. *ASCE-ASME J. Risk Uncertain. Eng. Syst. B Mech. Eng.*, 9(3):031106, 2023.

[185] R. Motoko and B. Dooley. Typhoon Hagibis slams into Japan, after landslides, floods and a quake. *The New York Times*, Oct. 2019. Available: [Online] https://www.nytimes.com/2019/10/12/world/asia/typhoon-hagibis.html.

[186] A. E. Motter and Y.-C. Lai. Cascade-based attacks on complex networks. *Phys. Rev. E*, 66(6):065102, 2002.

[187] A. E. Motter, S. A. Myers, M. Anghel, and T. Nishikawa. Spontaneous synchrony in power-grid networks. *Nature Phys.*, 9(3):191–197, 2013.

[188] G. Alanis-Lobato. CNM, MATLAB® Central File Exchange. Available: [Online] https://www.mathworks.com/matlabcentral/fileexchange/45734-cnm.

[189] M. F. Mukhtar, Z. Abal Abas, A. S. Baharuddin, M. N. Norizan, W. F. W. W. Fakhruddin, W. Minato, A. H. A. Rasib, Z. Z. Abidin, A. F. N. A. Rahman, and S. H. H. Anuar. Integrating local and global information to identify influential nodes in complex networks. *Sci. Rep.*, 13(1):11411, 2023.

[190] National Development and Reform Commission, National Energy Administration, National Data Bureau. Action plan for accelerating the construction of new-type power system, 2024. Available: [Online] https://www.gov.cn/zhengce/zhengceku/202408/P020240806793811358886.pdf.

[191] M. E. Newman. The structure and function of complex networks. *SIAM Rev.*, 45(2):167–256, 2003.

[192] M. E. Newman. *Networks: An Introduction.* London: Oxford University Press, 2018.

[193] M. E. Newman and M. Girvan. Finding and evaluating community structure in networks. *Phys. Rev. E.*, 69(2):026113, 2004.

[194] S. A. Nezam Sarmadi, A. S. Dobakhshari, S. Azizi, and A. M. Ranjbar. A sectionalizing method in power system restoration based on WAMS. *IEEE Trans. Smart Grid*, 2(1):190–197, 2011.

[195] M. Noebels, R. Preece, and M. Panteli. AC cascading failure model for resilience analysis in power networks. *IEEE Systems J.*, 16(1):374–385, 2022.

[196] North American Electric Reliability Corporation. Remedial action scheme definition development, 2014.

[197] J. Novacheck, J. Sharp, M. Schwarz, P. Donohoo-Vallett, Z. Tzavelis, G. Buster, and M. Rossol. The evolving role of extreme weather events in the US power system with high levels of variable renewable energy. Technical report, National Renewable Energy Laboratory (NREL), Golden, CO, United States, 2021.

[198] Pacific Northwest National Laboratory. Artificial intelligence/machine learning technology in power system applications. technical report, U.S. Department of Energy, 2024. Available: [Online] https://www.pnnl.gov/main/publications/external/technical_reports/PNNL-35735.pdf.

[199] G. A. Pagani and M. Aiello. The power grid as a complex network: A survey. *Physica A*, 392(11):2688–2700, 2013.

[200] S. Pahwa, C. M. Scoglio, S. Das, and N. Schulz. Load-shedding strategies for preventing cascading failures in power grid. *Electr. Power Compon. Syst.*, 41:879 – 895, 2013.

[201] A. Pai. *Energy Function Analysis for Power System Stability.* New York, USA: Springer, 1989.

[202] S. Panda, S. Mohanty, P. K. Rout, B. K. Sahu, S. M. Parida, I. S. Samanta, M. Bajaj, M. Piecha, V. Blazek, and L. Prokop. A comprehensive review on demand side management and market design for

renewable energy support and integration. *Energy Rep.*, 10:2228–2250, 2023.

[203] J. Parkinson. Hurricane Michael leaves 'unimaginable' destruction, 2018. Available: [Online] https://www.bbc.com/news/uk-43789747.

[204] R. Parshani, S. V. Buldyrev, and S. Havlin. Interdependent networks: Reducing the coupling strength leads to a change from a first to second order percolation transition. *Phys. Rev. Lett.*, 105:048701, 2010.

[205] R. Parshani, C. Rozenblat, D. Ietri, C. Ducruet, and S. Havlin. Inter-similarity between coupled networks. *Europhys. Lett.*, 92(6):68002, 2010.

[206] N. T. Pashiri, S. Azad, and M. T. Ameli. Distribution system resilience improvement against hurricanes: optimal line hardening according to reconfiguration potentials. *J. Eng.*, 2024(3):e12364, 2024.

[207] S. Paul, A. Poudyal, S. Poudel, A. Dubey, and Z. Wang. Resilience assessment and planning in power distribution systems: Past and future considerations. *Renew. Sustain. Energy Rev.*, 189:113991, 2024.

[208] Q. Peng, Q. Jiang, Y. Yang, T. Liu, H. Wang, and F. Blaabjerg. On the stability of power electronics-dominated systems: Challenges and potential solutions. *IEEE Trans. Ind. Appl.*, 55(6):7657–7670, 2019.

[209] R. Pfitzner, K. Turitsyn, and M. Chertkov. Controlled tripping of overheated lines mitigates power outages. *arXiv preprint:1104.4558*, 2011.

[210] A. G. Phadke and J. S. Thorp. *Computer Relaying for Power Systems.* New York, USA: Wiley, 2009.

[211] B. L. Preston, S. N. Backhaus, M. Ewers, J. A. Phillips, C. Silva-Monroy, J. Dagle, A. Tarditi, J. Looney, and T. King. Resilience of the us electricity system: A multi-hazard perspective. US Department of Energy Office of Policy. Washington, DC, 2016.

[212] T. Ptak, J. Brooks, and R. Stock. A systematic review and typology of power outage literature: Critical infrastructure, climate change and social impacts. *Renew. Sustain. Energy Rev.*, 218:115778, 2025.

[213] W. Quattrociocchi, G. Caldarelli, and A. Scala. Self-healing networks: Redundancy and structure. *PloS ONE*, 9(2):e87986, 2014.

[214] A. Raffin, A. Hill, A. Gleave, A. Kanervisto, M. Ernestus, and N. Dormann. Stable-baselines3: Reliable reinforcement learning implementations. *J. Mach. Learn. Res.*, 22(268):1–8, 2021.

[215] K. M. J. Rahman, M. M. Munnee, and S. Khan. Largest blackouts around the world: Trends and data analyses. In *Proceedings of the IEEE International WIE Conference on Electrical and Computer Engineering*, Pune, India, pp. 155–159, 2016.

[216] M. Rahnamay-Naeini, Z. Wang, N. Ghani, A. Mammoli, and M. M. Hayat. Stochastic analysis of cascading-failure dynamics in power grids. *IEEE Trans. Power Syst.*, 29(4):1767–1779, Jan. 2014.

[217] R. Rajkumar, I. Lee, L. Sha, and J. Stankovic. Cyber-physical systems: The next computing revolution. In *Proceedings of the Design Automation Conference*, pp. 731–736, 2010.

[218] H. Ren, I. Dobson, and B. A. Carreras. Long-term effect of the $n-1$ criterion on cascading line outages in an evolving power transmission grid. *IEEE Trans. Power Syst.*, 23(3):1217–1225, 2008.

[219] J. Renju. Australia to rescue thousands as Sydney faces worst floods in 60 years. *Reuters*, 2021. Available: [Online] https://www.reuters.com/business/environment/australia-evacuate-thousands-sydney-faces-worst-floods-60-years-2021-03-22/.

[220] B. Rolston, J. Gentle, Z. Priest, J. Laundergan, and B. Smith. North American electric reliability corporation critical infrastructure protection-related content from EnerNex. Technical report, Idaho National Laboratory (INL), Idaho Falls, ID United States, May 2024.

[221] V. Rosato, S. Bologna, and F. Tiriticco. Topological properties of high-voltage electrical transmission networks. *Electric Power Syst. Res.*, 77(2):99–105, 2007.

[222] F. Rosenblatt. Principles of neurodynamics. perceptrons and the theory of brain mechanisms. Technical report, Cornell Aeronautical Lab Inc Buffalo, NY, United States, 1961.

[223] D. E. Rumelhart, G. E. Hinton, and R. J. Williams. Learning internal representations by error propagation. Technical report, California Univ San Diego La Jolla Inst for Cognitive Science, 1985.

[224] G. Sabidussi. The centrality index of a graph. *Psychometrika*, 31(4):581–603, 1966.

[225] A. Saltelli. Sensitivity analysis for importance assessment. *Risk Anal.*, 22(3):579–590, 2002.

[226] A. Scala, S. Pahwa, and C. Scoglio. Cascade failures from distributed generation in power grids. *arXiv preprint arXiv:1209.3733*, 2012.

[227] A. Scherb, L. Garre, and D. Straub. Evaluating component importance and reliability of power transmission networks subject to windstorms: Methodology and application to the nordic grid. *Reliab. Eng. Syst. Saf.*, 191:106517, 2019.

[228] C. M. Schneider, A. A. Moreira, J. S. Andrade, S. Havlin, and H. J. Herrmann. Mitigation of malicious attacks on networks. *Proc. Natl. Acad. Sci.*, 108(10):3838–3841, 2011.

[229] J. Schulman, F. Wolski, P. Dhariwal, A. Radford, and O. Klimov. Proximal policy optimization algorithms. *arXiv preprint arXiv:1707.06347*, 2017.

[230] Y. Serizawa, E. Ohba, and M. Kurono. Present and future ICT infrastructures for a smarter grid in Japan. In *Proceedings of the 2010 Innovative Smart Grid Technologies (ISGT)*, Gaithersburg, MD, USA, pp. 1–5, 2010.

[231] J. Shair, H. Li, J. Hu, and X. Xie. Power system stability issues, classifications and research prospects in the context of high-penetration of renewables and power electronics. *Renew. Sustain. Energy Rev.*, 145:111111, 2021.

[232] X. Shan, X. Zhang, and C. K. Tse. Identifying sets of critical components that affect the resilience of power networks. *IEEE Trans. Circuits Syst. II Express Briefs*, 69(12):4859–4863, 2022.

[233] C. Shang, J. Gao, H. Liu, and F. Liu. Short-term load forecasting based on PSO-KFCM daily load curve clustering and CNN-LSTM model. *IEEE Access*, 9:50344–50357, 2021.

[234] Shenzhen Bus Group and China Southern Grid. Electric bus grid integration program, 2025. Available: [Online] https://fgw.sz.gov.cn/gkmlpt/content/12/12254/post_12254427.html, 2025.

[235] A. Shokri Gazafroudi, F. Neumann, and T. Brown. Topology-based approximations for n-1 contingency constraints in power transmission networks. *Int. J. Electr. Power Energy Syst.*, 137:107702, 2022.

[236] Y. Shu, G. Chen, J. He, and F. Zhang. Building a new electric power system based on new energy sources. *Strateg. Stud. CAE*, 23(6):61–69, 2021.

[237] C. Silva, P. Faria, Z. Vale, and J. Corchado. Demand response performance and uncertainty: A systematic literature review. *Energy Strategy Rev.*, 41:100857, 2022.

[238] D. Silver, et al. Mastering the game of Go without human knowledge. *Nature*, 550(7676):354–359, 2017.

[239] O. Smith, O. Cattell, E. Farcot, R. D. O'Dea, and K. I. Hopcraft. The effect of renewable energy incorporation on power grid stability and resilience. *Sci. Adv.*, 8(9):eabj6734, 2022.

[240] J. Song, M. Cheah-Mane, E. Prieto-Araujo, J. Amorós, and O. Gomis-Bellmunt. Grid equivalent representation of power systems with penetration of power electronics. *IEEE Trans. Power Del.*, 38(4):2742–2756, 2023.

[241] S. Sridhar, A. Hahn, and M. Govindarasu. Cyber-physical system security for the electric power grid. *Proc. IEEE*, 100(1):210–224, 2012.

[242] A. Srivastava, T. Morris, T. Ernster, C. Vellaithurai, S. Pan, and U. Adhikari. Modeling cyber-physical vulnerability of the smart grid with incomplete information. *IEEE Trans. Smart Grid*, 4(1):235–244, 2013.

[243] A. Staid. North American energy resilience model. Technical report, Sandia National Laboratory, Albuquerque, NM, United States, 2021.

[244] M. Starke, B. Xiao, P. K. Bhowmik, S. L. Campbell, M. Chinthavali, B. R. Dean, R. Sree Krishna Moorthy, M. Smith, and A. Thapa. Agent-based distributed energy resources for supporting intelligence at the grid edge. *IEEE J. Emerg. Sel. Top. Power Electron.*, 3(1):69–78, 2022.

[245] State Grid Corporation of China. Digital twin platform for urban substations. SGCC News Release, 2023. Available: [Online] https://jsnews.jschina.com.cn/sz/a/202303/t20230302_3172598.shtml.

[246] J. Stürmer, A. Plietzsch, T. Vogt, F. Hellmann, J. Kurths, C. Otto, K. Frieler, and M. Anvari. Increasing the resilience of the texas power grid against extreme storms by hardening critical lines. *Nat. Energy*, 9(5):1–10, 2024.

[247] Q. Su, C. Chen, and J. Li. Trendrank method for evaluating the importance of power grid nodes considering information network. *IET Gener. Transm. Distrib.*, 17(3):539–550, 2023.

[248] K. Sun. Complex networks theory: A new method of research in power grid. In *Proceedings of the IEEE Transmission and Distribution Conference and Exhibition Asia Pacific*, pp. 1–6, Dalian, China, 2005.

[249] C. W. Taylor, F. R. Nassief, and R. L. Cresap. Northwest power pool transient stability and load shedding controls for generation-load imbalances. *IEEE Trans. Power Appar. Syst.*, PAS-100(7):3486–3495, 1981.

[250] D. A. Tortorelli and P. Michaleris. Design sensitivity analysis: Overview and review. *Inv. Prob. Eng.*, 1(1):71–105, 1994.

[251] C. K. Tse, M. Huang, X. Zhang, D. Liu, and X. L. Li. Circuits and systems issues in power electronics penetrated power grid. *IEEE Open J. Circuits Syst.*, 1:140–156, 2020.

[252] US-Canada System Outage Task Force. Final report on the August 14th blackout in the United States and Canada. Technical report, US Dept. Energy and National Res. Canada, 2004.

[253] U.S. Department of Energy. Cybersecurity research, development, and demonstration (RD&D) for energy delivery systems, 2021. Available: [Online] https://www.energy.gov/ceser/cybersecurity-research-development-and-demonstration-rdd-energy-delivery-systems.

[254] T. Van Cutsem and C. Vournas. *Voltage Stability of Electric Power Systems.* Berlin, Germany : Springer, 2007.

[255] Z. Wan, G. Wang, Y. Yang, and S. Shi. SKM: scalable key management for advanced metering infrastructure in smart grids. *IEEE Trans. Ind. Electron.*, 61(12):7055–7066, 2014.

[256] C. Wang, Y. Dong, P. Sun, and Y. Lu. Cascading failure path prediction based on association rules in cyber-physical active distribution networks. In *proceedings of the 20th IEEE International Conference on Software Quality, Reliability and Security Companion (QRS-C 2020)*, Macau, China, pp. 458–464, 2020.

[257] G. Wang, J. Guo, S. Ma, X. Zhang, Q. Guo, S. Fan, and H. Xu. Data-driven transient stability assessment with sparse PMU sampling and online self-check function. *CSEE J. Power Energy Syst.*, 1–10, 2022.

[258] J. W. Wang and L. L. Rong. Cascade-based attack vulnerability on the US power grid. *Safety Sci.*, 47(10):1332–1336, 2009.

[259] K. Wang, R. Zhang, H. Yuan, Y. Pei, W. Zhang, and H. Xu. Construction and training of cloud computing-based smart grid operation risk early warning model. In *Proceedings of International Conference on Machine Learning, Big Data Analytics, Internet of Things (IoT) Security, and Privacy*, pp. 761–768, Cham, 2022. Springer International Publishing.

[260] L. Wang, X. Xie, Q. Jiang, H. Liu, Y. Li, and H. Liu. Investigation of SSR in practical DFIG-based wind farms connected to a series-compensated power system. *IEEE Trans. Power Syst.*, 30(5):2772–2779, 2015.

[261] S. Wang, Q. Dong, J. Zhang, J. Sun, X. Gu, and C. Chen. Robustness assessment of power network with renewable energy. *Electric Power Syst. Res.*, 217:109138, 2023.

[262] S. Wang, W. Lv, J. Zhang, S. Luan, C. Chen, and X. Gu. Method of power network critical nodes identification and robustness enhancement based on a cooperative framework. *Reliab. Eng. Syst. Saf.*, 207:107313, 2021.

[263] W. Wang, Y. Xu, and M. Khanna. A survey on the communication architectures in smart grid. *Comput. Netw.*, 55(15):3604–3629, 2011.

[264] X. Wang and F. Blaabjerg. Harmonic stability in power electronic-based power systems: Concept, modeling, and analysis. *IEEE Trans. Smart Grid*, 10(3):2858–2870, 2019.

[265] X. F. Wang and G. Chen. Complex networks: small-world, scale-free and beyond. *IEEE Circuits Syst. Mag.*, 3(1):6–20, 2003.

[266] Y. Wang, Z. Lin, X. Liang, W. Xu, Q. Yang, and G. Yan. On modeling of electrical cyber-physical systems considering cyber security. *Front. Inf. Technol. Electron. Eng.*, 17(5):465–478, 2016.

[267] Y. Wang, B. Wang, and M. Liang. A smart power system cascading failure accident prediction model based on probability analysis. In *Proceedings of the International Conference on Power System Technology*, Hangzhou, China, pp. 1–6, 2010.

[268] Z. Wang, M. Rahnamay-Naeini, J. M. Abreu, R. A. Shuvro, P. Das, A. A. Mammoli, N. Ghani, and M. M. Hayat. Impacts of operators' behavior on reliability of power grids during cascading failures. *IEEE Trans. Power Syst.*, 33(6):6013–6024, 2018.

[269] Z. Wang, A. Scaglione, and R. J. Thomas. Electrical centrality measures for electric power grid vulnerability analysis. In *Proceedings of the IEEE Conference on Decision and Control*, Atlanta, GA, USA, pp. 5792–5797, 2010.

[270] Z. Wang, A. Scaglione, and R. J. Thomas. A markov-transition model for cascading failures in power grids. In *Proceedings of the Hawaii International Conference on System Sciences*, Maui, HI, USA, pp. 2115–2124, 2012.

[271] Z. Wang, R. J. Thomas, and A. Scaglione. Generating random topology power grids. In *Proceedings of the Hawaii International Conference on System Sciences*, Waikoloa, HI, USA, pp. 183–183, 2008.

[272] D. J. Watts and S. H. Strogatz. Collective dynamics of small-world networks. *Nature*, 393(6684):440–442, 1998.

[273] F. Wenli, L. Zhigang, H. Ping, and M. Shengwei. Cascading failure model in power grids using the complex network theory. *IET Gener. Transm. Distrib.*, 10(15):3940–3949, 2016.

[274] A. Widuto. AI and the energy sector. Technical report, European Parliamentary Research Service, 2025. Available: [Online] https://www.europarl.europa.eu/RegData/etudes/BRIE/2025/775859/EPRS_BRI(2025)775859_EN.pdf.

[275] J. Winkler, L. Duenas-Osorio, R. Stein, and D. Subramanian. Performance assessment of topologically diverse power systems subjected to hurricane events. *Reliab. Eng. System Safety*, 95(4):323–336, 2010.

[276] C. J. Wong and M. D. Miller. *Guidelines for Electrical Transmission Line Structural Loading.* Reston: American Society of Civil Engineers, 2009.

[277] A. J. Wood, B. F. Wollenberg, and G. B. Sheblé. *Power Generation, Operation, and Control.* New York, USA: Wiley, 2013.

[278] J. Wu, C. K. Tse, F. C. Lau, and I. W. Ho. Analysis of communication network performance from a complex network perspective. *IEEE Trans. Circuits Syst. I Regul. Pap.*, 60(12):3303–3316, 2013.

[279] X. Wu, D. Wu, and E. Modiano. Predicting failure cascades in large scale power systems via the influence model framework. *IEEE Trans. Power Syst.*, 36(5):4778–4790, 2021.

[280] H. Xie, L. Tang, H. Zhu, X. Cheng, and Z. Bie. Robustness assessment and enhancement of deep reinforcement learning-enabled load restoration for distribution systems. *Reliab. Eng. Syst. Saf.*, 237:109340, 2023.

[281] S.-H. Xiong, Y.-J. Xin, Z.-S. Chen, R. M. Rodríguez, S.-H. Feng, L. Martínez, and W. Pedrycz. Two improved N-two-stage K-means clustering aggregation algorithmic paradigms for HFLTS possibility distributions. *Inf. Fusion*, 100:101964, 2023.

[282] H. Xu, J. Zhang, J. Yang, and L. Lun. Identifying important nodes in complex networks based on multiattribute evaluation. *Math. Probl. Eng.*, 2018, 2018.

[283] Y. Yan, Y. Qian, H. Sharif, and D. Tipper. A survey on smart grid communication infrastructures: Motivations, requirements and challenges. *IEEE Commun. Surveys Tuts.*, 15(1):5–20, 2013.

[284] Z. Yan and Y. Xu. A multi-agent deep reinforcement learning method for cooperative load frequency control of a multi-area power system. *IEEE Trans. Power Syst.*, 35(6):4599–4608, 2020.

[285] J. Yang, C. K. Tse, M. Huang, and X. Fu. Homoclinic bifurcation of a grid-forming voltage source converter. *IEEE Trans. Power Electron.*, 36(11):13176–13187, 2021.

[286] J. Yang, C. K. Tse, and D. Liu. Sub-synchronous oscillations and transient stability of islanded microgrid. *IEEE Trans. Power Syst.*, 38(4):3760–3774, 2023.

[287] S. Yang, W. Chen, X. Zhang, and W. Yang. A graph-based method for vulnerability analysis of renewable energy integrated power systems to cascading failures. *Reliab. Eng. Syst. Saf.*, 207:107354, 2021.

[288] X. Yang, D. Song, X. Tang, X. Song, P. Xu, and Z. Li. Practicable cascading failure process prediction analysis scheme and system development based on multiple information sources. In *Proceedings of the International Conference on Power System Technology*, Guangzhou, China, pp. 260–265, 2018.

[289] Y. Yang, T. Nishikawa, and A. E. Motter. Small vulnerable sets determine large network cascades in power grids. *Science*, 358(6365):1436902, 2017.

[290] R. Yao, S. Huang, K. Sun, F. Liu, X. Zhang, S. Mei, W. Wei, and L. Ding. Risk assessment of multi-timescale cascading outages based on markovian tree search. *IEEE Trans. Power Syst.*, 32(4):2887–2900, 2016.

[291] M. Youssef, C. Scoglio, and S. Pahwa. Robustness measure for power grids with respect to cascading failures. In *Proceedings of the International Workshop on Modeling, Analysis, and Control of Complex Networks*, San Francisco, CA, USA, pp. 45–49, 2011.

[292] X. Yu and Y. Xue. Smart grids: A cyber–physical systems perspective. *Proc. IEEE*, 104(5):1058–1070, 2016.

[293] X. Zha, Y. Liu, and M. Huang. Resilient power converter: A grid-connected converter with disturbance/attack resiliency via multi-timescale current limiting scheme. *IEEE J. Emerg. Sel. Topics Circuits Syst.*, 11(1):59–68, 2021.

[294] C. Zhan, C. K. Tse, and M. Small. A general stochastic model for studying time evolution of transition networks. *Physica A*, 464:198–210, 2016.

[295] B. Zhang, M. Wang, and W. Su. Reliability analysis of power systems integrated with high-penetration of power converters. *IEEE Trans. Power Syst.*, 36(3):1998–2009, 2021.

[296] L. Zhang, B. Zhu, and Y. Wang. Identification of vulnerable lines in power grids with wind power integration based on topological potential. *Electric Power Syst. Res.*, 234:110593, 2024.

[297] X. Zhang, J. Guo, T. Wang, S. Zeng, S. Ma, and G. Wu. Identifying critical elements to enhance the power grid resilience. In *Preceedings of the IEEE International Symposium of Circuits and Systems*, Seville, Spain, pp. 1–5, 2020.

[298] X. Zhang, D. Liu, H. Tu, and C. K. Tse. An integrated modeling framework for cascading failure study and robustness assessment of cyber-coupled power grids. *Reliab. Eng. Syst. Saf.*, 226:108654, 2022.

[299] X. Zhang, D. Liu, C. Zhan, and C. K. Tse. Effects of cyber coupling on cascading failures in power systems. *IEEE J. Emerg. Sel. Top. Circ. Syst.*, 6(2):228–238, 2017.

[300] X. Zhang and C. K. Tse. Assessment of robustness of power systems from a network perspective. *IEEE J. Emerging Sel. Top. Circuits Syst.*, 5(3):456–464, 2015.

[301] X. Zhang, H. Tu, J. Guo, S. Ma, Z. Li, Y. Xia, and C. K. Tse. Braess paradox and double-loop optimization method to enhance power grid resilience. *Reliab. Eng. Syst. Saf.*, 215:107913, 2021.

[302] X. Zhang, C. Zhan, and C. K. Tse. Modeling the dynamics of cascading failures in power systems. *IEEE J. Emerging Sel. Top. Circuits Syst.*, 7(2):192–204, 2017.

[303] Y. Zhang, M. E. Raoufat, and K. Tomsovic. *Remedial Action Schemes and Defense Systems.* Hoboken, USA: Wiley, 2016.

[304] Z. Zhang, R. Yao, S. Huang, Y. Chen, S. Mei, and K. Sun. An online search method for representative risky fault chains based on reinforcement learning and knowledge transfer. *IEEE Trans. Power Syst.*, 35(3):1856–1867, 2020.

[305] J. Zhao, M. Huang, H. Yan, C. K. Tse, and X. Zha. Nonlinear and transient stability analysis of phase-locked loops in grid-connected converters. *IEEE Trans. Power Electron.*, 36(1):1018–1029, 2021.

[306] B. Zhu and S. Sastry. SCADA-specific intrusion detection/prevention systems: A survey and taxonomy. In *Proceedings of the International Workshop on System Safety and Security*, Stockholm, Sweden, pp. 1–16, 2010.

[307] Y. Zhu, J. Yan, Y. Sun, and H. He. Revealing cascading failure vulnerability in power grids using risk-graph. *IEEE Trans. Parallel Distrib. Syst.*, 25(12):3274–3284, 2014.

[308] Y. Zhu, Y. Zhou, W. Wei, and L. Zhang. Real-time cascading failure risk evaluation with high penetration of renewable energy based on a graph convolutional network. *IEEE Trans. Power Syst.*, 38(5):4122–4133, 2023.

[309] R. D. Zimmerman, C. E. Murillo-Sánchez, and R. J. Thomas. MATPOWER: Steady-state operations, planning and analysis tools for power systems research and education. *IEEE Trans. Power Syst.*, 26(1):12–19, 2011.

[310] E. Zio, L. Golea, and G. Sansavini. Optimizing protections against cascades in network systems: A modified binary differential evolution algorithm. *Reliab. Eng. Syst. Saf.*, 103:72–83, 2012.

# Index